The Economics of Transportation Systems: A Reference for Practitioners

Kara M. Kockelman
T. Donna Chen
Katie A. Larsen
Brice G. Nichols

The Economics of Transportation Systems: A Reference for Practitioners

Published by www.createspace.com.
CreateSpace is a DBA of On-Demand Publishing LLC, part of the Amazon group of companies.

Initial work to assemble this Reference was funded by a grant from the Texas Department of Transportation, awarded to the University of Texas at Austin's Center for Transportation Research. The deliverable associated with the grant is found at http://www.utexas.edu/research/ctr/pdf_reports/0_6628_P1.pdf .

Printed and bound in the United States of America.

ISBN-13: 978-0615714790

FIRST EDITION

Preface

This Reference was borne of a two-year Texas Department of Transportation (TxDOT) research project, with multiple work tasks. Our mission was to develop a comprehensive yet reasonably concise transportation economics text for department personnel to appreciate the economic implications of their work and assess such impacts for more optimal transportation planning, design, policy-making, and budgeting decisions. While many transportation planners and engineers feel unfamiliar with economic principles, their job duties frequently involve economic decisions. These decision processes may appear casual in nature, and rely heavily on past practices, like pre-determined agency budget splits and a preference for asphalt pavements over cement concretes. Nonetheless, they are rooted in economic considerations and consequences. Fundamentally, travel is an economic activity and good economic judgment is vital to transportation planning and engineering.

What was initially an encyclopedic and academic effort, to be comprehensive and often conceptually abstract, took a detour as we realized that such an approach simply did not suit the needs of practitioners. While fundamental economic concepts remain key in this Reference, our book targets an audience seeking an *applied* understanding of transportation economics, through frequent examples and case study descriptions.

Many of the Reference's core concepts derive from an extensive set of course notes, developed by the first author to teach a rigorous graduate-level engineering course in Transportation Economics at the University of Texas at Austin. However, focus group meetings and interviews with TxDOT staff revealed a strong preference for relating all contents to common activities of the DOT. Essentially, the busy practitioner needed the Reference to emphasize opportunities for incorporating economic concepts into existing practices, rather than the meaningful but typically mathematical abstractions found in economic textbooks. To serve readers long removed from college coursework and seeking news they can use in their professional activities, accessible language is used throughout the Reference. The organization relies on a Terms, Concepts, and Examples structure, with margin callouts of content highlights. Further details and nuance are addressed in the "In-Depth Look" sections, at the end of each chapter, for those interested in digging deeper.

Open-ended interviews with TxDOT staff helped identify the topics of greatest interest for our intended audience. Additionally, surveys of outside peer reviewers (Ken Cervenka of the Federal Transit Authority, Patrick DeCorla-Souza of the Federal Highway Administration, Dr. Gerald de Jong of the University of Leeds, Dr. Robin Lindsey of the University of British Columbia, Todd Litman of the Victoria Transport Policy Institute, Chris Porter from Cambridge Systematics, Dr. Erik Verhoef of VU University Amsterdam, and Chris Williges of the System Metrics Group) directed the content of detailed case studies in the Reference's final chapter. To ease one's transition into this lengthy Reference, an Overview chapter provides a roadmap to all contents, emphasizing their connection and their ties to specific case studies.

While keeping the Reference concise and practical was a primary goal, we sought extensive outside review, by other experts, from around the world, to help avoid gaps in content and identify relevant literature. Nuances in specific topics were clarified via consultation with expert economists, like Dr. Kenneth Small (on the topic of welfare assessment), Dr. Glen Weisbrod (on double-counting), Dr. Erik Verhoef (on the significance of network effects in complementary and competing links and corridors), Ken Cervenka (on transportation social equity issues), and Dr. Robin Lindsey (on distinguishing schedule delay from schedule unreliability/uncertainty). Dr. Leigh Boske served as a formal project advisor, offering support on economic impact analysis. Freight transport economist Robert Harrison provided insights on costs and benefits. Todd Litman and Dr. Gerard de Jong made important suggestions (such as emphasizing accessibility over simple mobility, and clearly including freight transport and logistics topics).

Draft chapters were reviewed by topic experts across the U.S. and Canada. We are grateful for all the substantive feedback of Todd Litman, Dr. Patrick DeCorla-Souza, and Dr. Michael Lahr. In response to such feedback, a summary table of all transport benefits and costs was added to Chapter 2 (Transportation Costs and Benefits), and a recommended list of statistical modeling software programs was added to Chapter 8 (Econometrics for Data Analysis). Lists of transportation applications and associated data sets were also added to Chapter 9 (Data Sets), to underscore the value in comprehensive and consistent transportation-related data for performance evaluation and planning.

Many others also played key roles in making this Reference a reality. TxDOT's financial support was critical for thousands of hours that produced most of the current manuscript. In particular, we would like to thank TxDOT's Dr. Duncan Stewart, Matt MacGregor, and Ron Hagquist for their proactive feedback through every step of the original Reference's development. We also must thank all those who made contributions to the Reference's initial contents. Chapter 6 (Project Evaluation) has its origins in Dr. Fernanda Leite's lecture notes for her Project Management and Economics course at U.T. Austin, which Sami Kolahdoozan developed into an initial draft. We also are grateful to Sami for initial writing towards Chapter 9 (Data Sets) and Chapter 5 (Investment and Financing). We are grateful to Dr. Yiyi Wang, for her contributions of content in Chapter 8 (Econometrics for Data Analysis), the Reference's most technically rigorous chapter. We also want to acknowledge Jay Chmilewski, an undergraduate intern who enhanced Chapter 3 (Regulation and Competition), with examples in railroad, airlines, and public transit under the direction of Dr. Boske, and who created the Wordle graphics for chapter title pages. For the Reference's user-friendly formatting and many instances of greater readability, we thank our primary editor, Maureen Kelly at U.T.'s Center for Transportation Research. Maureen has the expertise and eye for creating hundreds of pages of accessible content in the engineering and economics lexicon. Lastly, we want to thank the editorial assistance of Annette Perrone and Stefanie Pascacio, who helped keep this work on track, for over two years of intensive activity.

We all hope the final content and contribution are of use to readers from around the globe, in practice and research, application and education. Please do not hesitate to let us know as new topics and case studies come to mind, for inclusion in future versions of this work.

In dedication to our families & collaborators,

Kara Kockelman, Donna Chen, Katie Larsen, and Brice Nichols

Table of Contents

Introduction and Overview

"I am an engineer, so I never use economics—do I?"

Transportation planners and engineers often feel unfamiliar with economic principles, and some assume that economics does not apply to their job duties. In practice, most transportation professionals can regularly employ economic concepts and techniques for decision-making—and many do, albeit unconsciously. Due to a variety of time and data constraints, many transportation practitioners' decision-making processes are not formally documented and emerge via "engineering judgment." However casual in nature, the wisdom behind such judgment comes from past experiences and is rooted in economic considerations and consequences. In fact, many rules of thumb for transportation investment and policy arose from economic backgrounds.

Consider this example: due to pavement aging and regular use, many rural state and county roads are in need of rehabilitation or reconstruction. Should transportation agencies install more expensive but longer lasting concrete pavements or rely on less expensive asphalt overlays? The rule of thumb is to go with asphalt, for a variety of reasons, but a definitive answer is not simple. If strict near-term *budget constraints* did not exist, the decision presumably would be based on a *life-cycle cost analysis*, used to reveal the solution that yields the lowest annual equivalent cost or maximum *net present benefit* over a long-term horizon, reflecting *risk and uncertainty* in flow volumes, materials prices, vehicle sizes, and other economic indicators. In the face of tight budgets, immediate *tradeoffs* loom. Asphalt pavements may be favored simply to ensure a consistent level of pavement quality across the district under limited funding conditions, while emphasizing *equity* in funds disbursement—thus covering more funding requests in a given year. However, if certain FM roads carry significantly more truck traffic, and some are in areas with high levels of black clay (which causes premature distress on asphalt pavement and so requires higher maintenance costs), should these roads be candidates for concrete pavements? What if such a consideration requires some lighter-traffic roads to be maintained less frequently? What is the cost passed onto the users of the lighter-traffic roads who may experience slower *travel times* and increased *vehicle repair and maintenance costs*? This common topic is rife with economic considerations.

Fortunately, a wide variety of tools is available to help transportation professionals address these common but fundamentally complex questions with more confidence than a rule of thumb offers. All of the following questions also apply. Have you ever had to ponder one or more of these?

- How much should contractors be charged for project schedule delays?
- How should DOTs prioritize capacity-expansion, maintenance, and operations projects?
- With limited funding, should DOTs focus on implementing multiple smaller projects, or allocate a significant amount to relatively few larger projects?
- Should a new highway include or exclude frontage roads? What are the monetary and other costs associated with constructing these frontage roads relative to the benefits they provide?

- Should right-of-way (ROW) acquisition for a new-build project include room for a future passenger rail corridor (or other future connecting facilities)? What is the likelihood of rail implementation (or construction of future connections) compared to the uncertainty of future ROW acquisition cost?
- Should a speed limit be raised (to save travel time) or lowered (to guard against severe crashes and increase energy efficiency)? What speed changes (and times savings) can we expect from drivers, and how do all costs and benefits compare?
- If adding a relief route attracts new development (e.g., a big-box retailer) to the bypass frontage, but the competition closes several smaller shops in the city's historic downtown, what is the overall economic impact to the city? And to the region?
- What is better for DOT budgets, the environment, and travelers: gas taxes, vehicle-miles-travelled (VMT) fees, or tolls by time of day and location?

These are just a few of the questions where successful solutions are improved by an economic understanding. This Reference is designed to introduce transportation practitioners to the underlying economic realities of their profession. Ultimately, good engineering judgment, which is vital to defensible and optimal decision-making, relies in large part on good economic judgment.

Economics as a Tool for Transportation Decision-making

From travel time savings to job creation (both direct and indirect), income growth to property value changes, motor vehicle crashes to air quality and noise impacts, and microeconomic choices to macroeconomic shifts, transportation policies and investments carry great weight. Where formally assembled data is available, economic analysis tools allow decision-makers to comprehensively evaluate projects. For large projects with significant costs and many others closely scrutinized by the public, practitioners feel more confident about decisions with "numbers to back them up." Even when data are lacking and/or decision impacts are minor, a basic understanding of various economic principles will aid transportation professionals in anticipating the direction and general magnitude of project (and policy) effects. Such understanding helps identify key project impacts and leads to more educated and robust decision-making. An understanding of current and future data needs also helps engineers and planners identify—and remedy—important data limitations for enhancements in future decision-making. For example, unsafe curves on two-lane highways with high fatality rates can be prioritized on the basis of these crashes' very high economic and social costs, as discussed in Chapter 1, Costs and Benefits of Transportation. And the geo-coding of network design databases can be prioritized to better map to police-report crash information systems, ensuring more accurate crash counts by segment for statistical regression applications (as discussed in Chapter 8, Econometrics for Data Analysis.)

An additional motivation for DOT staff to become familiar with economic analysis tools is the trend of evolving federal mandates that require economic impact analysis and comprehensive quantification of transportation costs and benefits. For example, the USDOT Transportation Investment Generating Economic Recovery (TIGER) Discretionary Grant Program required that applicants monetize project benefits in the categories of livability, economic competitiveness, safety, state of good repair, and sustainability (as discussed in Chapter 5, Investment and Financing). Such mandates will motivate agency staff to understand and apply economic techniques in order to pursue funding through various federal channels. Even if such analyses are not performed in-house, an understanding of basic economic analysis principles helps staff members critically review outsourced analyses and more appropriately guide consultants' activities. This Reference seeks to enable such understanding while enhancing a wide variety of DOT staff activities. Following is a sneak preview of key concepts covered.

Reference Overview

For those who want a strong sense of the Reference without reading it from beginning to end, this section summarizes the chapters while demonstrating how topics relate across chapters. This overview also lists sample transportation considerations that each chapter addresses, pointing readers to specific chapters for more details on topics of greatest interest to them.

Chapter 1: Costs and Benefits of Transportation

Without question, transportation plays a vital role in human interactions. Before delving into analytical techniques, transportation engineers and planners should be able to comprehensively tabulate the various benefits and costs associated with transportation decisions, as these costs and benefits are fundamental to project and policy valuations, both economic and otherwise. The Reference's first chapter lays this foundation.

Chapter 1 covers the estimation of **capital costs** and **operating costs**, which may be **fixed** or **variable**. In the **short run**, capital costs such as construction and design are considered fixed, while operating costs such as maintenance and traffic management are considered variable. However, in the **long run**, all transportation facilities will eventually need replacement or major rehabilitation, and capital costs can also be considered variable costs, as Chapter 1 explains. The chapter also differentiates between **marginal** and **average** costs, which are key concepts for optimal supply, demand, and cost allocation decisions. For example, a new four-lane highway is unlikely to cost twice as much as a two-lane highway (in the same location) because the marginal cost of adding lanes is likely to be less than that of the first and second lanes, thanks to **economies of scale**.

Chapter 1 also introduces **opportunity costs** and **indirect internal costs**, which can be difficult to observe and/or quantify, but can be monetized based on choice behaviors. For example, no physical monetary exchange occurs when a driver is stuck in traffic on a congested highway, yet that delay results in missed economic (and other) opportunities. **Willingness-to-pay** (WTP) measures allow planners to estimate the **value of travel time** (VOTT), which may be the monetary value a salesperson places on spending an extra hour pitching a product or a parent places on spending at his/her child's soccer game (as opposed to sitting in traffic). Moreover, travelers are concerned about **travel time reliability**. A truck driver on the way to an important or time-constrained delivery can better anticipate and prepare for a consistent 30-minute travel time than one that averages 25 minutes, but regularly varies between 15 and 50 minutes.

Another critical, yet sometimes overlooked, benefit-cost component emerges from the concept of externalities, or **external costs** to society. Transportation system users impact the safety and well-being of others via these developments: traffic crashes; the effect of mobile emissions on air quality; traffic noise; and effects on wetlands, groundwater quality, endangered species, and other wildlife habitats. Chapter 1 discusses how these external impacts can be—and often are—valued, as compared to the **internal costs and benefits** of transportation projects and policies.

Some example questions that Chapter 1 addresses are as follows:

- What factors into the marginal and average ownership and operating costs ofvehicles?
- Why do airline, trucking, and shipping industries rely on "hub-and-spoke" networks? What is their economic advantage?
- How are the VOTT and value of reliability (VOR) estimated, and how do these compare?
- What are the economic and other benefits of a 50% crash reduction when a dangerous intersection is realigned for better visibility?
- How do the benefits of a noise wall compare to its construction costs?

Chapter 2: Pricing of Transportation Services

While Chapter 1 emphasizes the significant benefits and costs involved in providing and using transportation systems, Chapter 2 examines how these costs can best be allocated. In other words, **who should pay for transportation services and how?** Transportation pricing refers to fees (and incentives) incurred by travelers, including transit fares, cargo fees, fuel taxes, tolls, parking fees, vehicle registration fees, and insurance payments.

A key Chapter 2 concept is the notion of **consumer surplus**, or the difference between the maximum price a traveler is willing to pay for a good or service and the price he/she actually pays. While providing transportation free of charge is infeasible, tolling or taxing in pursuit of maximum profit contradicts the goal of putting public interests first. In theory, **social welfare** is maximized when **marginal cost pricing** is used, which is when users pay the equivalent of their added cost to the system (in terms of delays for those who follow, crash costs they may be responsible for but don't pay for, emissions that others will be breathing, and so forth). For example, a **flat-rate tolled road** cannot moderate congestion as well as a **dynamically/variably priced** road. Variable tolls are generally designed to fall (to zero, potentially) when traffic is light (because external delay costs are negligible then) and rise when traffic is heavy (because added delays are quite high under near-capacity conditions). In theory, a dynamically priced road can be priced so that level of service (LOS) F never occurs. But with only incomplete information available to drivers, unpredictable events (such as traffic crashes), and a lack of substitutable alternative routes and modes (and imperfect pricing of such alternatives), even facilities with the best applications of congestion pricing can occasionally experience congestion.

In addition to fuel taxes and vehicle registration fees, Chapter 2 describes other pricing strategies to achieve fuller cost recovery and better reflect user costs. These include **congestion pricing**, **highway cost allocation** methods, and **VMT fees**. The chapter also examines **equity** issues that can arise from transportation pricing (and other) policies, such as impacts to specific socio-economic groups and/or people with special mobility needs.

Example questions that Chapter 2 addresses are as follows:

- How much should commercial trucks pay per mile of freeway driving, versus light-duty vehicles?
- How does a fuel tax differ from a mileage-based user fee? What considerations determine an optimal VMT fee?
- How can tolling be deployed to prevent congestion in the presence of non-tolled alternative routes?
- What freight pricing mechanisms can regulate truck travel demand on busy urban corridors?
- How do the equity impacts of gas taxes compare to those of VMT fees?
- What are common administrative costs to implement automated tolling?

Chapter 3: Regulation and Competition

Due to their tremendous importance and complexity, transportation markets are subject to various forms of regulation. As transportation supply and travel demand have evolved over time, policies and regulations have developed accordingly. **Economic, safety, environmental, and social regulations** are set by multiple government entities to enhance procedures and behaviors, and these regulations impact market outcomes—including **competition** across modes and within modes (e.g., airlines and railroads).

Environmental and safety regulations can affect every transportation market participant, from car manufacturers (who must abide by **fuel economy standards** and **vehicle safety requirements**) to transportation agencies (who must pursue many kinds of **environmental impact studies** in order to receive federal transportation funding). **Wage regulations** establish elevated minimum-pay rates for different types of labor and tend to raise the cost of federally funded construction projects.

Competition between operators depends upon many factors, including the nature of demand (e.g., local vs. inter-regional) and technology (e.g., high speed vs. low speed, shared vs. exclusive ROW, electric vs. conventional vehicles, and online reservations vs. first come-first served settings). Risk and uncertainty also vary by mode and setting. As a result, regulatory policies vary across sectors (e.g., air vs. rail), operator types (e.g., private vs. public carriers), vehicle types (e.g., passenger cars vs. motorcycles), times of day (e.g., nighttime vs. daytime speed limits), and so forth. In some cases, **deregulation**, or the removal of government mandates, has allowed more efficient operation of transport markets. Chapter 3 provides examples of deregulation in the railroad and airline industries that have resulted in growth of freight and long-distance passenger travel. On the other hand, increased regulation through government ownership of previously private transit systems has preserved a system whose service arguably suffers from too little competition.

Example questions that Chapter 3 tackles are as follows:

- What are some cost-control strategies to increase competition among bidding contractors for transportation construction projects?

- What are the carbon emissions and energy implications of Corporate Average Fuel Economy (CAFE) standards?
- What safety benefits do air bag, car seat, and seat belt laws provide?
- What impacts did trucking's deregulation have on U.S. highways?

Chapter 4: Transportation, Movement, and Location

Chapter 4 describes interactions between transport and **location choices**, **land values**, **wages**, and **economic development**. The "chicken and egg" relationship between system provision and land use decisions is discussed in the context of distinguishing **accessibility** from **mobility**. Transportation engineers have long quantified the travel-time savings benefits of network improvements. However, travel is a **derived demand**, a by-product of the need to work, shop, visit with others, and so forth. Other than the occasional joy ride or Formula One race, travel itself is not the desired activity. Getting from point A to point B only matters if point B is a quality destination that offers the traveler the satisfaction of a more **direct demand** (for labor, food, human interaction, etc.). While the quest for mobility looks solely at travel times, speeds, and distances, accessibility also considers the quality of the destinations, which tends to increase with higher land use intensities.

While activity site locations tend to drive travel patterns in the short term, in the longer term transportation infrastructure pricing and provision shape urban form. In addition to impacting land use decisions, transportation policies and investments impact land values, the prices of goods and services, and wages. Chapter 4 describes how transportation influences site accessibility and thus business and household **location choices**. Because access is valued by businesses and households, transportation investments (and speed limits and tolling policies) impact land values via their **bid-rent curves**.

Chapter 4 presents estimates of property-value impacts from many case studies of rail transit and highway investments. Such impacts are generally context-specific, with economically thriving communities already experiencing population and employment growth tending to benefit the most from transportation investments. Studies of **urban wage gradients** find that employees with higher commute costs require higher wages, creating wage differences between urban-center workers and those in suburban employment centers, with their typically lower commute costs. Chapter 4 examines key features of the following example questions:

- Why do businesses tend to cluster by industry (e.g., Silicon Valley and Wall Street)?
- How do transportation investments impact nearby land values?
- How do raised medians and other access management strategies impact local sales, by business type?

Chapter 5: Transportation Investment and Financing

Transportation agencies face budget constraints and must set funding priorities. Chapter 5 describes U.S. road, bridge, and railway infrastructure conditions and investment needs. Transportation financing strategies are presented in the context of traditional revenue sources and other, more innovative financing methods. Traditional sources include state and federal funding, via motor fuel taxes, bond proceeds, tolls, state motor fuel taxes, vehicle registration fees, along with local funding via property and sales taxes.

Funding shortfalls have motivated the pursuit of other financing methods. Chapter 5 summarizes key features of federal **Section 129** loans, **TIFIA** loans, and **TIGER/MAP-21** grants, along with opportunities such as state infrastructure banks (**SIBs**), private activity bonds (**PABs**), tax increment financing (**TIF**), and public-private partnerships (**PPPs**). Each method has advantages and disadvantages, in specific contexts, and example projects financed under each method are provided throughout the chapter.

Chapter 5 content addresses a range of questions, including the following:

- What types of transportation projects have utilized TIFIA financing?
- What is the federal grant application process, and what kind of projects do the programs target?
- From a DOT's perspective, what are the advantages and disadvantages of design-build projects? How do features compare between a new-build project versus a concession or lease on an existing facility?

Chapter 6: Project Evaluation

Chapter 6 describes two different approaches to project evaluation and selection: traditional **engineering economy**-based techniques and **multicriteria analysis** (**MCA**) methods. This chapter defines and then demonstrates how to use **discount rates**, **internal rates of return** (**IRRs**), **payback periods**, **breakeven analysis**, and other techniques. The chapter outlines typical steps in a comprehensive **cost-benefit analysis** (**CBA**) to reflect agency costs, user benefits and costs, and externalities, as introduced in Chapter 1. Chapter 6 also discusses **constrained optimization**, an important tool for maximizing total benefits under budget and other constraints (or, for example, minimizing costs subject to supply and demand constraints). Such techniques can be quickly applied using MS Excel functions.

Unlike traditional optimization techniques, where all outcomes are characterized in a single metric (such as dollars), MCA can reflect a host of non-quantifiable considerations, such as environmental justice and public support, thus allowing project rankings to be reasonably calculated across multiple criteria in various dimensions. Chapter 6 presents a Kansas DOT MCA application utilizing weighted numeric scores for project selection to highlight the flexibility (and potential interpretation issues) of such tools. The simplest form of MCA, **simple additive weighting** (**SAW**), is discussed in detail and illustrated with an example. **Data envelopment analysis** (**DEA**), a widely used decision-making model to compare relative performance of units within systems, is also covered.

Chapter 6's final section introduces **sensitivity analysis**, which helps decision-makers identify the degree to which analysis outputs (such as net present values, project rankings, and traffic flow predictions) are affected by changes in inputs. For example, link performance function parameters and population and job growth rates are critical assumptions in travel demand models that can greatly impact future years' traffic predictions (and project benefit-cost ratios).

Sensitivity analysis can quantify the range of likely decision outcomes, helping decision-makers guard against uncertainty and risk.

Example questions that Chapter 6 helps address are as follows:

- What are the steps for a proper CBA?
- What is the difference between real interest rates and nominal interest rates?
- What are the costs and benefits associated with a bridge replacement project, and how can future benefits be compared to present costs?
- How can the user benefits and construction and rehabilitation costs of flexible and rigid pavement alternatives be compared using life-cycle cost analysis (LCCA)?
- What is the optimal way to select a set of top projects under agency budgets and other constraints?

Chapter 7: Economic Impact Analysis of Transportation Investments

In addition to anticipating direct costs and direct benefits of policies and projects, transportation professionals are interested in a wide range of less direct economic impacts. Chapter 7 opens with a discussion of why **economic impact analyses (EIAs)** are conducted, describing regulatory requirements such as **environmental impact statements** and public information and planning needs. While a CBA can reveal project alternatives that maximize net benefits, an EIA attempts to anticipate wage, employment, sales, and related impacts.

Economic indicators such as spending by households and businesses, employment levels, wages, business sales, tax revenues, exports and imports, and capital investment expenditures can be the focus of an EIA. Transportation investment has the potential to affect a variety of economic indicators (via time savings and price changes), and these changes can be categorized as **generative** or **redistributive impacts**. As Chapter 7 described, generative impacts produce a net economic gain, while redistributive impacts essentially shift economic activity from one area to another, netting zero economic gain.

Chapter 7 also introduces **input-output** and **general equilibrium models** of the economy. These models' foundations are presented alongside model strengths and limitations, helping identify a variety of critical issues that relate to all economic analysis methods. An in-depth section offers a deeper look at the common issue of **double-counting** economic impacts, identifying a dozen types of EIA-based double-counting errors.

Chapter 7 helps address questions like these:

- How do EIA and CBA differ?
- What is the difference between economic value and economic impact?
- How can engineers anticipate the multiplicative effects of transportation spending?
- What is double-counting and how can it be avoided?

Chapter 8: Econometrics for Data Analysis

This chapter is a departure from project-focused evaluation and impact analysis. It tackles the fundamentals of transportation data analysis, characterizing mathematical relationships across wide samples of data points. **Econometric analysis** involves advanced statistical models to help practitioners analyze a variety of transportation data and discern the interactions and relationships between various variables (in order to pursue more optimal policies and investments, while predicting future trends). The results of these flexible statistical models are used to predict ROW acquisition costs, mode and route choices in the presence of tolls, the impact of gas prices on VMT, the effects of increased speed limits on crash counts and injury severities, the impact of household characteristics and income on vehicle ownership choices, and much more.

Chapter 8 details the steps taken for econometric analysis, including **data selection**, **model specification**, and **parameter estimation** (including use of MS Excel's Regression command). Crucial to the understanding of econometrics is the realization of how different types of data affect choice of model and estimation methods. The chapter introduces a wide variety of **continuous** (both linear and non-linear) and **discrete choice** model specifications to illustrate the models and methods that are most appropriate for different data types and research questions. Example specifications include ordinary versus **feasible generalized least squares**, use of **instrumental variables** and seemingly unrelated regression systems, and **multinomial logit** versus **ordered probit models**.

The ultimate goal of econometric regression analysis is to determine the explanatory variables that impact the response variable—and to what extent. Chapter 8's final sections describe how to determine **statistical versus practical significance**.

Chapter 9: Data Sets and Chapter 10: Case Studies

The Reference's last two chapters identify data resources and describe real-world transportation applications of various economic methods and tools discussed in the main chapters. The **Data Sets** chapter lists example transportation applications for a wide variety of publicly available data sets, and mentions some data collection trends (e.g., use of GPS and Bluetooth-based cell-phone data).

Many economics concepts apply simultaneously in specific transportation project and policy contexts. The Reference's **Case Studies** chapter details several applications that demonstrate this interconnectedness, while tying such concepts to real-world settings. The featured case studies include a benefit-cost evaluation of a New Jersey DOT highway extension project, the local sales impacts of bypasses for small to mid-sized Texas communities, an economic impact assessment of various congestion pricing scenarios in the Dallas-Fort Worth Metroplex, and an estimation of ROW acquisition costs.

All together, these case studies, analysis methods, and transportation economic fundamentals reveal a world of concepts and tools that should strengthen the practice of transportation engineering, planning, and policymaking. Ultimately, travel is an economic activity, and DOT decisions impact our quality of life in a number of significant and measurable ways. To ignore such opportunities is imprudent. Resource constraints and public interest should not allow a lack of economic understanding to continue. The motivation is clear, and the opportunities to incorporate economic ideas in the practice of transportation planning and implementation are endless. *Here's to better decision-making and public communication!*

Chapter 1: Costs and Benefits of Transportation

1.1 Introduction

Transportation constitutes nearly 17% of household expenditures (BLS 2011), 30% of U.S. greenhouse gas (GHG) emissions (Greene 2011), and 70% of domestic petroleum consumption (EIA 2013). In a world of limited resources, volatile materials prices, energy security issues, and multiple environmental concerns, it is imperative to understand and accurately model how transportation investments and policies impact stakeholders and society.. Economic practice and theory require familiarity with a variety of costs and benefits. So do transportation engineering and planning practice. This chapter presents the most common cost and benefit concepts encountered in transportation economics. While transportation can provide many hard to quantify benefits such as access to new markets, customers, health care, and education, a transportation project's or policy's economic value can be estimated by anticipating its quantifiable costs and benefits. Table 1.1 lists the typical costs and benefits considered in this chapter's transportation context.

Figure 1.1 illustrates that all costs can be categorized as internal or external costs, and this chapter describes their inter-relationships. Further, economists regularly refer to variable and fixed costs, and Figure 1.1 also illustrates this complementary relationship, with dashed lines around capital and operating costs indicating that these can be either variable or fixed in different contexts. The sum total of fixed and variable costs equals the total costs. The concepts of marginal and average costs derive from the total costs. Section 1.2 explains the role of these cost concepts in transportation economics.

Key Terms

- ✓ **TC:** total cost
- ✓ **AC:** average cost
- ✓ **MC:** marginal cost
- ✓ **VOR:** value of reliability

For applications of costs and benefits concepts, see Chapter 6 for cost-benefit analysis (CBA) and Chapter 7 for economic impact analysis (EIA).

Table 1.1: Typical Costs and Benefits Used in CBA, EIA, and Other Project Evaluation Methods

Potential Costs (and Benefits) of Transportation Projects and Policies	Description	Examples
Capital	One-time design and construction costs	A new facility's capital costs include planning, preliminary engineering, project design, environmental impact analysis, right-of-way (ROW) acquisition, construction, equipment purchases, etc.
Operating	Recurring operations, maintenance, and rehabilitation costs	Typical highway operating costs include traffic management, crash- or weather-related repair and cleanup, equipment (vehicles, traffic signals, signs), utilities, resurfacing (but not reconstruction), etc.
Vehicle	Vehicle ownership and maintenance costs such as fuel, tire replacement, insurance, etc.	Pavement resurfacing improves road conditions and reduces vehicle wear and maintenance costs.
Travel Time	Lost time and productivity	Implementation of signal timing coordination on an arterial street enables faster travel times and reduces delay.
Travel Time Reliability	Variance of schedule uncertainty	Dynamically priced high-occupancy/ toll (HOT) lane keeps travel speeds close to free flow speed and reduces variability in travel time.
Safety	Number, severity, and cost of crashes	Addition of rumble strips reduces the number of crashes related to driver fatigue.
Emissions	Health and other costs of vehicle-produced pollution due to changes in travel speeds, distances, times of day, fuels, and vehicle types	Fleet conversion from diesel to compressed natural gas vehicles reduces emissions.
Noise	Discomfort and property value loss due to increased traffic noise	Construction of a sound wall between a freeway facility and an adjacent neighborhood reduces traffic noise.
Ecological Impacts	Travel's impacts on wildlife habitat, water flow, and water quality	Planned roadway alignment runs through an endangered species habitat, impeding animal movements through the area.

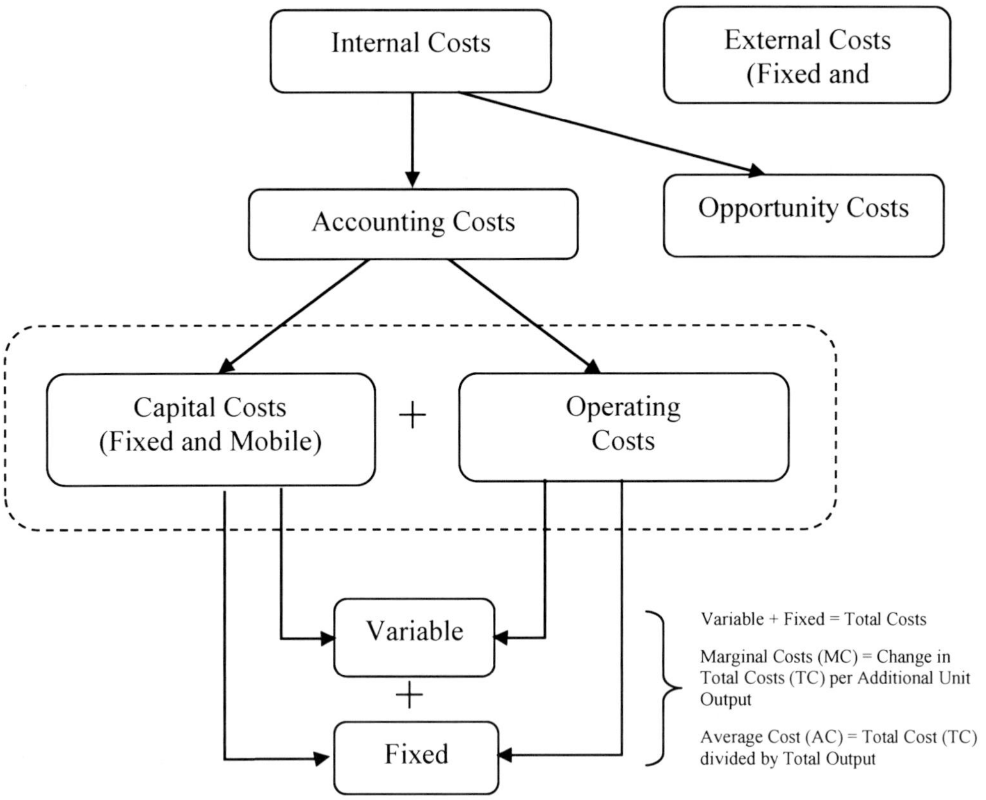

Figure 1.1: Cost Concept Relationships

1.2 Internal Costs and Benefits

Broadly speaking, **internal costs** are those borne by system operators and/or users while **external costs** are those borne by non-users. Internal costs include construction, maintenance, operation, and road user costs (e.g., fuel and registration fees). External costs, such as noise and air pollution costs borne by members of the community other than the transportation system users, are discussed later in this chapter.

In computing internal costs, an appreciation for the difference between real accounting costs and opportunity costs is necessary because the former are real, experienced, monetary costs, while the latter are potential, missed benefits, as described further below.

Accounting Costs

Accounting costs refer to transactions when real monetary changes occur. These costs are represented in traditional engineering project cost estimates—including but not limited to construction, maintenance, and operation costs. These costs can occur once, as in the case of initial project costs (like right-of-way (ROW) acquisition and design), or be regular and continuous, as in the case of operations, maintenance, and rehabilitation. The distinctive accounting-cost concepts of capital, operating, marginal, average, fixed, variable, short-run and long-run costs are presented in the following sections.

Capital vs. Operating Costs

In transportation, **capital costs** typically refer to *fixed capital costs* for facilities and *mobile capital costs* for vehicles. **Operating costs** are incurred for goods and services used to maintain and operate a facility, vehicle, or service. The benefit counterpart to operating costs is *operating revenues* (as well as *savings* in operating costs). Accounting techniques effectively merge capital and operating costs by including depreciation of capital goods within operating expenses. Categories of operating costs are essentially the same across all modes, whereas capital costs differ, as noted in Table 1.2.

Mode	Capital Costs	Operating Costs
Automobiles	Vehicle, roadway, traffic signals	Fuel, labor, maintenance, supplies
Railroads	Rail cars, tracks, stations, rail yards, signal systems	
Airlines	Airports, traffic control systems	
Ships	Ports, ships	

Table 1.2: Comparison of Capital & Operating Costs for Different Transportation Modes

Estimating Capital Costs

Transportation planners and engineers must assess a number of capital costs when estimating project expenditures during the planning phase. In addition to standard construction costs, funding must be allocated for project design, environmental process activities, and ROW acquisition (FHWA 2007). The Federal Highway Administration (FHWA) also recommends that a 5–10% contingency be included in projects to account for unforeseen changes, though a contingency of up to 15% may be used for projects particularly susceptible to "scope creep." Low-risk projects must have well-defined scopes and schedules and properly identified risks and uncertainties (FHWA 2007). For projects entailing substantial risk and/or uncertainty, the FHWA (2007) recommends that ranges be used for early estimates. For example, a planning stage estimate could state project cost as $15 million, with a 95% confidence interval or probable range of $14 million to $18 million.

Two basic procedure types are used to estimate project construction costs: detailed bid item estimates and broader conceptual estimates. *Bid item estimates* are conducted by estimating quantities and prices of specific items that contractors will bid on. An engineer may estimate, for example, that a project will require 300 linear feet of 36 in. diameter corrugated metal pipe (CMP) as one bid item. The engineer must then determine the price per foot of 36 in. CMP. He or she may do this by reviewing historical projects that used 36 in. CMP, emphasizing recent installations in nearby and similar locations with similar quantities of 36 in. CMP and similar project sizes. The FHWA cautions against using historical bid prices unless the projects are for similar work and similarly sized. As an alternative or supplement to estimating based on historical bid prices, the engineer may estimate the price that a potential contractor will purchase 36 in. CMP for and anticipate the labor and equipment costs required to install it. Regardless of whether historical bid prices or materials and labor are used to generate estimates, engineering judgment should be used for all assumptions. However, detailed bid item estimates are often time consuming and the individual quantities may be hard to obtain before substantial design work has been completed.

Conceptual cost estimates are typically appropriate for sketch planning because they require a lower level of detail. A conceptual estimate is similar to a bid item estimate in that quantities of specific items are estimated and assigned a cost per unit. The difference lies in the magnitude and number of the quantities being estimated. A large project would look at the number and characteristics of new lane miles, on and off ramps, signalized intersections, bridges, and other *major* items. The costs of smaller items are figured into the larger ones. For example, a conceptual estimate would estimate the cost of a "50-foot by 3-mile resurfacing using 6 inches of asphalt treated base and 4 inches of asphalt" instead of estimating individual quantities and prices for tons of asphalt, tons of asphalt-treated base course, square feet of pavement removal, concrete manhole collars, and other items that may be measured in a bid item estimate. For example, generic cost-per-mile models developed by the Florida Department of Transportation estimated the 2012 cost of new construction of a two-lane divided urban interstate with median, barrier walls, and full inside and outside shoulders to be $8.9 million per mile, and the cost of a rural arterial widening from two lanes to four lanes with shoulders to be $2.0 million per mile. The FHWA recommends that certain items such as traffic control, environment mitigation, and utility relocation be estimated separately from the base project and itemized within the final estimate.

Resource
The FHWA recommends that planners consult their ROW division staff when generating cost estimates. These individuals have the greatest expertise and will generate the most accurate ROW cost estimates.

Design and ROW costs must also be anticipated. Design is often calculated as a percentage of project costs. Agencies typically assign a default design cost percentage, then adjust it up or down based on the project's complexity. Also, design cost percentages for very small projects are generally higher because various administrative overhead costs typically apply, regardless of project size. ROW costs can be estimated during the planning process by first estimating the amount and location of new ROW required and then ascertaining whether any structures will be taken (e.g., parking lots and billboards). General property values (per square foot or acre) can then be applied for an early estimate of acquisition costs. Regression models using past ROW acquisition cost data can also be developed, and their parameter estimates used, as discussed in the Case Studies section of this Reference.

Estimating Operating Costs

Changes in operating costs may be determined by using standard values for expected operation and maintenance costs. Agencies often estimate standard maintenance costs for each foot-mile or lane-mile of pavement. Items such as road signs, guardrail, luminaire heads, and signal indications may have design lives shorter than the project life, so their replacement should be factored into maintenance costs. For projects with substantial new electrical features, daily electricity demands for illumination, signals, and intelligent transportation system applications must also be included.

Concept Example: Operating Costs
Not all projects result in increased operations and maintenance (O&M) costs. For example, a roundabout replacing a traffic signal should result in lower overall O&M costs. Intersections with roundabouts require less electricity than signalized intersections (although intersection illumination is still required). Furthermore, signal cabinets, signal indications, vehicle detectors, and pedestrian pushbuttons do not have to be provided and then replaced (as components wear out). Nevertheless, most transportation investments add more pavement, guardrail, crash cushions, electroliers, signs, bridges, signals, and/or other features that can increase maintenance costs instead of lowering them.

In general and in the short run, capital costs tend to be considered **fixed costs**, while operating costs tend to be considered **variable costs**. As discussed in a later section of this chapter, all costs are variable in the long run, though the time horizon for which that is true varies by project type. All facilities eventually need to be replaced or updated, and the costs for doing so can be considered variable costs over the long term, or fixed capital costs at/near the time of their application.

Marginal vs. Average Costs

As noted in Table 1.3, **marginal cost (MC)** refers to the change in total cost when the quantity produced changes by one unit (or an infinitesimal unit), whereas **average cost** refers to total cost divided by the total number of units produced (such as seat-miles delivered by an bus operator, square feet of pavement laid by a contractor, and miles travelled by a commercial truck fleet). To get a sense of this difference, consider how the total cost of a new four-lane highway is unlikely to be twice as much as that of a two-lane highway in the same location. Essentially, marginal costs tend to fall with project size, so the marginal cost of the third and fourth lanes is likely to be less than that of the first and even second lanes, thanks to economies of scale in bringing equipment and workers out to a job site, ordering materials, managing the inspections, and so forth. Nevertheless, the *average* cost per lane of the new highway is the total cost divided by four.

Table 1.3 also defines two types of marginal costs (point versus incremental) and compares them to the definition of average costs. Point marginal cost calculations can come from taking derivatives (if a continuous mathematical function exists for total costs), while incremental MC calculations reflect simpler (but somewhat less elegant) mathematics.

Table 1.3: Marginal and Average Cost Comparison

Costs	**How To Calculate (TC= Total Cost; Q= Total Output)**	**Interpretation**
Point Marginal Costs (MC)	MC = dTC/dQ (derivative of TC with respect to Q) Example: TC=\$200+4Q → dTC/dQ = \$4/unit	The instantaneous slope of the TC function, relative to output (Q). This represents the infinitesimal change in total costs for an infinitesimal change in total output (often a function of output level).
Arc Marginal Cost (incremental)	$\text{ARC MC} = \frac{(TC_2 - TC_1) - TC_1}{Q_2 - Q_1}$ Example: (\$600-\$400)/(100 units - 50 units) = \$4/unit	Normalized change in costs for a specific, discrete change in total output.
Average Cost (AC)	AC = TC/Q Example: \$600/100 units = \$6/unit average cost	Also called unit cost, AC is total cost divided by total output.

Transportation Application: Marginal Vehicle Ownership and Operating Costs

Chapter 6 discusses depreciation of vehicle capital ownership costs and use of interest rates for standard discounting techniques.

For automobile travel, marginal vehicle operation and ownership costs (such as tire wear and vehicle depreciation) are private costs borne by the users. The American Automobile Association estimated the cost of gas, maintenance, insurance, licensing, financing, and registration to be in the range of \$0.47 to \$0.72 per mile for a sedan, with a base cost of \$0.179 per mile for gas and maintenance in 2010 dollars. Other estimates place operating costs (not including ownership and insurance) at \$0.21 per mile (Polzin et al. 2008), or \$0.173 per passenger-car mile, \$0.217 per pickup truck, van, or SUV mile, and \$0.49 per commercial-truck mile (Barnes & Langworthy 2003). However, depreciation is an inevitable cost of asset ownership, at an estimated \$0.062 per added mile of passenger car use (Barnes & Langworthy 2003). In other words, greater use means faster depreciation of this asset.

Average Costs and Economies of Scale

Average costs will vary as production expands. Economies of scale and scope, as well as constraints on production, depend on how average costs change in response to the expansion of production. Average costs help explain why increasing ship sizes have limits and why major airlines tend to rely on hub and spoke systems.

Economy of scale (EOS) is an important concept in any production process, including the provision of transportation. EOS indicates what happens to average costs when scaling up a transportation project or process. A transportation service, facility, or organization can experience either (rising) *economies of scale, constant economies of scale*, or *diseconomies of scale*, depending on how average costs change as output (such as seat-miles or VMT) increases, as described below and shown in Table 1.4. A related term, **returns to scale** (RTS), is also common in economics and often used interchangeably with EOS. However, RTS refers to production or output scaling, rather than cost scaling, as described below.

Economies of Scale vs. Returns to Scale

Economies of scale indicate average *cost* savings (per unit) as output/production increases, whereas **returns to scale** indicate the factor change in *production* in response to a factor increase in all inputs. When input prices are constant and do not change with the firm or agency's purchase decisions, EOS and RTS are essentially equivalent (i.e., increasing RTS will occur with EOS and vice versa).

Table 1.4 presents the three cases when EOS and RTS align in direction of change (increasing, constant, and decreasing), when input prices are fixed. Returns to scale are determined by analyzing a production function, which relates the maximum possible quantity of output with a given quantity of inputs. Economies of scale are determined by analyzing the average cost curve.

When a transportation organization's purchasing power lowers the price of inputs or its heavy demands increase input prices, EOS and RTS may no longer move in the same direction. For example, an organization that experiences an increasing returns to scale in production (a doubling of inputs [like workforce, fuel, and equipment] more than doubles outputs) could potentially have diseconomies of scale (e.g., input prices increase so much that average costs rise following doubled production).

Table 1.4: Economies of Scale and Return to Scale Cases when Input Prices are Constant

Case	Economies of Scale Interpretation	Returns to Scale Interpretation
Increasing Economies of Scale or Increasing Returns to Scale	$MC(Q) < AC(Q)$ The average cost (per unit) decreases as output increases. *Example:* Average input costs per seat on Amtrak decrease as seat-miles increase.	Increasing all inputs by same proportion results in a more-than-proportional increase in the level of output. *Example:* Number of Amtrak seat-miles increases by 60% when all inputs increase by 40%.
No Economies of Scale or Constant Returns to Scale	$MC(Q) = AC(Q)$ The average cost (per unit) stays the same as output increases. *Example:* Average input costs per seat stay the same as seat-	Increasing all inputs by same proportion results in the same proportional increase in the level of output. *Example:* Number of seat-miles increases by 50% when all inputs

	miles increase.	increase by 50%.
Diseconomies of Scale or Decreasing Returns to Scale	MC(Q) > AC(Q) The average cost (per unit) increases as output increases. *Example:* Average input costs per seat increase as seat-miles increase.	Increasing all inputs by same proportion results in a less-than-proportional increase in the level of output. *Example:* Number of seat-miles increases by 10% when inputs increase by 20%.

Diminishing Returns vs. Decreasing Returns to Scale

Diminishing returns sounds similar to **decreasing returns**, but the former only applies to increases in one input or factor of production at a time (rather than a factor increase in all inputs). For example, increasing the number of employees at a toll booth will eventually result in diminishing returns because only so many employees can be in a toll booth. The marginal or added number of vehicles served (the output) per added worker will fall and eventually hit zero (and possibly turn negative) because the numbers of toll booths, lanes, and queued vehicles (three other key inputs) did not increase. If all inputs—number of employees, tollbooths, waiting vehicles, and lanes—are increased by the same factor, one would generally expect a proportional rise in output, and thus constant RTS. While constant RTS is most common in practice, and often assumed by economists, some production processes show rising and falling RTS, depending on details of how inputs interact.

Transportation Application: Airline Hub-and-Spoke Operations

The airline, trucking, and shipping industries regularly rely on hub-and-spoke operations to exploit the economies of using larger vehicles (and fuller vehicles, via more appropriately sized-to-load vehicles or higher load factors) to transport more passengers or goods at a lower average cost (and with more frequent trip scheduling, which travelers and shippers greatly appreciate). Trans-shipment points (like the Dallas-Fort Worth airport and Chicago rail yard) allow carriers to consolidate goods or passengers into larger vehicles (or longer trains) for routes with the demand to support the consolidation.

Hub and spoke networks are designed with larger vehicles for high-demand routes (e.g., San Francisco to New York City) and smaller vehicles for lower-demand routes (e.g., Sacramento to San Francisco) to economize on average costs. Large and small vehicles exchange passengers or goods at the hubs.

U.S. airlines were clearly moving towards hub and spoke structures by the mid-1980s because of economies of vehicle size. Figure 1.2 shows visually the change from direct service to hub and spoke for cities around Atlanta, Georgia (with Table 1.5 providing corresponding airport codes). Because hub and spoke designs tend to increase total travel mile (by reducing the number of direct trips, from one's origin non-stop to one's destination), many airlines have more than one U.S. hub, as shown in Table 1.6's listing of out-of-state hubs. Hub and spoke operations do carry the following added costs:

- Increased operating costs due to mileage increases,
- Additional terminal costs for passengers,
- Additional take-off, landing, and operation costs, and
- Potential delays for passengers and goods if hub operations cease or slow down temporarily (e.g., due to weather).

Airports themselves do not offer clear economies of scale, so the advantage of hub and spoke lies in vehicle size and scheduling frequency for passengers, not airport size. For larger market pairs, of course, non-stop service endures, thanks to high demand (such as San Francisco to Chicago and New York City to Washington, D.C.).

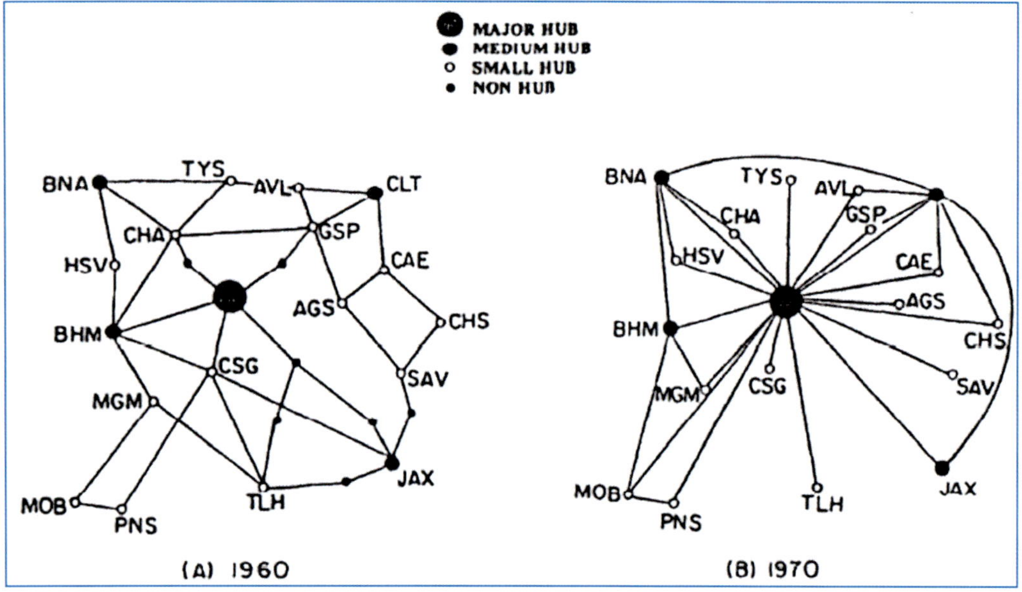

Figure 1.2: The Switch to Hub-and-Spoke Airline Operations around Atlanta, Georgia (see Table 1.5 for Airport Code Locations) (Source: Kanafani & Ghobrial 1985)

Table 1.5: Airport Codes and Corresponding Locations for Figure 1.2

Airport Code	Airport Location	Airport Code	Airport Location
BNA	Nashville, TN	PNS	Pensacola, FL
TYS	Knoxville, TN	CSG	Columbus, GA
CHA	Chattanooga, TN	AGS	Augusta, GA
HSV	Huntsville, AL	GSP	Greensville/Spartanburg, SC
BHM	Birmingham, AL	CLT	Charlotte, NC
AVL	Henderson, NC	CAE	Columbia, SC
MGM	Montgomery, AL	CHS	Charleston, SC
MOB	Mobile, AL	SAV	Savannah, GA
JAX	Jacksonville, FL	TLH	Tallahassee, FL

Total Costs: Fixed vs. Variable, and Short Run vs. Long Run

As described earlier in this chapter, **fixed costs** are costs that do not change with the level of output, such as vehicle miles traveled (VMT) along a new highway. In other words, fixed costs are independent of the output: they must be paid even if no output is being produced. Capital infrastructure, such as railroad track and highway lanes, is typically considered a fixed cost, although some components can be variable costs, due to variations in the number of tracks needed, runways used, gates at an airport, and/or lanes along a highway.

Variable costs are costs that <u>do change</u> (at least somewhat) with the level of output, including fuel used, system operators paid, and vehicle maintenance fees. Variable costs may be avoidable; lowered production can mean lower variable costs.

The **total cost** of producing goods and services is the sum of all fixed and variable costs, as shown in Figure 1.3. Fixed and variable costs are often differentiated by one's timeframe of reference. For instance, a cost can be fixed in the **short run** but variable in the **long run**. Essentially, there are no fixed costs in the long run.

Figure 1.3: Variable and Fixed Costs

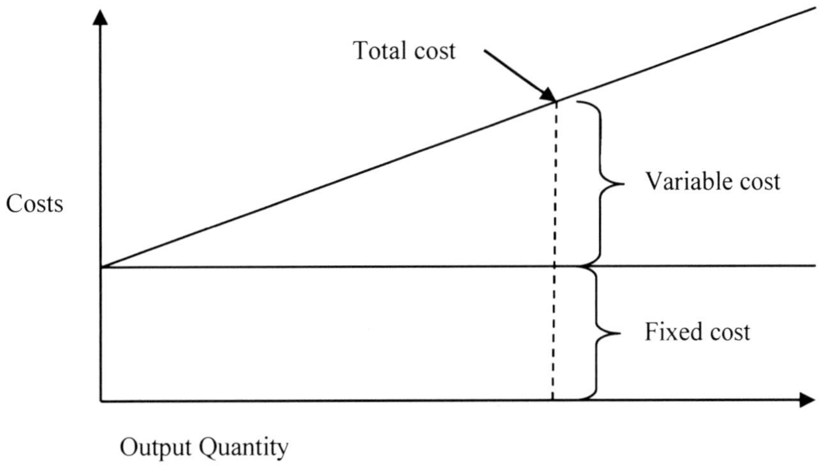

Table 1.6 summarizes several key variable and fixed costs associated with highway transportation (Small & Verhoef 2007), as borne directly by highway users (consumers) and highway providers (producers). This table has other columns and rows, as added in Section 1.3 of this chapter, relating to external costs, borne by others.

Table 1.6: Variable and Fixed Short-Run Costs of Automobile Travel ($ per Vehicle-Mile) (Small & Verhoef 2007)

Variable Costs	Average Private Cost ($/mile)
(1) Operating and maintenance	$0.141/mile
(2) Vehicle capital	0.170
(3) Travel time	0.303
(4) Schedule delay and unreliability	0.093
(5) Crashes	0.117
(6) Government services	0.005
Fixed Costs	

(8) Roadway	0.016
(9) Parking	0.007
Average Total Costs (not including externalities)	$0.852/mile

*All costs measured in US$ per vehicle-mile in 2005 prices.

Concept Example: Fleet Vehicle Operating Costs

Researchers have modeled fleet operating costs as both fixed and variable, with fixed costs including vehicle depreciation, financing, and insurance, and variable costs including vehicle repairs, maintenance, and fuel.

Opportunity Costs

Opportunity cost is the lost value/benefit of an investment or activity's "next-best" (foregone) alternative. These benefits may be monetary or non-monetary in nature (such as lost time or other valued impacts from the missed opportunity). One way to think about opportunity costs is in the context of shadow prices, which give the "cost" (or reduced objective function value) of a constraint on one's choices.

Another type of opportunity cost is implicit cost, expressed in transportation terms as the value of travel time (described later in this section).

Shadow Prices

In a broad sense, **shadow prices** are used to measure changes in a situation's outlook resulting from "any marginal change in the availability of commodities or factors of production" (Squire 1975, p. 26).

If there are costs set outside of the market, such as with labor or other supply contracts, then the accounting costs need to be adjusted to become ***shadow prices*** that indicate what would have been paid, or not paid, for the labor and supplies. Without the adjustment, the opportunity cost will not reflect the economic cost in the market of the transportation project or policy.

For our purposes, shadow prices determine how an *objective value* (some measure of benefit or cost) is impacted by altering some constraining factor by one unit. (More technically, the shadow price is equal to the Lagrange multiplier on a given constraint at the point of optimal solution.)

Chapter 6 provides more details of shadow pricing.

Concept Example: Shadow Price

In transportation, the objective may be user benefits that are to be maximized, but are constrained by factors like the number of busses available in a transit fleet. The shadow price of an additional bus in the fleet is the change in total user benefits that follows an increase in the fleet. Similarly, shadow prices can be applied to a minimization problem: manipulating a constraint on total costs, such as reducing a maintenance project's roadway resurfacing time by one day, will result in total cost reductions. Even if reducing rehabilitation time on a road by one full day is not realistically possible, the shadow price provides insight into the value of this additional time.

Implicit Costs

Essentially, **implicit cost** is the value of an owned asset, whether physical (such as a machine) or temporal (such as a person's time). Such assets could be used in other, productive pursuits (such as a truck serving another freight trip or a person enjoying more time with family). A DOT's accounting expenditures do not capture the opportunity costs of using goods, labor, or time that the DOT does not pay for directly, such as when it uses a machine or property that it purchased in the past. The DOT's implicit cost is the cost of the machine at present market prices. Another example is when a person is delayed on a congested roadway; the traveler's implicit cost is his/her value of time. An agency's facilities and the land they occupy can be considered implicit costs, because the current use of office or land space is not available for renting or selling to others.

Transportation Application: Travel Time and Reliability

The value of travel time (VOTT) is a type of implicit opportunity cost. Everything else constant (such a total number of trips to and from various activity sites), a decrease in travel time (an increase in travel time savings) reduces a project's opportunity cost of the project; and an increase increases the project's opportunity cost. Figure 1.4 breaks out the components of VOTT.

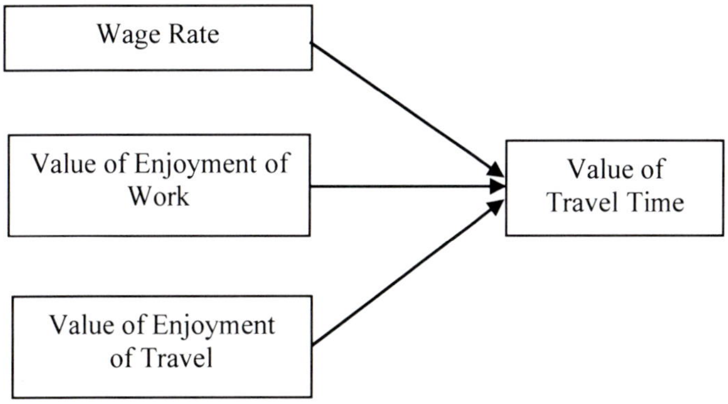

Figure 1.4: Components of VOTT

VOTT is the amount of money a traveler is willing to pay for time savings. In more technical terms, VOTT is the marginal rate of substitution of money for travel time (while keeping a traveler's overall happiness or utility constants). The variation in VOTT and value of reliability (VOR, discussed below) across a population of travelers can significantly affect project evaluations because some

Resource

An extensive body of research focuses on valuing the value and implicit cost of travel time. See Zamparini and Reggiani's 2007 literature survey.

travelers are willing to pay more than others to save travel time (thanks to having more income or tighter time constraints, for example).

Estimating the Value of Travel Time

Estimates of VOTT for individual travelers vary widely, depending on where and how the data are collected and the methods of analyzing the data. One method of collecting data regarding VOTT is to use traveler survey data to estimate how much a traveler is willing to pay for time savings. For example, *revealed-preference surveys* ask what travelers actually do and pay, while *stated-preference surveys* ask what they would most likely do and pay. Thanks to various data regression techniques, analysts can hone in on distributions of VOTT across survey respondents.

See the In-Depth Look section at this chapter's end for other VOTT methods and estimates.

Because VOTT estimates emerge from willingness-to-pay (WTP) considerations, higher-income individuals regularly exhibit higher VOTTs than lower-income individuals. To recognize differences among travelers, more than one estimate of VOTT can be used when assigning trips to the network for determining routes and then a project's implicit costs and benefits. Using a variety of traveler classes is especially important when anticipating the impacts of tolling policies. For instance, higher-income travelers may be more able and willing to pay a toll at congested times of day. If everyone had the same VOTT, most variable tolling policies (to reflect congestion externalities) would generally fail to improve social welfare. Thanks to variation or "heterogeneity" in the traveling population, thoughtfully designed variable toll policies can improve overall traveler welfare (Verhoef & Small 1999).

In their guidance for economic analysis of federal transportation projects and policies, the USDOT recommends hourly values of travel time (in 2000 U.S. $) of $10.60–$21.20 for local surface modes of travel, $14.80–$21.20 for intercity travel, and $18.80 for truck driver travel, both local and long distance (USDOT 2003).

Of course, a traveler's VOTT varies by other details of the trip as well. If one is late to an important meeting or the airport, VOTTs can run very high, even for low-income travelers.

Reliability of Travel Times

See the In-Depth Look section at this chapter's end for more VOR methods and estimates.

In addition to persons valuing the (average) time they spend traveling, they also value *reliability*, the lack of uncertainty in travel times. Transportation projects that increase reliability offer an important benefit; those that decrease reliability create a cost.

Most theories explaining aversion to unreliable travel are based on costs of unexpected arrival times at work, which are greater for being late than for being early. Estimates of the VOR, as with VOTT, vary from $10.20 to $32 per hour of late or early arrival (Brownstone & Small 2005). VOR may be estimated with WTP values determined from revealed- or stated-preference surveys.

In transportation settings, implicit costs can be the majority of travel's economic costs. Thus, assessing delay's implicit costs when planning transportation projects and policies is an essential calculation.

Opportunity Cost Example: Cost of Delay

Delays can increase a project's cost considerably, during both planning and construction. These additional costs may not always be obvious, but quantifying delay costs is useful in prioritizing resources for allocation.

Perhaps the most apparent cost of transportation project delays is the cost of redirecting traffic around construction zones. The Kentucky Transportation Center identified three distinct categories of costs associated with construction: vehicle operating costs, user delay costs, and crash costs.

- *Vehicle operating costs* are simply the physical costs involved with redirecting traffic, including the cost of additional fuel consumption from vehicles changing speeds, idling, or taking longer detours. While these costs may be the smallest portion of project delay costs, they can be computed more exactly.
- *User delay costs*, the cumulative value of the additional time road users must spend to detour from a construction area, are more difficult to quantify. Research suggests values of approximately $16 per automobile-hour and $28 per truck-hour (2010 dollars). While some debate exists about these exact values, the total cost will remain high even if smaller values are assumed (Rister & Graves 2002).
- *Crash costs* involve the average cost of a crash and the increased likelihood of a crash occurring due to construction. This cost, though significant, varies widely from one project to another.

While all three of these costs will be present in any construction project, delays can greatly increase the total cost by extending the time-period over which they act.

Another way in which project delay costs accrue is in the opportunity cost of waiting. When a project is put off for future construction, the problems the construction intended to alleviate remain, and their costs accrue.

Example

Costs of Delay

The major at-grade railroad junction in Fort Worth, near Tower 55, results in sizable delay costs. For several years, plans have been made to separate the massive east/west and north/south freight flows so that both may continue unimpeded. Presently, however, the tower still represents a bottleneck in the state's rail network. "Due to the volume of trains, each train must come to a complete stop prior to crossing. The average wait time is 15 minutes, with 90-minute delays during peak period. Long freight trains with lengthy wait times at Tower 55 are responsible for several negative impacts to the region," notes the North Texas Council of Governments (NCTCOG 2011). Postponing the start of this project not only means additional costs for freight companies from the delays, but also greater environmental costs (air and noise pollution) caused by keeping trains (and then waiting cars and trucks) at a stand-still in the DFW Metroplex.

For these reasons, contractors are sometimes granted early-finish bonuses (for completing a facility prior to its originally scheduled completion date). For example, the contract for an emergency bridge replacement near Toyah, Texas included a $10,000 per day early-finish bonus for the contractor (Smith 2011). Extreme rainfall in the rural region had caused the IH-20 bridge to fail, prompting a detour 20 miles longer than the original route, and demanding immediate DOT and contractor actions.

Another serious cost of delaying a project is the uncertainty it invites. If the price of raw materials rises significantly, overall project costs can become unmanageable. For example, the cost of constructing the Oakland Bay Bridge in California more than doubled (from $2.6 billion to around $6 billion), largely due to an increase in the price of steel (Diesenhouse 2005). The lead-time to buy materials also changed from 3 to 8 months, causing further delays. The effect of the sudden shortage in steel caused many projects across the country to either temporarily halt or stop altogether. Obviously, this cost can be computed accurately only in retrospect, but planners must remain mindful of its possibility.

Traditionally, the costs of delaying a project are not considered when choosing between different alternatives. However, these costs can be useful in determining the priority ranking, especially because putting something off for the future can actually be far more expensive than doing it now.

1.3 External Costs and Benefits

The provision and pursuit of transportation regularly entails **external costs** and sometimes external benefits, as borne by those *not* making the travel or transport-infrastructure-provision decision. These so-called "externalities" include the annoyance of highway noise, harm to adjacent properties and bystanders during crashes, and the visual and health impacts of air and water pollution. Reductions in these external costs (due to provision of sound walls, low-noise pavements, cleaner vehicles, and safer roadways) can be considered external benefits.

External costs and benefits are reflected in comprehensive cost-benefit analyses to estimate the overall economic value and social welfare impacts of transportation projects and policies.

Table 1.7 presents an extended version of Table 1.6 to show not just variable vs. fixed costs, but privately borne (internal) vs. total social costs (external) during a typical commute trip during a peak time of day on a congested U.S. network. This table also shows marginal versus average costs for adding one more mile to one's trip, versus dividing total commute cost by total commute VMT. It is interesting to see how high travel time costs are in comparison to other costs, and how high the marginal costs of time and (un)reliability are. Vehicle capital, O&M, and crash costs are also quite high. At the end of the day, one passenger-vehicle-mile costs travelers and society at large on the order of $1, with external costs (social minus private costs) accounting for about 30% of the total.

Type of Cost	Private (Internal)	Social (Internal + External)	
	Average	Average	Marginal
Variable Costs			
Costs borne mainly by highway users			
(1) Operating and maintenance	$0.141/mile	0.141	0.141
(2) Vehicle capital	0.170	0.170	0.170
(3) Travel time	0.303	0.303	0.388
(4) Schedule delay and unreliability	0.093	0.093	0.172
Costs borne substantially by non-users			
(5) Crashes	0.117	0.140	0.178
(6) Government services	0.005	0.019	0.019
(7) Environmental externalities	0	0.016	0.016
Fixed Costs			
(8) Roadway	0.016	0.056	
(9) Parking	0.007	0.281	
Total Costs	$0.852/mile	$1.219/mile	

Table 1.7: Variable and Fixed Short-Run Costs of Automobile Travel (Small & Verhoef 2007)

Transportation Application: Safety Impacts

Motor vehicle use affects not only the road, air, and water systems, but also the safety of those in the vehicles and nearby. Transportation projects or polices can either increase or decrease the number of crashes occurring, and their severity, resulting in costs or benefits to travelers and society at large. Vehicle owners buy insurance to cover many crash costs, but auto insurance is typically a flat fee, rather than per-mile driven, and many costs are not recovered from insurers (such as delays in lanes blocked by crashed vehicles, deployment of emergency personnel, and the pain and suffering of crash victims). Crash cost allocation is complex due to the nature of shared responsibility in a crash. According to the USDOT, just "about one-quarter of the cost of crashes is paid directly by those involved, while society in general pays the rest," suggesting that external costs are substantial. In 2000, driver-funded insurance companies paid about half the cost of all U.S. highway crashes (BTS 2003). Figure 1.5 breaks down payments for motor vehicle crashes by source.

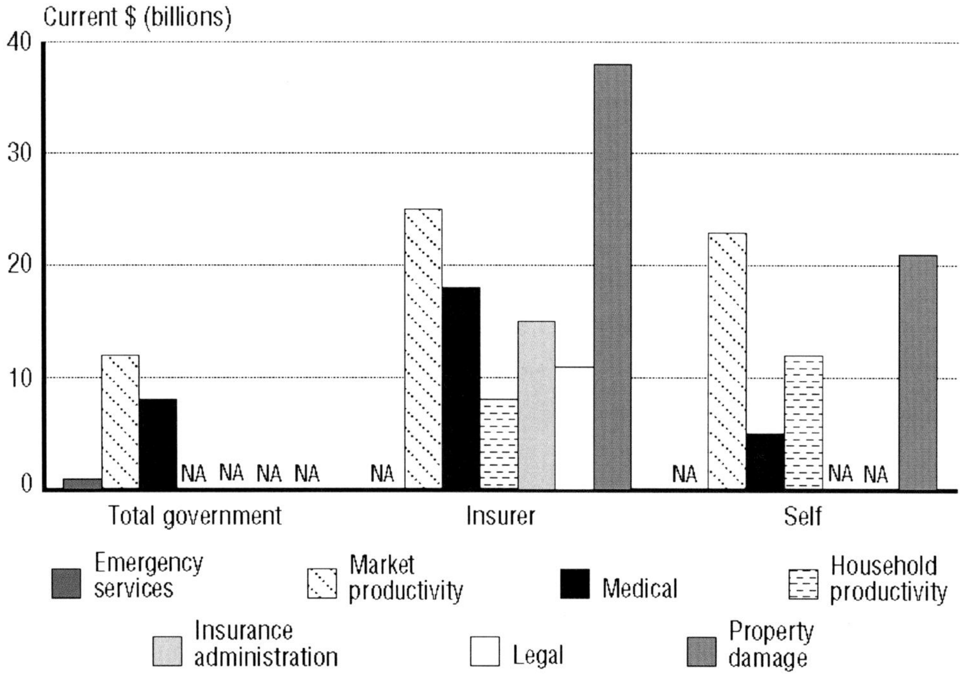

Figure 1.5: Estimated Sources of Payment for Motor Vehicle Crashes (2000 US$) (Source: BTS 2003)

The National Highway Traffic Safety Administration's *Economic Impacts of Motor Vehicle Crashes 2000* (Blincoe et al. 2002) provides rigorously estimated crash costs. Average total market costs of lost productivity, medical services, travel delay and property damage per crash range from $2,762 per no-injury crash to $977,208 per fatal crash. These values include market costs, such as lost productivity, medical services, travel delay, and property damage, but they do not include non-market factors, such as the value of life, pain and suffering, or values based on WTP in order to avoid collisions.

Other crash valuations may consider WTP measures, which are based on the price that a person is willing to pay for a marginal increase in safety. A person may be willing to pay millions of dollars to save his or her own life, but the implied value of life based on reducing the probability of serious injury or death via air-bag purchase and installation tends to be lower—and highly variable. For example, one study by the European Conference Ministers of Transport (2000) found that a person's past experience of a crash increased his/her average stated WTP by a hundred-fold (i.e., by 10,000%). Furthermore, WTP was found to vary with respondent age and household income, with those near 40 years of age placing the highest value on human life. Table 1.8 presents values based on WTP from National Safety Council estimates.

Table 1.8: WTP-Based Motor Vehicle Crash Cost Estimates (Source: NSC 2010)

Crash Severity	**WTP** (Per Injured Person)
Death	$4,200,000
Incapacitating Injury	$214,200

Non-Incapacitating Injury	$54,700
Possible Injury	$26,000
Property Damage Only	$2,400 (no injury)

Note that Table 1.8's crash costs are per injured person, rather than per crash. Oftentimes, more than one person is injured during a crash, so analysts intending to use crash costs based on WTP measures should account for the total number of persons injured. The California Benefit/Cost model assumes an average of 1.15 fatalities per fatal injury crash and an average of 1.49 injuries per injury crash (Caltrans 2010).

Transportation Application: Air Quality Impacts

Mobile source emissions, such as carbon dioxide (CO_2), oxides of nitrogen (NO_x), reactive organic gases or volatile organic compounds (ROG and VOC), fine particulate matter (PM2.5 and PM10), carbon monoxide (CO), and various other hazardous air pollutants, constitute a major fraction of human-caused emissions and are responsible for air quality concerns in hundreds of U.S. cities and counties.

Emissions costs can be broadly associated with the costs of damages caused by the pollutants to human, plant, and animal health, as well as damage to buildings and ecosystems. Another influential factor is whether to focus on the cost to control or reduce such emissions, rather than compensate for harm done.

Each pollutant carries different costs, and human health costs are dominated by particulate matter. The oxides of nitrogen and sulfur, NO_x and SO_2, form acid rain, which has been associated with ecosystem damage as well as degradation of structure exteriors and the built environment. Ozone exposure in humans is associated with breathing difficulty, asthma, airway, and lung inflammation and lung damage. Ozone deposits on plants reduce the efficiency of photosynthesis and have contributed to 90% of air-pollution-associated U.S. crop losses (Murphy & Delucchi 1998). Carbon dioxide (CO_2), a greenhouse gas, contributes to climate change.

A second variable affecting emission costs is their location, or more specifically, the potential for exposure of humans, plants, animals, and structures. Meteorological conditions as well as other factors (such as activity patterns) significantly influence the impact that the geographic location of emissions will have on human exposure and health outcomes.

Two additional, somewhat subjective variables impact emissions costs:

1. The value placed on human life and health. No universally accepted value has been determined, but many studies now include the economic (and opportunity) costs of lost productivity.
2. The range of impacts to natural resources and the built environment.

Models like the EPA's MOBILE 6.2 can quantify the costs of air pollution impacts by assessing the change in quantity and mix of emissions when a transportation project effects changes to vehicle-miles, vehicle-hours, and vehicle-trips. Emissions rates depend on transportation facility type (freeway, arterial, local road, or ramp), vehicle speed, year of analysis, vehicle type, and age. Once the emission rates are estimated, a dollar value can be applied per unit of emissions to assess air pollution costs and benefits (McCubbin & Delucchi 1996 and Mailbach et al. 2008). Emissions costs have been estimated (in 2010 \$) as \$2,900 to \$5,800 per ton of hydrocarbon, \$70 to \$140 per ton of CO, \$620 to \$5,600 per ton of NOx, \$620 to \$6,400 per ton of SO_2, and \$9,300 to \$830,000 per ton of PM2.5. As alluded to earlier, these costs depend on population density (and thus level of human exposure) and meteorology, as well as the local population's wealth and income (and thus WTP to protect one's lungs, views, and property).

Transportation Application: Noise Impacts

Traffic noise is a nuisance. Beyond that, sustained exposure to traffic noise can cause hearing impairment and sleep disturbance, increasing stress levels in those living and working nearby. In addition to human health impacts, studies indicate that traffic noise reduces the price of nearby homes (Gamble et al. 1974, Lewis et al. 1997, Bowes & Ihlanfeldt 2001).

Chapter 8 discusses hedonic price models, used to estimate traffic noise levels.

Numerous factors influence levels of traffic noise, including vehicle type, engine type, traffic speed, pavement type and condition, and presence of noise barriers. For examples, passenger cars are generally quieter than buses and motorcycles. Within passenger cars, electric and hybrid vehicles are generally quieter than gasoline and diesel vehicles. Due to the wide range of inputs affecting noise, it is difficult to quantify noise costs for projects in which the change in traffic noise is small. However, for larger projects in which traffic noise changes are significant, hedonic price models can estimate the effects of traffic noise on property values. One study recommends a depreciation of 0.5% in property value per decibel increase in traffic noise above 50 dB. A 2009 online review of a variety of other research suggests the following variations by vehicle type:

- Automobile noise costs 1.3¢ per mile on urban roads and 0.7¢ per mile on rural roads, on average.
- Electric car noise costs 0.4¢ per mile on urban roads and 0.4¢ per mile on rural roads.
- Motorcycle noise costs 13.2¢ per mile on urban roads and 6.6¢ per mile on rural roads.

Transportation Application: Ecological Impacts

Wildlife habitat and water quality can be impacted (and mitigated) by transportation improvements in many ways. Habitat disruptions (particularly those of endangered species) are difficult to monetize and are typically mitigated on a case-by-case basis according to level of impact.

Vehicles deposit rubber, oil, and other polluting particles on pavements. Rain washes these pollutants over impervious roadway surfaces into nearby areas, causing groundwater and/or wetlands contamination. Additional new impervious surfaces make the ground less permeable and increases runoff. Hydrologic impact models can predict increased storm water management costs. As with noise impacts, hedonic price models can also be used to monetize water quality impacts.

1.4 Summary

This chapter discusses and quantifies key transportation cost and benefit concepts. While some costs are relatively easy to observe and anticipate, including fixed capital costs (such as initial construction expenses), many are difficult to observe or quantify, but are very significant for economic evaluations. Unseen costs include opportunity costs and indirect costs, which lead to concepts like shadow prices and the significance of travel time estimation, reliability, and traveler welfare.

Many concepts in this chapter have overlapping definitions and applications. For instance, *capital* and *operating costs* tend to be *fixed* or *variable*, and be *short-run* or *long-run* costs depending on the analyst's timeframe of reference (with all costs and benefits being variable in the long run). Capital costs tend to represent fixed initial costs that may take place at specific points in time (especially at a project's start), while operating costs include ongoing expenses from repeated maintenance, fuel, and labor use.

The idea of *fixed* versus *variable* costs and benefits relates to *short-run* versus *long-run* horizons Other key concepts for project costing and economic impact assessments include *marginal* versus *average costs*, *economies of scale*, and *returns to scale*.

An appreciation of *external costs and benefits* is critical to understanding the environmental, delay-related, and safety impacts of transportation activities. To keep things in perspective, consider that Americans spend roughly $1 trillion per year in travel time, and lose on the order of $300 billion per year to highway crashes, along with roughly $100 billion per year to congestion (from time and fuel losses) and $50 billion to vehicle emissions. Such metrics help transportation planners, engineers and policymakers gain a sense of how important transportation is to economic systems, quality of life, and the long-term health and safety of current and future generations. Optimal decision-making in the transportation arena requires a comprehensive and long-term perspective on a variety of project and policy alternatives.

1.5 An In-Depth Look

Estimating Value of Travel Time (VOTT)

A common method for VOTT estimation involves collecting travel time and travel cost data for various travelers' mode options and their chosen alternative(s). A logit choice model is calibrated from such data, and the ratio of its utility function's parameters on travel time and cost gives the VOTT estimate (VOTT = [marginal utility of time]/[marginal utility of money] = [$/minute] or [$/hour]).

A wide range of VOTTs has been calculated by various researchers:

- Levinson and Smalkoski (2003) estimated U.S. heavy-duty truck VOTTs to average $49.42 per hour.
- Brownstone and Small (2005) estimated those of morning commuters in the Los Angeles area along routes SR-95 and I-15 to generally lie between $20 and $40 per hour, using revealed preference surveys. These values are nearly two times greater than the hypothetical values emerging from stated preference surveys.
- The Oregon DOT (2006) estimated VOTTs of $16.31 per hour for the average auto, $20.35 per hour for light trucks, and $29.50 per hour for heavy trucks.
- Schrank and Lomax (2009) used $15.47 per hour for autos and $102 per hour for commercial vehicles in their regional congestion cost calculations.
- Litman (2009) recommends that paid travel be valued at 150% of the wage rate, commuting and congested travel be valued at 50% of the wage rate for drivers' and 35% of passengers' wage rates, uncongested travel be valued at 25% of wage rate, and pleasurable travel be valued at $0/hour.
- Zamparini and Reggiani (2007) conducted a meta-analysis of VOTT studies for Europe and North America and estimated VOTT to average 82% of the wage rate, with an average of 68% for North American travelers. They estimated travel time valuations at 55% of the wage rate for commuting, 146% for business travel, and 60% for other activities.

See Chapter 8 for details on utility functions and logit choice models.

Estimating Value of Reliability (VOR)

Unreliability is typically measured as the standard deviation in travel times (from day to day, and minute to minute) on a set route at a given time of day. Accurately estimating the VOR and transportation projects and policies' reliability impacts is a fairly new research topic in transportation. Reliability measurement requires detailed data because many observations over several days of speed and traffic settings are needed to assess travel time variability on a roadway or route.

Various VOR estimates have been developed, including the following:

- Using data for travel on the variably tolled SR-91 in Southern California, Brownstone and Small (2005) estimated travelers' VOR to be \$12 to \$32 per hour of standard deviation in arrival time, or roughly 95% to 145% of the corresponding VOTT on those links (the VOTT was between \$20 and \$40 per hour). They also estimated a much higher VOR for women than men—roughly twice as high. They hypothesized that because women have more child-care responsibilities; their schedule flexibility is limited, making them more likely to choose the variably tolled SR-91 road.
- Litman's (2009) literature review suggests a VOR range of \$10.20 to \$15.60 per hour of standard deviation in arrival time.

Cost Functions and Returns to Scale in Production

Determining returns to scale involves an analysis of a transportation sector's production functions. **Production functions** are simply equations for predicting the quantity of output (e.g. ton miles of truck-cargo transport and person-miles of plane travel) as a function of all inputs' quantities, x_i (where i indexes input type, such as fuel, employee hours, and vehicles). **Cost functions** are similar to production functions in that they predict the cost of production as a function of the output (Q) and the prices of all inputs, where p_i or w_i is the (unit) price or wage of input x_i.

Production functions can take on a variety of forms, as Table 1.9 illustrates.

Table 1.9: Examples of Production Function Forms

Production Functions	General Form (where x_i=quantity of input i; α_n= indicates how input i affects the output, y)	Parameters to Estimate
Constant Elasticity of Substitution (CES)	$y = \left(\alpha_1 x_1^p + \alpha_2 x_2^p + \dots + \alpha_n x_n\right)^{1/p}$	"p" is the elasticity of substitution parameter. The actual elasticity of substitution equals $1/(1-p)$.
Leontief (production function only)	$y = (x_1\alpha_1,\ x_2\alpha_2, \dots,\ x_n\,\alpha_n)$	"α_i" is the amount of industry output i needed for production of one unit of y
Cobb-Douglas	$y = Ax_1^{\alpha_1} x_2^{\alpha_2} \dots x_n^{\alpha_n}$	"A" indicates the general scale of production. "α_i" are elasticities of production with respect to each input.

1.6 References

Barnes, G. & Langworthy, P. (2003) Per Mile Costs of Operating Automobiles and Trucks. Humphrey Institute of Public Affairs, University of Minnesota. Available online at http://www.hhh.umn.edu/centers/slp/pdf/reports_papers/per_mile_costs.pdf.

Bates, J., Polak, J., Jones, P., & Cook, A. (2001) The Valuation of Reliability for Personal Travel, *Transportation Research Part E*, Vol. 37, 191–229.

Blincoe, L., Seay, A., Zaloshnja, E., Miller, T., Romano, E., Luchter, S., & Spicer, R. (2002) The Economic Impact of Motor Vehicle Crashes 2000. National Highway Traffic Safety Administration.

Bowes, D. R. & Ihlanfeldt, K. R. (2001) Identifying the Impacts of Rail Transit Stations on Residential Property Values. *Journal of Urban Economics* 50, 1–25.

Boyer, K. D. (1998) Principles of Transportation Economics. New York: Addison Wesley Longman.

Brownstone, D. & Small, K. A. (2005) Valuing Time and Reliability: Assessing the Evidence from Road Pricing Demonstrations. Department of Economics, University of California, Irvine.

Bureau of Labor Statistics (2011) Consumer Expenditure Survey, Bureau of Labor Statistics.

California Department of Transportation (Caltrans) (2010) Life Cycle Benefit Cost Analysis Model (Cal B-C). Available online at http://www.dot.ca.gov/hq/tpp/offices/ote/benefit.html.

Cohn, E. (1992) Returns to Scale and Economies of Scale Revisited. *Journal of Economic Education* 23(2),123–24.

Diesenhouse, S. (2005) Commercial Real Estate; Rising Steel Prices Force Changes in Construction Plans. New York Times. available online at http://query.nytimes.com/gst/fullpage.html?res=9905EEDF133EF930A25757C0A9639C8B63&pagewanted=1.

Energy Information Administration (2013) Monthly Energy Review, April 2013.

European Road Safety Observatory (2006) Cost-Benefit Analysis. European Commision, Directorate-General Transport and Energy. Available online at http://www.dacota-project.eu/Links/erso/knowledge/Fixed/08_measures/cost_benefit_analysis.pdf.

Florida Department of Transportation (2012) Generic Cost per Mile Models. Available online at ftp://ftp.dot.state.fl.us/LTS/CO/Estimates/CPM/summary.pdf

Gamble, H., Sauerlender, O., & Langley, J. (1974) Adverse and Beneficial Effects of Highways on Residential Property Values. *Transportation Research Record* 508, 37–48.

Gelles, G. M. & Mitchell, D. W. (1996) Returns to Scale and Economies of Scale: Further Observations. *Journal of Economic Education* 27(4), 259–261.

Gomez-Ibanez, J., Tye, W. B., & Winston, C. (1999) Essays in Transportation Economics and Policy: a Handbok in Honor of John R. Meyer. Washington, D.C. The Brookings Institution.

Greeene, D.L. and Plotkin, S.E. (2011). *Reducing Greenhouse Gas Emissions from U.S. Transportation*. Pew Center on Climate Change. Available online at: http://www.c2es.org/docUploads/reducing-transportation-ghg.pdf

Heiner, J. and Kockelman, K. (2005). Costs of Right-of-Way Acquisition: Methods and Models for Estimation. Journal of Transportation Engineering, 131(3), 193–204.

Intergovernmental Panel on Climate Change (IPCC) (2007) Climate Change 2007: Synthesis Report. Contribution of Working Groups I, II, and III to the Fourth Assessment Report of the Intergovernmental Panel on Climate Change. Geneva, Switzerland. IPCC. Available online at http://www.ipcc.ch/publications_and_data/ar4/syr/en/main.html.

Janke, B., Gulliver, J. S., & Wilson, B.N. (2011) *Development of Techniques to Quantify Effective Impervious Cover*. Center for Transportation Studies, University of Minnesota. Available online at: www.cts.umn.edu/Publications/ResearchReports/reportdetail.html?id=2058.

Kanafani, A. & Ghobrial, A. A. (1985) Airline Hubbing—Some Implications for Airport Economics, *Transportation Research Part A* 19(1), 15–27.

Levinson, D. & Smalkoski, B. (2003) Value of Time for Commercial Vehicle Operators in Minnesota. TRB International Symposium on Road Pricing, University of Minnesota. Available online at http://nexus.umn.edu/Papers/TruckValueOfTime.pdf.

Lewis, C. A., Buffington, J. L., Vadali, S. R., & Goodwin, R. E. (1997) Land Value and Land Use Effects of Elevated, Depressed, and At-Grade Level Freeways in Texas. TxDOT Research Report 1327–2.

Litman, T. (2009) Transportation Cost and Benefit Analysis (2nd ed.). Victoria Transport Policy Institute. Available online at http://www.vtpi.org/tca/.

Maibach, M., Schreter, C., Sutter, D., van Essen, H. P., Boon, B. H., Smokers, R., Schroten, A., Doll, C., Pawlowska, B., & Bak, M. (2008) Handbook on Estimation of External Costs in the Transport Sector, Delft, the Netherlands. Available online at http://ec.europa.eu/transport/sustainable/doc/2008_costs_handbook.pdf.

McCubbin, D. & Delucchi, M. (1996) The Social Cost of Health Effects of Motor-Vehicle Air Pollution. Institute for Transportation Studies, University of California, Davis. Available online at http://www.its.ucdavis.edu/publications/1996/UCD-ITS-RR-96-03(11).pdf.

Murphy, J. & Delucchi, M. (1998) A Review of the Literature on the Social Cost of Motor Vehicle Use in the United States. *Journal of Transportation and Statistics*, Vol. 1, No. 1, Bureau of Transportation Statistics, January 1998, 15–42.

National Safety Council (NSC) (2010) Estimating the Costs of Unintentional Injuries. Statistics Department, National Safety Council, Itasca, Illinois, and Children's Safety Network, Economics and Insurance Resource Center, Institute for Research and Evaluation. Available online at http://www.nsc.org/news_resources/injury_and_death_statistics/Pages/EstimatingtheCostsofUnintentionalInjuries.aspx.

North Central Texas Council of Governments (NCTCOG) (2011) Streamlined Project Delivery—Tower 55 Projects. Available online at http://www.nctcog.org/trans/spd/freightrail/tower55/index.asp.

Oregon Department of Transportation (2006) The Value of Travel Time: Estimates of the Hourly Value of Time for Vehicles in Oregon 2005. Available online at http://www.oregon.gov/ODOT/TD/TP/docs/publications/ValueTravelTime2005.pdf.

Pearce, D.W. & Markandya, A. (1989) *Environmental Policy Benefits: Monetary Valuation.* Organisation for Economic Co-operation and Development.

Pike, R. W. (2001) Optimization for Engineering Systems. Louisiana State University. Available online at http://mpri.lsu.edu/bookindex.html.

Polzin, S., Xuehao C., & Raman, V.S. (2008) Exploration of a Shift in Household Transportation Spending from Vehicles to Public Transportation, Center for Urban Transportation Research. Available online at http://www.nctr.usf.edu/pdf/77722.pdf.

Risbey, T., de Silva, H., & Tong, A. (2007) Road Crash Cost Estimation: A Proposal Incorporating a Decade of Conceptual and Empirical Developments. Bureau of Transport and Regional Economics, Department of Transport and Regional Services, Australian Government. Available online at http://www.bitre.gov.au/publications/96/Files/ATRF07_Road_Crash_Cost.pdf.

Rister, B. & Graves, C. (2002) The Cost of Construction Delays and Traffic Controls for Life-Cycle Cost Analysis of Pavements. Kentucky Transportation Center. available online at http://www.ktc.uky.edu/Reports/KTC_02_07_SPR197_99_SPR218_00_1F.pdf.

Samuelson, P. A. & Nordhaus, W. D. (2004) Economics (18th ed). New York, New York: The McGraw-Hill Campanies, Inc.

Schrank, D. and Lomax, T. (2009) 2009 Urban Mobility Report. Texas Transportation Institute. Available online at http://tti.tamu.edu/documents/mobility_report_2009_wappx.pdf.

Small, K. & Verhoef, E. (2007) The Economics of Urban Transportation. New York, Routledge.

Small, K. & Yan, J. (2001) The Value of "Value Pricing" Roads: Second-Best Pricing and Product Differentiation. *Journal of Urban Economics*, 310–336.

Small, K. A. (1992) Urban Transportation Economics. Harwood Academic Publishers.

Smith, S. (2011). Personal communication, November 21, 2011.

Squire, L. & Van Der Tak, H. G. (1975) Economic Analysis of Projects. International Bank for Reconstruction and Development, Washington, D.C.: Johns Hopkins University Press.

Texas Department of Transportation (2011). Value of Time and Road User Costs. Memo to District Engineers, May 11, 2011.

Truett, L. J. & Truett, D. B. (1990) Regions of the Production Function, Returns, and Economies of Scale: Further Considerations. *Journal of Economic Education* 21(4), 4113–19.

US Department of Transportation (USDOT) (2003) Revised Departmental Guidance: Valuation of Travel Time in Economic Analysis. Office of the Secretary of Transportation, U.S. Department of Transportation. Available online at http://ostpxweb.dot.gov/policy/Data/VOTrevision1_2-11-03.pdf.

US Department of Transportation, Bureau of Transportation Statistics (BTS) (2003) *Transportation Statistics Annual Report*. Available at: http://www.bts.gov/publications/transportation_statistics_annual_report/2003/html/chapter_02/economic_impacts_of_motor_vehicle_crashes.html.

US Department of Transportation, Federal Highway Administration (FHWA) (2007) Major Project Program Cost Estimating. Available online at: http://www.fhwa.dot.gov/ipd/pdfs/project_delivery/major_project_cost_guidance.pdf.

US Environmental Protection Agency (EPA) (2009) Ozone and Your Health. Available online at http://www.epa.gov/airnow/ozone-c.pdf.

US Environmental Protection Agency (EPA) (2011) 2011 U.S. Greenhouse Gas Inventory Report. United States Environmental Protection Agency. Available online at http://www.epa.gov/climatechange/emissions/downloads11/US-GHG-Inventory-2011-Complete_Report.pdf.

US Environmental Protection Agency (EPA) (2012) What is Acid Rain? Available online at http://www.epa.gov/acidrain/what/index.html.

Verhoef, E. & Small, K. (1999) Product Differentiation on Roads: Second-Best Congestion Pricing with Heterogeneity under Public and Private Ownership. Department of Spatial Economics, Free University Amerstam, and Department of Economics, University of California at Irvine. Available online at http://www.tinbergen.nl/discussionpapers/99066.pdf.

Wang, M. Q., Santini, D.J., & Warinner, S.A. (1995) Monetary Values of Air Pollutants in Various U.S. Regions. *Transportation Research Record 1475*, 33–41.

Welter, D., Ates, M., Loftus-Otway, L., Matthews, R., & Harrison, R. (2009) Estimating Texas Motor Vehicle Operating Costs. Center for Transportation Research, University of Texas at Austin. Available online at http://www.utexas.edu/research/ctr/pdf_reports/0_5974_1.pdf.

Zamparini, L. & Reggiani, A. (2007) Meta Analysis of the Value of Travel Time Savings: A Transatlantic Perspective in Passenger Transport. Springer Science & Business Media, LLC. Available online at http://www.springerlink.com/content/h663q51u448078x1/fulltext.pdf.

Chapter 2. Pricing of Transportation Services

2.1 Introduction

A crucial topic in understanding transportation markets, pricing is perhaps the most widely researched area within transportation economics. Transportation pricing refers to fees and financial incentives incurred by travelers, including (but not limited to) transit fares, cargo fees, fuel taxes, variable and flat rate tolls, parking fees, vehicle registration fees, and insurance payments. When the price of vehicles, parking, fuel, and transit fare change, travel activity patterns are affected.

Pivoting off of Chapter 1's discussion of the costs and benefits of transport, this chapter explores the interaction of demand and supply, congestion and costs, and tolls and revenues. While air and freight modes rely largely on private markets to balance supply and demand, roadways are provided largely as a public service and rely on fuel taxes as the primary source of funding. Travelers are accustomed to paying a premium for airfare during peak holiday travel periods and paying extra for parking during special events, when demand is high. Yet the concept of tolls that rise and fall with traffic demand is relatively new. As the gap between fuel tax revenues and road system provision and maintenance costs widens, tolling is becoming a popular consideration (for example, IH 10 high-occupancy toll [HOT] lanes in Houston, SR 91 Express lanes in Orange County, VMT-based pricing in Oregon, etc.). Furthermore, pricing strategies can help society internalize the negative externalities of congestion, air pollution, noise, and crashes.

Key Terms

- ✓ **ROH:** rule of half
- ✓ **CP:** congestion pricing
- ✓ **VOTT:** value of travel time
- ✓ **SRMC:** short-run marginal cost
- ✓ **LRMC:** long-run marginal cost

2.2 What is an Optimal Price?

The appropriateness of a pricing strategy depends on the goal at hand. Is it to keep traffic flowing at 60 mph or to recover roadway maintenance and bridge replacement costs for the next 10 years? Is it to incentivize the purchase of hybrid and electric vehicles or to reduce the number of high emissions vehicles in urban areas? A scheme designed to maximize the profit of a private business in an unregulated environment differs greatly from one designed to maximize social welfare. In order to pick a good tax or toll strategy, agencies must define their objectives. Simplifying the case to a single mode, optimal pricing can still be viewed from multiple perspectives.

In theory, each trip provides each user with some net benefit known as consumer surplus. **Consumer surplus** is the difference between the maximum price a consumer is willing to pay and the price he or she actually pays. The sum of all realized net benefits is the market's overall consumer surplus. In markets with lots of consumers (travelers), consumer surplus can be represented by the area under the demand curve and above the market price, as shown in Figure 2.1 and computed using the integral of the demand curve out to the quantity (number of trips) consumed and above the cost of the trip in Equation 2.1.

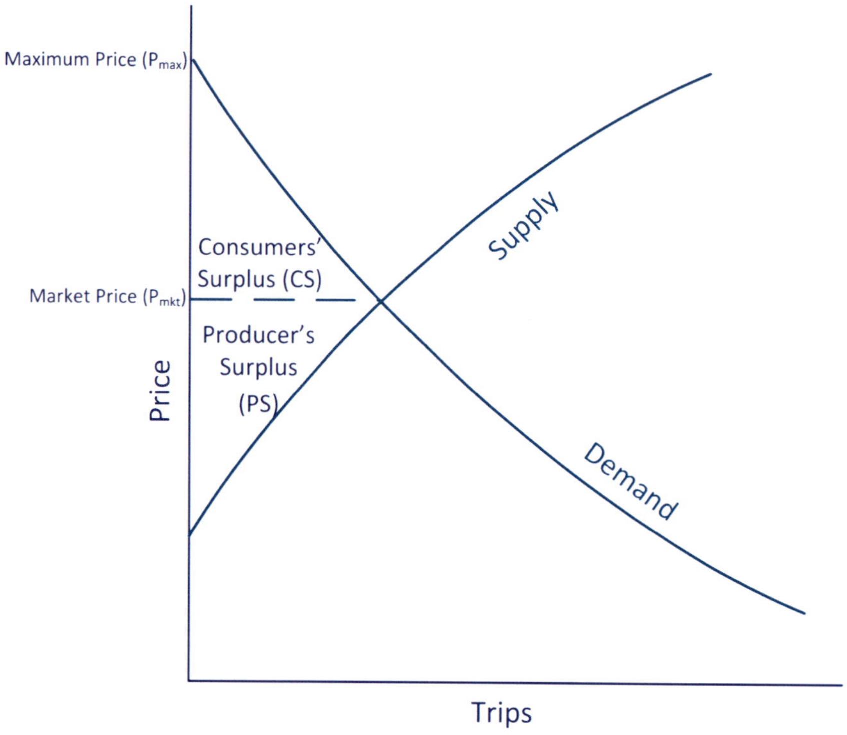

Figure 2.1: Surplus and Demand and Consumer Surplus

$$CS = \int_{Pmkt}^{Pmax} D(P)dP \tag{2.1}$$

Concept Example: Consumer Surplus

Assume vehicle operating and fuel costs are on average \$40 for a trip between Austin and San Antonio, and average value of travel time is \$12/hour for the 1.5-hour trip. Then the total cost per traveler for the trip can be approximated as

$$\$40 + (\$12 \times 1.5) = \$58/traveler$$

If, on average, the trip between Austin and San Antonio is worth \$60 to the traveler, then the \$2 difference between traveler cost and traveler's valuation of the trip is an individual consumer's surplus. If there are 40,000 travelers who make this trip on IH 35 each day, then the total consumer surplus (net travelers' benefit) would be worth \$80,000.

Producer surplus, on the other hand, is the amount of benefit that producers gain by selling at a higher market price than the least price they would be willing to sell for. As seen in Figure 2.1, consumer surplus is maximized when the price is zero, while producer surplus is maximized when the price is at a maximum. The actual market price generally falls between these two extremes, and the sum of consumer surplus and producer surplus is called **social welfare**.

Transportation Application: Traveler Welfare

Traveler welfare, sometimes called consumer or firm surplus, depends on direct user costs (such as fuel, tolls, fares, and vehicle maintenance), travel times (an opportunity cost), and the base attractiveness of various choices (time of day, destination, and mode). Savings in direct costs are direct benefits.

Estimation of Welfare Value

Changes in traveler welfare can be estimated using the Rule of Half (RoH). Figure 2.2 shows the gain in traveler welfare (surplus) when travel cost falls. The benefit to existing users equals Area 2, while the benefit to new users equals Area 1. The RoH assumes a linear demand function between before and after demand points for each origin-destination pair.

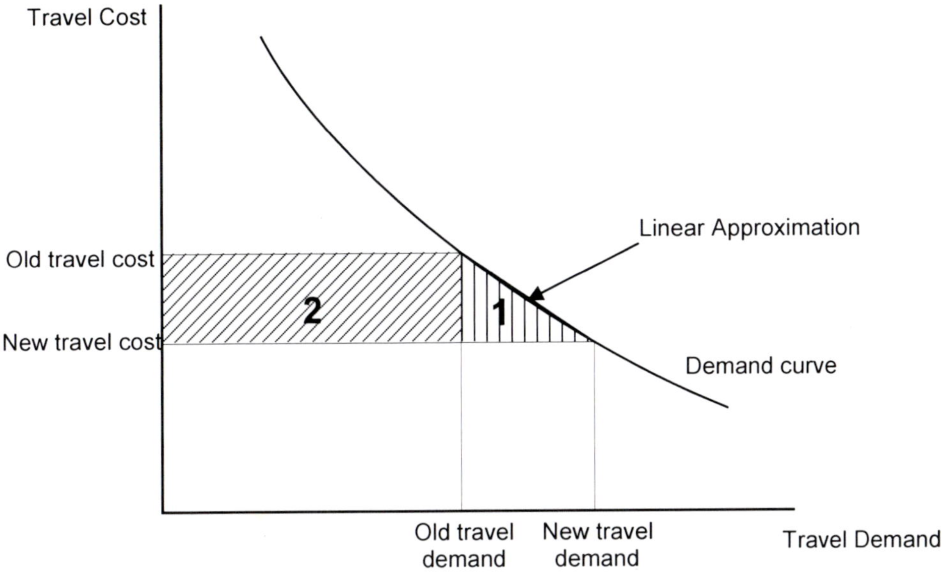

Figure 2.2: Rule-of-Half Showing Changes in Consumer Surplus (in Shaded Areas) as Travel Price Falls (and Demand Rises)

Maximizing Profit

Most private businesses wish to maximize profits, defined as the difference between total revenue (*TR*) and total cost (*TC*), as Equation 2.2 shows.

$$Profit = TR - TC = P_{mkt} \cdot Y - TC \quad (2.2)$$

Here P_{mkt} is the market price and Y is the quantity of goods sold, like vehicles per day on a tolled road or seats on an airplane between El Paso and Dallas-Fort Worth. Profit is maximized when the increase in total revenue generated by serving one additional trip is equal to the increase in total cost due to serving that one additional trip, or when marginal revenue (*MR*) is equal to the marginal cost (*MC*), as demonstrated in Equation 2.3.

See Chapter 1 for a complete discussion total cost, marginal cost, and average cost.

$$Max\ Profit: \frac{d(P_{mkt} \cdot Y)}{dY} - \frac{d(TC)}{dY} = P + \frac{dP}{dY}Y - MC = MR - MC = 0 \quad (2.3)$$

Because the demand for transportation is generally somewhat elastic, the higher the price charged, the fewer trips the consumers will make ($\frac{dP}{dY} < 0$), which translates into optimal profit-maximizing price marginal cost.

Maximizing Social Benefits

From a transportation agency's point of view, it is neither ideal to operate free of charge (transportation facilities not rationed by pricing will be rationed by congestion) nor to charge for maximum profit (as the goal of public institution is one of serving the community's best interests). Public agencies are to put their customers (the public at large) first. In theory, maximizing net social benefits ensures that transportation system resources are used optimally. Agency revenues (a benefit) offset user charges (a cost), and total social welfare (SW) is computed as shown in Equation 2.4:

$$SW = Profit + CS = TR - TC + CS \qquad (2.4)$$

By the law of diminishing marginal utility, the more trips a traveler makes between destinations A and B, the less he or she is willing to pay for each additional trip between those destinations. Across a market of potential consumers, the same downward sloping demand situation emerges, as marginal social benefit (or willingness to purchase) falls with quantity purchased. The optimal quantity served occurs when social welfare is at maximum, or when marginal social cost rises up to effectively cancel the marginal social benefit of the last unit purchased (i.e., that with the lowest marginal social benefit served), as expressed in Equation 2.5.

$$Max\ SW: \frac{d(P_{mkt} \cdot Y)}{dY} - \frac{d(TC)}{dY} + \frac{d(CS)}{dY} = P + \frac{dP}{dY}Y - MC - \frac{dP}{dY}Y = P - MC = 0 \qquad (2.5)$$

Marginal cost pricing allows for an economically efficient system, where social benefits are maximized.

Thus, the optimal price to maximize social welfare equals the marginal cost of producing the last unit sold/served. In effect, **marginal cost pricing** ensures there is no subsidy or waste in transportation service provision and transportation service use.

In general, marginal cost pricing for maximum social benefit (P_{SB}) falls between maximum consumer surplus pricing ($P_{CS}=0$) and maximum profit pricing (P_P), as Figure 2.3 depicts.

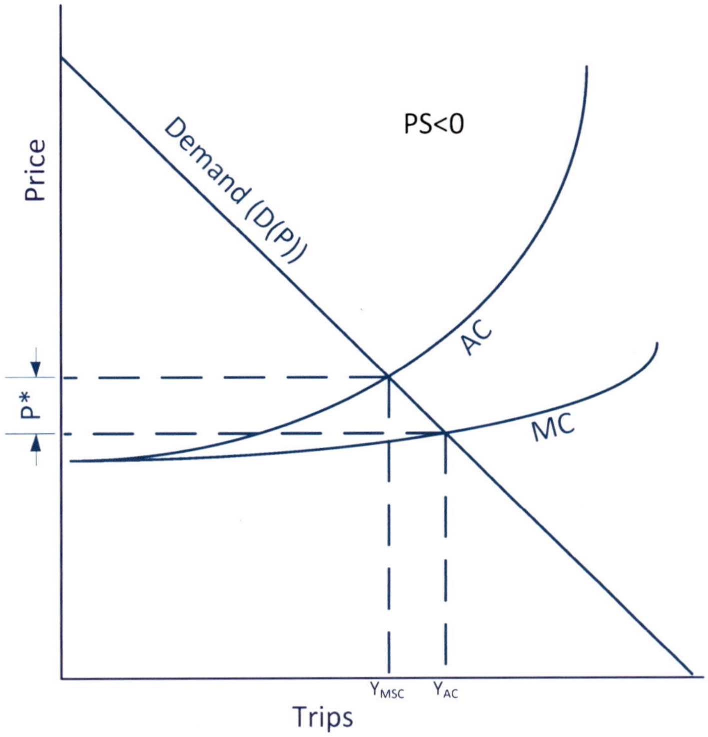

Figure 2.3: Marginal Cost Pricing

Yield Management

Due to the high capital cost of transportation investments, average costs tend to be higher than marginal costs, and so marginal cost pricing will result in financial losses for the operator. Fixed costs need to be recovered, which requires average-cost pricing, subsidies, or something called *yield management*. Frequently practiced by the airline industry, yield management involves price discrimination, as discussed below.

Yield management can involve various forms of legal price discrimination.

Price Discrimination

Price discrimination is the practice of a service provider varying the price of the same service for different users, provided that they cover their marginal costs. Generally, three kinds of price discrimination are related in economic textbooks:

- **First-degree price discrimination** varies the price based on each individual user's willingness or ability to pay. This is rarely practiced in reality.
- **Second-degree price discrimination** varies the price based on the quantity sold. In transportation, this translates to price differentiation by either the number of trips or the length of the journey. For example, regular transit users can buy an unlimited use card each month or year.
- **Third-degree price discrimination** varies the price based on the segment of the market or group of consumers. This is the most common form of price discrimination in transportation. Examples include first

class versus economy class fares, express versus regular bus services, and senior citizens' and student discounts.

Boyer (1998) described the near-perfect monopoly of Michigan's Mackinac Bridge, the only surface transportation route that connects the peninsulas. To recover the 1956 construction cost of $150 million, Boyer suggested different toll strategies that exemplify third-degree price discrimination:

1. Set a high standard toll targeted towards tourists and a discounted toll for local residents with residency certification.
2. Charge higher tolls on weekends (when tourists are more likely to use the bridge) and lower tolls on weekdays.
3. Offer lower tolls at off-peak periods to capture users.
4. Discount tolls on a particular day of the week, advertised locally in advance.
5. Distribute coupons through local media sources to capture price-sensitive residents.

Dynamic Yield Management

Dynamic yield management is similar to price discrimination in that it results in variable pricing. However, it is more of a scarce resource allocation strategy, whereas price discrimination emphasizes revenue maximization.

Concept Example: Dynamic Yield Management

Consider the example of airlines raising the prices on a flight as the number of remaining seats falls. The airline then allocates scarce remaining capacity to those with a higher willingness to pay. Airlines also employ minor product differentiation in offering first, business, and coach class fares. Because first-class service costs an airline more than coach class service, this fare structure differs from strict price discrimination (where the same product is priced differently to distinct consumer groups). Dynamic yield management arguably provides a more efficient market structure because firms offering a variety of services can offer lower unit prices than those specializing in a single service.

By allowing operators with high fixed costs (like high speed rail and airline operators) to operate at a financially feasible point, dynamic yield management retains a service that may otherwise be financially infeasible. The availability of the transportation service is still beneficial to all travelers, even if some pay more than others. The presence of competing carriers in all markets helps prevent monopolized pricing and excessive profits.

2.3 Roadway Pricing

Unlike most markets where price-setting firms provide all the economic resources (and thus bear all costs), transportation costs are shared among system users, the operators/owners, and non-users. For example, the total cost of highway transportation can be expressed as Equation 2.6:

$$TC = Y \cdot AC_{User} + Y \cdot T \cdot VOTT + FC_{Gov} + VC_{Gov} + SC_{Crash} + SC_{Env} \quad (2.6)$$

where $Y \cdot AC_{User}$ represents vehicle ownership and operating costs (AC_{User}) for all travelers (Y), costs such as gas, maintenance, insurance, licensing, financing, and registration, but excluding tolls. Here, $Y \cdot T \cdot VOTT$ represents the travel time costs, a function of travel time per trip (T) and value of travel time ($VOTT$), for all travelers (Y).

- FC_{Gov} represents the fixed costs of roadway infrastructure for the government agency or other highway owner. Like other cost components, the capital cost of building roads varies significantly based on geography and roadway-specific features.
- VC_{Gov} represents the variable costs of roadway infrastructure for the owner, including pavement maintenance, law enforcement, and emergency management.
- SC_{Crash} represents the social costs borne by both users and non-users from crash risk.
- SC_{Env} represents the social costs borne by both users and non-users from air, noise, water, and other environmental impacts.

Short-Run Marginal Cost Pricing

In short-run marginal cost (SRMC) pricing, all costs in Equation 2.6 are considered except the infrastructure cost, which is considered a fixed cost in the short term. Thus, the corresponding SRMC for each user can be expressed as Equation 2.7:

$$SRMC = MC_{User} + MT \cdot VOTT + MR \cdot VOR + MC_{Gov} + MSC_{Crash} + MSC_{Env} \quad (2.7)$$

Each component of the SRMC equation is discussed below.

See Chapter 1 for a detailed discussion of fixed versus variable costs, plus ways to estimate the MCs of transportation.

User Vehicle Operating Costs (MC_{User})

For automobile travel, marginal vehicle operation and ownership costs (MC_{User})—such as tire wear and vehicle depreciation—are private costs borne by the users. Vehicle capital ownership depreciation and interest costs can be calculated by standard discounting techniques, as discussed in Chapter 6.

Travel Time Costs

The marginal cost of travel time ($MT \cdot VOTT$) accounts for travel time costs borne by the marginal traveler and any delays he or she imposes on others. The term MT represents the change in total system travel time due to the addition of one more vehicle, which depends on existing congestion levels in the system. The Bureau of Public Roads (BPR) function relating travel time to the volume-capacity ratio is typically used to estimate changes in average travel time on individual links in the roadway system $(T_f - T_o)$. The overall change in link TT is thus $T_f \cdot V_f - T_o \cdot V_o$, as shown in Equation 2.8.

$$T_f = T_o(1+\propto \left(\frac{V}{C}\right)^5) \quad (2.8)$$

For efficient or optimal travel decisions, users imposing delays on others should, in theory, pay for such costs, in addition to their own travel time costs. One researcher makes the following point:

> *(E)ven though each user is simultaneously and instantly providing and consuming [his/her] own time...[he/she] is demanding an increase or decrease of other economic resources, namely, other individuals' time. Unless that change in demanded resources is paid or rewarded for, the market will be not be socially efficient (Jara-Diaz 2007, p.117).*

Wide variations are present in individual values of travel time (VOTTs). Because the value of travel savings typically is tied to wealth and wage rates, high-income individuals regularly exhibit VOTTs higher than those of lower-income individuals.

See Chapter 1 for details on estimating VOTT and VOR.

Schedule Delay Costs

In addition to placing value on travel time itself, individuals value certainty in information or expectations and being on schedule. Schedule-delay costs, like travel time costs, can be substantial. Such uncertainty can be estimated as a convex function of volume-capacity ratios, similar to a shifted version of the link performance function, as follows in Equation 2.9:

$$r_a = r_a^0\left(1 + \sigma\left(\gamma + \frac{v_a}{c_a}\right)^\tau\right) \quad (2.9)$$

where r_a^0 is the free-flow travel time variance of link a, and σ, γ and τ are function parameters. Fagnant et al. (2010) estimated these parameters to be $\sigma = 2.3$, $\gamma = 0.7$, and $\tau = 8.4$ from data for 2- to 5-mile-long freeway segments in Atlanta, Los Angeles, Seattle, and Minneapolis. Similar to calculations for travel time costs, the travel time unreliability multiplied by users' value of reliability (VOR) determines the total system (un)reliability costs.

Provider Service Costs

Marginal provider costs (MC_P) include maintenance and operation costs. Small and Verhoef (2007) estimated this cost to be \$0.019 per vehicle-mile using 2006 FHWA data, including law enforcement, crash response and system administration. The majority of MC_P is dominated by highway maintenance expenses, which are traditionally covered by state and local governments through title, license, and registration fees and subsidies from general tax revenues.

See Chapter 1 for estimates of marginal safety and environmental costs.

Safety and Environmental Costs

Crash costs (MSC_{Crash}) and environmental costs (MSC_{Env}) are external costs resulting from increases in crash risk and impacts to air, water, noise, and other ecological elements.

Long-Run Marginal Pricing

Though the idea of social welfare optimization using marginal cost pricing is well favored, there exists much debate about whether SRMC or long-run marginal cost (LRMC) pricing is more appropriate. SRMC does not account for the capital cost of roadway construction, which has significant implications for cost recovery and congestion. Potentially, the shorter the time period under consideration, the lower the SRMC, which can lead to minimal charges for road use that are ineffective in containing excess demand. However, Verhoef (2000) examined three models of long-run marginal *external* cost pricing involving the factors listed above and concluded that in the long run, SRMC pricing still provides economic efficiency by effectively controlling demand.

See Chapter 1 for a discussion of fixed vs. variable costs.

To account for capital costs, LRMC treats capacity either as a continuous variable or a discrete variable measurable in units such as lanes. In the short run, these costs are considered fixed and can be incorporated in road tolls based on average cost. Small and Verhoef (2007) used 2005 US Department of Commerce data to estimate the average capital cost at $0.056 per vehicle-mile for urban passenger vehicles. They accounted for depreciation of the entire U.S. highway capital stock assuming a 7% interest rate and 20-year infrastructure lifespan.

Static vs. Dynamic Pricing

The optimal tolling approaches discussed thus far assume flat-rate tolling or static pricing, appropriate for congestion levels that stay relatively constant through an extended period (such as the peak). In cases of more dynamic bottlenecking due to construction, crashes, or other extraordinary events, dynamic pricing strategies are more effective in substantially changing traveler behavior.

Consider the case of the single pure bottleneck, where no delays arise if inflow (V) to the system lies below capacity (C). The rate of queue, when a queue exists, and free-flow travel time through the bottleneck is zero. Under optimal dynamic tolling, no queue should exist and the bottleneck should flow at capacity until demand falls below C. The optimal toll influences the inflow (Q_i) and replaces delay cost. This dynamic bottleneck model implies that rescheduling departure times at trip origins as a result of roadway pricing can significantly improve system efficiency. Small and Verhoef (2007) discuss a triangle toll schedule using piecewise linear functions to toll optimally based on the expected travel time through the bottleneck. However, shifting schedules to accommodate optimal departure times may be infeasible without large investments in system administration.

Second-Best Pricing

First-best, or socially optimal, pricing requires perfect information on travel times and tolls and perfectly competitive markets, which cannot realistically be achieved. Travelers are generally not aware of all alternative routes and cannot anticipate a variable toll or congestion conditions before beginning their trip. A first-best pricing scheme assumes that all other competing modes and routes are priced at (optimal) marginal cost, which is rarely the case in transportation. When marginal cost pricing for delays imposed on others is applied on all roadways in a network, the externality of congestion is fully internalized and the socially optimal, or first-best, scenario is achieved. In reality, first-best pricing roadway is infeasible due to the existence of non-tolled roads near tolled facilities and the availability of competing modes of transportation (e.g., transit and bike) that are not priced at marginal cost.

Marginal cost pricing, or first-best pricing, is not optimal when users don't have perfect information and transportation markets are not perfectly competitive.

Additionally, while social welfare optimization through the principle of marginal cost pricing can be modeled "perfectly" in a conceptual network, designing a first-best pricing scheme for real-world applications faces many constraints. The model assumes access to perfect information on prices for travelers and costs for modelers and those setting the tolls. As one researcher states (Verhoef 2000, p. 319), "such a situation can only be realized if one would apply some 'Big Brother' type of electronic road charges, using very sophisticated technologies that can monitor the actual emissions, the place and time of driving, the driving style, and the prevailing traffic conditions; and that allows the regulator to adjust the charge accordingly." Aside from the political controversy that such a pricing scheme would surely cause, present-day technologies are still far from making the idea feasible in practice. Even with advanced technologies, pricing systems still have capital and operations costs that may offset many (and sometimes all) of the benefits achieved through rationalizing traffic flows and reducing delays.

Other issues even technology cannot resolve. For example, drivers will always have imperfect information due to uncertain events (such as traffic crashes) and lack perfectly substitutable routes and modes (and the available modes may be imperfectly priced). Marginal cost pricing is not optimal when drivers do not know all their options when choosing and/or altering their routes (such as all the local streets that could bypass a bottleneck, or all the variable tolls at any point in time).

When first-best pricing cannot be achieved due to market imperfections (including imperfect information), second-best pricing schemes cost-effectively are considered. **Second-best pricing** strives to maximize traveler (or societal) welfare based on marginal cost pricing principles while reflecting technological, political, and financial constraints. Many different second-best pricing strategies exist, but this discussion focuses on two simplified cases to illustrate the concept.

Case 1: Not All Links in a Network Are Tolled

Consider the case of two completely comparable routes in parallel, one tolled (T) and one untolled (U). In a perfect market, both routes are tolled and the number of trips on each will be roughly the same ($Q_U = Q_T = Q_{equilibrium}$). However, because one route is not tolled, many travelers will shift to the untolled route, due to its lower out-of-pocket costs so that $Q_U > Q_{equilibrium} > Q_T$. Marginal social cost of the tolled route ($P_T^* = TMSC_T - AC_T$) only captures the benefit of those trips on T and not the spillover trips now on U. Thus, the second-best toll for T in this scenario is shown in Equation 2.10:

$$P_T^{**} = (TMSC_T - AC_T) - (TMSC_U - AC_U)\left(\frac{Q_U - Q_{equilibrium}}{Q_T - Q_{equilibrium}}\right) > P_T^* \quad (2.10)$$

Here, the term $\frac{Q_U - Q_{equilibrium}}{Q_T - Q_{equilibrium}}$ represents the number of trips added to U per trip removed from T. When demand is elastic (responsive to tolls), this term will be less than one as some trips removed from T will no longer be made on either road. This term is also negative ($Q_T < Q_{equilibrium}$), so the second-best toll is the P_T^* plus a fraction of the marginal social cost price of U (representing recapture of the benefits of the trips lost by T to U).

The welfare gain generated by tolling just one route at P_T^{**} is greater than that generated by tolling P_T^* under the toll and untolled scenarios, but less than the welfare gained if *both* routes were tolled at marginal social cost and flow (first-best scheme). Additionally, the calculation of a second-best toll requires more information than the first-best scenario (demand and cost elasticities in addition to marginal external costs). Second-best tolling can result in more "mistakes" and less optimal toll settings, which can reduce traveler welfare.

A Verhoef and Small (2004) study of the efficiency of three express-lane pricing schemes found that relative welfare gain at P_T^{**} increased as the capacity of the express lane (as a fraction of total capacity of the highway) increased. This makes intuitive sense because zero express lane capacity corresponds to two parallel routes that are both untolled, while express-lane capacity of 100% corresponds to all routes, both tolled. For express lanes offering approximately one-third of the total capacity, the relative efficiency of the second-best toll (P_T^{**}) is about 30% of first best tolling. Surprisingly, they also found that if tolled capacity is less than 65% of total capacity, pricing at a "quasi first-best" P_T^* on the tolled route with an untolled parallel alternative actually harms travelers overall (Small and Verhoef 2007).

Deriving second-best pricing requires more information than first-best pricing.

The relatively simple case of the parallel routes demonstrates that the second pricing equation becomes more complex (requires more inputs) with more alternative routes, increasing the potential for error when strictly optimizing. However, with a well-calibrated travel demand model and appropriate measures of traveler benefits, second-best pricing can still be effective at managing travel demand. In general, expressways and major urban arterials with fewer substitutes make easier candidates for road pricing.

Case 2: Unable to Differentiate Price by User Group

As mentioned earlier, the costs each user imposes on other travelers varies with vehicle, driver, and traffic characteristics. Imagine the relatively simple case of two vehicle classes: those with high emissions (H) and those with low emissions (L), and $MEC_H > MEC_L$. When technology or political acceptance prohibits differentiation of pricing groups, a weighted average second-best toll (P_T) higher than $P_{L\text{-}MEC}$ and lower than $P_{H\text{-}MEC}$ can be assigned. In the case where the demand curve for low-emission vehicles and high-emission vehicles are identical, P_T is simply the average of $P_{L\text{-}MEC}$ and $P_{H\text{-}MEC}$. With the same P_T assigned to both vehicle classes, the low-emission vehicles are overpriced and the high-emission vehicles are underpriced, resulting in some welfare losses (represented by the shaded areas of the graphs in Figure 2.4) as compared to the first-best, group-specific toll scheme.

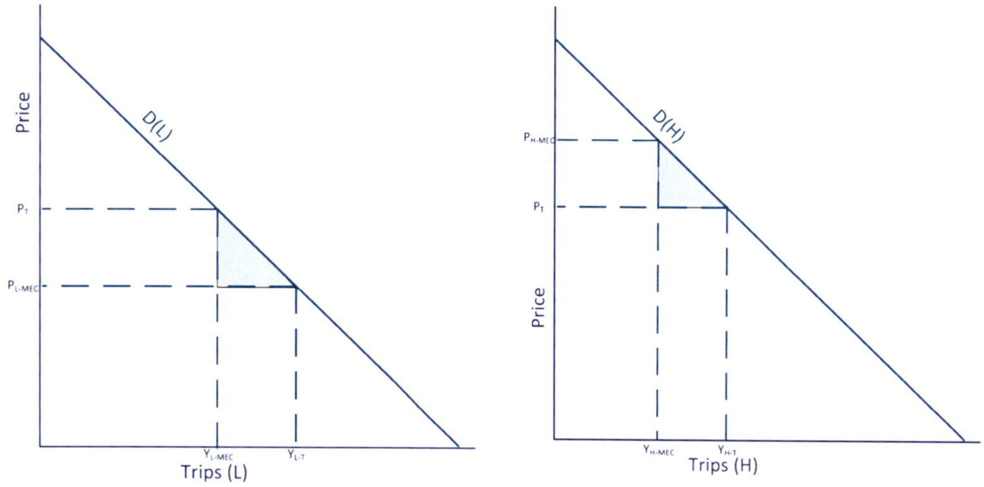

Figure 2.4: Second-Best Pricing With Two User Groups

Even though the two cases presented above are simplified, they highlight important characteristics about second-best pricing. Fundamentally, first-best is always better than second-best. The welfare gains of a second-best pricing scheme cannot surpass that of a first-best pricing scheme. In addition to not offering welfare-maximizing prices, placing identical tolls on both vehicle classes does not encourage drivers of high-emission vehicles to purchase lower emission vehicles. Moreover, second-best schemes require more information to formulate than first-best schemes, thus increasing the odds of imperfect pricing choice. However, second-best schemes can provide insight as to which factors are most influential in pricing and generally perform quite a bit better than pricing schemes that ignore indirect effects.

In addition to formal optimization methods, trial and error can also be used to determine second-best prices, such as the use of Safirova et al.'s (2007)

LUSTRE model for second-best tolling in the Washington DC area.

Implementing Roadway Pricing

As discussed in Chapter 5, most transportation revenues do not base user fees on marginal social-cost pricing. Automobile travel is currently underpriced because user fees do not account for most external costs. Traditional user fees include vehicle registration fees, which are independent of road use and fuel taxes. The flat-rate nature of federal and state fuel tax has not kept pace with increasing transportation capital and maintenance costs. Furthermore, as natural gas and electric vehicles (including hybrid electric vehicles) increase their market share, traditional fuel taxes will reflect smaller shares of actual road use. As fuel tax revenues fall, a move toward alternate funding instruments such as sales taxes and bonds (which are not related to roadway usage) pulls transportation finance farther away from marginal cost pricing, leading to a less efficient market in which demand does not reflect transportation costs. To better reflect user costs, the following pricing strategies are recommended by Litman (2011b):

- Achieve fuller cost recovery by increasing fuel taxes and tolls to reflect actual user costs.
- Utilize weight-distance fees to reflect roadway costs that are proportional to vehicle class as heavy vehicles cause more pavement wear and tear than light vehicles.
- Price roads variably to reflect congestion costs that vary with location, time, and vehicle type.
- Convert traditional fixed costs, such as insurance and registration fees, to variable costs based on annual usage.
- Consider new fuel taxes and emissions fees to reflect the environmental costs of driving.

See Chapter 5 for a detailed breakdown of transportation revenue and expenditure sources.

Congestion pricing and distance-based tolls seek to allocate travel costs based on usage. However, the public tends to perceive tolls as a new tax on existing, unpriced infrastructure. Researchers Podgorski and Kockelman (2005) conducted a survey of public perceptions of toll roads in Texas and found that Texans generally preferred to keep existing roads free of tolls, reduce tolls after recovering construction costs, charge higher tolls for trucks, and not implement congestion pricing. The survey also revealed that Texans who already commute on toll roads and during peak periods tended to support HOT lanes. Generally, frequent toll road users were more supportive of a wide range of transportation policies. Familiarity breeds new values and understanding.

The strategies of congestion pricing, allocation of joint and common costs, and VMT taxes for allocating cost based on usage are described in the following sections.

Congestion Pricing

Also known as *value pricing*, **congestion pricing (CP)** consists of pre-scheduled or truly dynamic tolls that vary in anticipation of demand or in response to actual demand, in order to avoid excessive congestion and delays. Utilizing the principle of supply and demand, CP manages congestion on roadways by adjusting the price (toll) to control the demand (traffic volume). VOTTs of added traffic vary based on trip type, household income, and other factors. The following example applies a uniform VOTT to illustrate a simple case of congestion tolling to account for the marginal cost of travel time.

Concept Example: Congestion Pricing

Given a 10-mile section of a 4-lane highway (8,000 veh/hr capacity) with a free-flow speed (FFS) of 60 mph, what is the optimal congestion price for the last or "marginal" vehicle entering? Assume the average VOTT is $12/hr.

The optimal congestion charge for the marginal vehicle is the added delay cost that the vehicle imposes on the entire system, which can be estimated using the BPR function discussed previously (Equation 2.8). Initial travel time (T_o) based on FFS is 10 minutes. Estimating that "C" level-of-service conditions (FFS) occur at 75% of capacity gives a value of 6,000 veh/hr. When the system has 7,999 vehicles, the average travel time is 14 minutes 44.30 seconds. When the system has 8,000 vehicles, the average travel time is 14 minutes 44.44 seconds. The difference seems minor until the time is summed up over all the vehicles in the system, showing an increased delay of 19 minutes for the whole system when the 8,000th vehicle enters—or with VOTT applied, about $4.

Dynamically priced lanes are generally designed to ensure reasonable flows and/or speeds subject to toll rate limits. For example, the IH 15 HOT lanes in San Diego have a maximum toll cap of $8 and SR 167 HOT lanes in Seattle have a maximum toll cap of $9 (DeCorla-Souza 2009). Presently, the U.S. has fewer than 10 HOT lanes with an explicit objective of moderating congestion via variable pricing. Several other locations have peak and off-peak pricing (e.g., New York bridges at $8 and $6), but the rate differentials tend to be very small (e.g., the New Jersey Turnpike, California's San Joaquin Toll Road, and Florida's Ft. Myers-Lee County route) with little to no effect on congestion (DeCorla-Souza 2009).

Examples

Dynamically Priced Lanes

In Southern California, pricing on SR 91 is pre-set, and thus not truly dynamic, but near-peak prices vary hourly, and by day of week and direction of travel. Friday afternoon eastbound travelers (between 3:00 and 4:00 p.m.) pay the highest tolls (at 97¢ per mile), while weekend fares are nearly flat (generally around 30¢ per mile between 8:00 a.m. and 10:00 p.m.). As with California's IH 15, SR 91 essentially aims for a "C" level-of-service target (Roth 2009, OCTA 2009), offering maximum flow at nearly free-flow-speed (zero-delay) conditions. California's SR 91 authority has been making small adjustments in rates almost every year, with little political fight, thanks to the size of the increments, regularity of past experiences, and clear existing policy (Samuel 2009). SR 91's policies appear to be the most public; monitoring for rate changes occurs on 12-week cycles (though, in practice, its toll schedules tend to change just once a year).

In contrast, California's IH 15 and Minneapolis's IH 394 offer truly dynamic pricing with capped rates. These caps do not change often, largely because they are set sufficiently high (Samuel 2009). Guidelines for changing these rates vary by location: while Minneapolis targets speeds of 50 to 55 mph along IH 394 (Roth 2009), the Acting Director of Denver's Colorado Tolling Enterprise has the authority to change IH 25 rates during certain times of day if needed, and the governing board can later adopt the rate increase in a concurring resolution.

Colorado's E-470 toll road is also governed by a board, and bond covenants have a toll rate structure built in, with periodic rate increases scheduled and subsequently approved by the board (Caitlin 2009).

Allocation of Joint and Common Costs

Allocating cost responsibility requires different methods for joint versus common costs and is a critical component in roadway tolls, shipping rates, and other modes of transportation.

Allocation of joint costs in a market with varying demand between points A and B, such as empty rail cars or trucks on many return trips, requires setting rates that reflect the different demand in each direction. Costs for "true joint products are produced in fixed proportions," which "means there can be no variability in costs making it logically impossible to specify the cost of, say, an outward journey when only the overall cost of the round trip is known" (Button 2010, pp. 80–81). Shipping rates along a low-demand direction (e.g., from B to A) tend to be less than the rates for the high-demand direction (e.g., A to B) in order to encourage the market to use the vehicles for the return trips. The joint costs are mostly allocated to the higher demand direction because marginal costs to serve return trips are relatively low.

Chapter 5 discusses traditional transportation revenue sources.

Allocation of common costs is an area of ongoing research and policy debate. Luskin et al. (2002) used four different cost allocation methods for allocating Texas highway costs to different vehicle classes. They concluded that fees and tax revenues from light vehicles (autos and pickup trucks) exceed their share of highway system costs (with revenue-to-cost ratios over 1), whereas combination trucks and buses do not (revenue-to-cost ratios less than 1). They also describe four different highway cost allocation methods and the desirable properties of such methods, as presented in Table 2.1 and Table 2.2.

Applied to our aggregated results, this criterion of fairness would lead to the conclusion that light vehicles—autos and particularly pickup trucks—are cross-subsidizing combination trucks and buses (Table 2.1).

Table 2.1: Desirable Properties of Highway Cost Allocation Methods (Source: Luskin et al. 2002)

Property	Description
Completeness	Highway costs are fully paid by highway users.
Rationality	Vehicle classes do not pay more than they would if they chose to be part of any smaller coalition of vehicle classes for which an exclusive facility is assumed available.
Marginality	Vehicle classes are charged at least enough to cover their marginal costs.
Consistency	Repeating the method gives consistent results.

Two of the approaches in Table 2.2 enjoy all four properties: the modified incremental approach and the generalized method. These were used by the Texas Department of Transportation in a 1994 allocation study, and are considered "superior methods" (Luskin et al. 2002). The incremental and proportional methods are considered more traditional methods.

Table 2.2: Highway Cost Allocation Methods (Source: Luskin et al. 2001)

Method	Description
Incremental Method	The cost of a highway facility designed for a class of vehicles is calculated. Then, in a process repeated until all vehicle classes have been assigned costs, the incremental cost of also serving the next class (by weight or some other measure) is determined and the added costs (marginal costs) are assigned to the added class. This method offers completeness, rationality, and marginality, but not consistency because cost allocation depends on the order of vehicle class additions (e.g., lightest to heaviest weight vs. heaviest to lightest weight).
Proportional Method	Vehicle classes are assigned the costs according to a measure of usage (such as vehicle-miles traveled and equivalent single-axle loads [ESALs]). This method offers completeness, but not rationality or marginality.
Modified Incremental Approach	Vehicle-miles traveled are typically used to proportionally divide the costs among classes of vehicles. First, non-overlapping cost portions attributable to each class are computed (e.g., costs associated with the particular vehicle class only). Then the overlapping cost portions for pairs or groups of classes are computed, followed by computation of overlapping costs for all the vehicle classes (e.g., costs common to all vehicle classes). The total costs allocated to one particular vehicle class are the non-overlapping costs plus the fraction of overlapping costs attributed to the vehicle class in each group of vehicle classes. This method satisfies all four properties.
Generalized Method	Based on the theory of cooperative games, the method uses a linear programming mathematical formulation to find a single point that allocates cost responsibilities to the vehicle classes. This method satisfies all four properties.

VMT with Taxes

Vehicle-mile or **VMT taxes** seek to more equitably charge for roadway usage based on distance (and ideally incorporating vehicle weight and emissions), rather than travel consumption taxes (which reflect fuel economy more than pavement damage and other costs.) Public services provided to vehicle users include policing, traffic lights, and emergency services, for which the costs are estimated to run about 1 to 4¢ per vehicle-mile. VMT taxes can also be priced to account for emissions or air quality impacts, habitat loss, stormwater management, and heat-island effects (Litman 2011b). Estimates of other external costs of light duty vehicles run about 2 to 5¢ per VMT. In contrast, current U.S. fuel taxes average about 2¢ per mile on a 20-mpg vehicle and 1¢ per mile on a 40-mpg hybrid electric vehicle (Litman 2011b).

The 2001 Oregon Legislature established a Road User Fee Task Force "to develop a design for revenue collection for Oregon's roads and highways that will replace the current system for revenue collection." The Oregon Department of Transportation (ODOT) conducted a 12-month study of two strategies for more efficiently collecting revenue: 1) replacing the gas tax with a mileage-based fee collected at gas stations, and 2) using this system to collect congestion charges. The study demonstrated the feasibility of collecting mileage fees at gas stations, and how different pricing zones can be established electronically with variable fees charged for driving in each zone at particular times of day. The pilot program tested two different kinds of mileage fees: a flat per-mile charge and a premium for travel in congested zones during peak hours. Despite paying a VMT fee equivalent to the current gas tax, the per-mile charge group reduced their VMT by about 12%; the congestion-fee group reduced peak period travel by 22% when compared to the per-mile group. Oregon's Road User Task Force recently recommended 2011 legislation for new fees on plug-in vehicle owners. Other states currently exploring VMT tax options include Alabama, California, Iowa, Indiana, Kentucky, Michigan, Minnesota, Utah, and Washington.

In 2011, The Netherlands tested a GPS-based VMT tax, which varied based on fuel efficiency, day, and route. Drivers were billed monthly, and proponents saw this system as a possible revenue generator in lieu of gas taxes and toll roads. Although politically unpopular, this approach is being considered by several other countries (Rosenthal 2011).

Freight Movements

Freight shipments play a critical role in the U.S. economy. Rail freight movements accounted for 15% of the value and 21% of the weight of total U.S. domestic freight activity while truck freight traveling on the Interstate system accounted for 62% of the value and 28% of the weight. Pipelines, maritime modes, and air carried the rest (BTS 2006). The FHWA estimates that domestic freight volumes will more than double between 2002 and 2035, with truck and air/truck combination modes expected to experience the fastest growth. Interstate and international deregulation over the last two decades has increased the number of U.S. operators.

Roadway pricing for freight is complicated by the geographic distribution of truck traffic. Highway corridors that experience the heaviest truck traffic run from Indiana to Illinois, Pennsylvania to New Jersey, Michigan to Ohio, and New Jersey to New York as illustrated by Figure 2.5's plot of the 2002 Freight Analysis Framework (FAF).

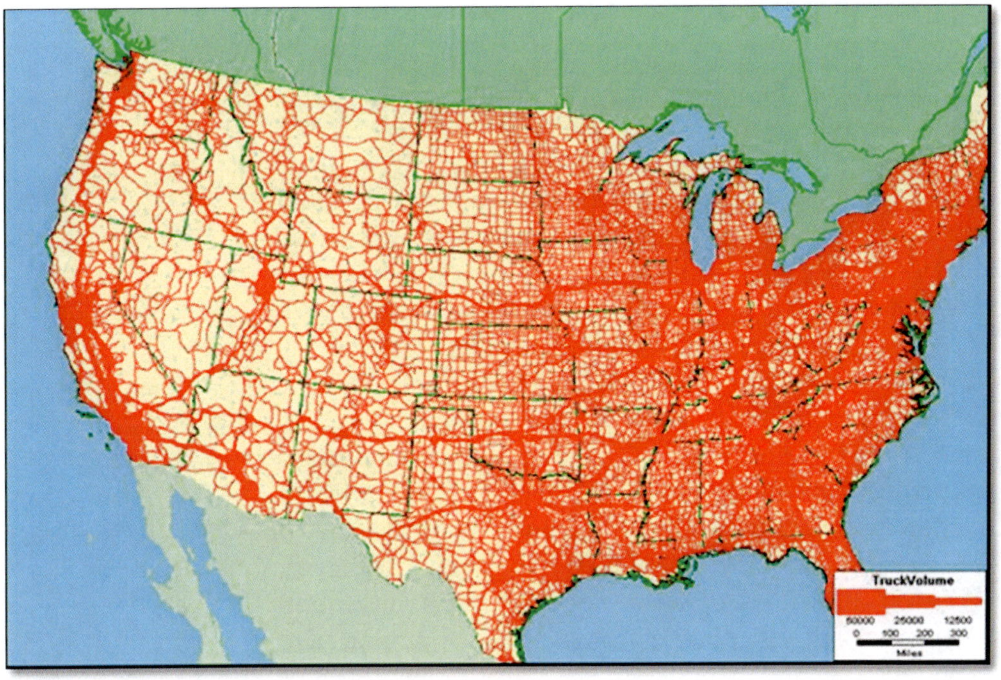

Figure 2.5: 2002 Truck Traffic Volume on U.S. Highway Network (Source: USDOT 2007)

Freight in Urban Areas

A 2008 American Transportation Research Institute survey approximates motor carrier marginal expenses at $83.68 per hour, with fuel and oil costs accounting for 39% of these costs; driver pay, benefits, and bonus payments account for 30%.

Due to high labor and equipment costs, the trucking industry is particularly sensitive to congestion. A 1999 NCHRP report surveyed truck operators and estimated values of travel time savings during congested conditions to range from $144 to $193 per hour and the value of schedule delay savings at $371 per hour (Small 1999). More recently, Texas A&M Transportation Institute estimated in 2007 the average cost of lost time and fuel for commercial vehicles in congestion to be $77.10 per vehicle-hour. Another study of operators estimated an average value of travel time to be $49.42 per truck-hour (Levinson & Smalkowski 2003). Additionally, trucks are high contributors of noise and air pollution, conditions already plaguing urban areas that will only worsen with increased truck traffic.

A study for freight demand management in the New York metropolitan area estimated the operator cost of off-hour deliveries to be about 30% less than delivering during regular business hours. Those carriers with fewer delivery stops were most inclined to participate in off-hour deliveries. The researchers estimated that tolls would be required to shift a significant amount of NY freight traffic to off-hours (Holguín-Veras et al. 2010).

All of these factors contribute to further incentives to price roadways for efficient use and reduce congestion for both commercial and non-commercial users in order to minimize delay, reliability, and transportation-related environmental costs. Some freight pricing mechanisms are discussed below.

Freight Pricing Mechanisms

The Tioga Group's (2011) extensive examination of freight-pricing options considered VMT fees, international trade fees, vehicle sales and excise taxes and fees, freight activity taxes and fees, highway tolls, and fuel-tax reforms. Each option was evaluated for technical and legal feasibility, institutional feasibility, revenue potential, cost, efficiency, environmental impacts, modal impacts, economic impacts, and political and public acceptance. The study concluded that waybill taxes and carbon taxes are not viable pricing mechanisms, at least not at the present time. But VMT fees based on distance, time, and location are one of the most technically viable options. Many trucks already report VMT (as required by the International Fuel Tax Agreement); thus, many fleets can implement distance-only VMT fees without GPS or cellular technology. But the compliance burden will still be great, particularly for smaller operators. VMT fees based on time and location require significantly higher incremental implementation, compliance, collection, and reinforcement costs. Time- and location-specific fees will also require GPS or other location technology, thus requiring longer implementation periods and facing more political and acceptance barriers, due to privacy concerns. However, time-of-day and location-based VMT fees provide a closer connection between congestion and roadway use.

Example

Freight Pricing Schemes in Europe

In 2001, Switzerland's flat-rate heavy-vehicle charge was replaced by a performance-related fee. All freight vehicles whose total weight exceeds 3.5 metric tons (about 7,700 pounds) are charged based on weight, and emissions rating. Austria followed suit in 2004 with a similar ETC system for trucks. Using satellite technology, Germany introduced a toll system for trucks in 2005. Operators have the choice of installing on-board units for automated tracking of truck movements or advance booking of truck routes (online or via computerized booking terminals).

2.4 Road Pricing's Impacts of Equity

The distinct roles that local, regional, state, and federal governments play in financing transport systems can lead to heated debates on how funds should be regulating and collected. Should revenues come from income, sales, and property taxes, or through user fees (such as tolls, transit fares, and fuel taxes)? Equity is a key criterion in this debate, for assessing the distribution of benefits and costs associated with transportation investments and policies. Litman (2011a) defines the following types of transportation equity:

- **Horizontal equity** measures whether individuals and groups considered equal in ability and need receive equal shares of resources and bear equal costs. Horizontally equitable public policies encourage cost-based pricing. This equal exchange of transportation spending versus revenue collections has been called *market equity*.

- **Vertical equity in income and social class** (also referred to as *social and environmental justice*) refers to impacts of transportation on different stockholders' incomes and socially disadvantaged groups. Even though transportation is a derived demand, it is essential in daily life and can be considered a right. Thus, policies favoring disadvantaged groups (such as tiered tax systems) are considered *progressive*, and policies that burden the disadvantaged (such as flat-tax systems) are called *regressive*. Policies that distribute funding to help bridge the accessibility and mobility gaps across different groups help improve vertical equity.

Addressing equity is complicated, in part because the goals of horizontal and vertical equity can be at odds in practice. While vertical equity favors subsidies for the disadvantaged, horizontal equity calls for economically efficient usage-based pricing. Equity is potentially achieved through economic efficiency, where users bear the costs they impose on society, unless a subsidy is justified. Because economically, socially, and/or physically disadvantaged people typically have less access to cars, policies that favor alternate modes help create vertical equity. Following are Litman's (2011a) suggestions for improving horizontal and vertical equity.

Measures to Improve Horizontal Equity

- Correct current planning biases that favor certain groups and specific modes (e.g., per capita funding and per trip funding tend to favor densely populated areas and the auto mode).
- Increase variable roadway user fees (tolls and fuel taxes) to reflect the actual cost of auto travel, which varies by time of day, location, and vehicle type.
- Price parking facilities and allow parking cash-outs for workers who choose cash over subsidized parking.
- Base vehicle insurance and registration fees on annual VMT.
- Consider environmental taxes and emissions fees so that drivers are accountable for the environmental externalities of auto use.

Measures to Improve Vertical Equity

- User-based pricing can be structured to favor economically, socially, and physically disadvantaged people (e.g., discounts and vouchers can be provided to those who qualify for low-income benefits).
- Reward alternative modes of travel by funding sidewalks, bike lanes,

and public transit.

- Support carsharing and bikesharing programs.
- Support multi-mode accessible land use patterns and locate public services in places accessible without a private car.

Congestion Pricing: Is It Equitable?

Public and political opposition to CP policies often arise based on vertical equity concerns. Because the traveler (and trip) population is heterogeneous, those with higher values of time tend to benefit the most from pricing (as time-savings values are more likely to exceed the toll). Referring back to Figure 2.2, even though the optimal price maximizes net welfare, users of trips between Y_{TMSC} and Y_{AC} have been "priced off" the toll road and are taking a less-preferred route or mode, or eliminating the trips altogether.

Example

A Case Study of Income Equity: Tolling vs. Taxation

Schweitzer and Taylor (2008) evaluated the income equity of a local option sales tax to fund roadways was evaluated against a tolling option for Orange County's SR 91 freeway. The local option sales tax was more popular politically, partly because the tax burden can be imposed at a low level across more people. However, shifting financing via tolls to a local sales tax was found to shift the cost burden away from middle-income travelers to those in low and high-income groups. In comparison to the sales tax, tolling was more progressive for lower-income households.

Redistribution of CP Revenue

Eliasson and Mattson (2006) examined three different revenue redistribution schemes of Stockholm's cordon charges. The study found that using CP revenues to lower value-added taxes benefitted high-income travelers the most and using the revenue for transit benefitted low-income travelers the most.

Although higher-income travelers use CP more often and bear most of the charges, a low-income traveler going by car in the peak direction during the peak hour is still more affected by the charges. Alternatives for these low-income travelers can be provided in the form of toll exemptions or rebates and discounted "lifeline" pricing based on income. For example, Kalmanje and Kockelman's (2004) credit-based congestion pricing (CBCP) grants drivers a monthly allowance of travel credits (typically monetized) to use on priced roads. The policy proposes drivers do not pay money out of pocket unless they exceed their allowance. Those spending less than their limit can use the credits later or exchange them for cash, bus passes, or other benefits. For drivers with special, socially desirable travel needs (e.g., welfare-to-work participants and single-parent low-income household heads), extra credits may be allotted. In essence, CBCP encourages travelers to budget their travel based on congestion. Similar to CBCP, DeCorla-Souza's (2000) Fair and

Chapter 10's Case Study 3 examines the economic impacts of CBCP.

Intertwined (FAIR) Lanes work on the basis of providing toll credits to those regularly using/needing the free lanes adjacent to tolled lanes. Accumulated credits allow for periodic free use of the tolled lanes.

Revenue redistribution, which can aid in bridging the gap in user benefits for different income groups, has also played a key role in increasing the political acceptability of CP. Finally, as populations and travel behaviors vary from region to region, the distributional impacts of pricing policies need to be evaluated on a city- and scheme-specific basis.

2.5 Summary

This chapter focuses on the concepts behind optimal transportation pricing. In other words, who should pay for transportation services and how? In order to implement appropriate pricing, decision-makers must first define their objectives. The concept of consumer surplus is critical in defining and solving for optimal marginal cost prices; system benefits are (theoretically) maximized when user prices are equal to marginal benefits received. Short-run marginal-cost pricing requires that tolls reflect vehicle operating costs, travel time costs, schedule delay costs, government service costs, motor vehicle crash costs, and vehicle emissions costs. In addition to these costs, long-run marginal-cost pricing also should reflect the capital costs of roadway construction. However, transportation markets do not offer perfect competition and the amount of information required to determine real-time, congestion-sensitive, and vehicle-specific pricing levels is unrealistic. This chapter describes two cases of second-best pricing, to maximize benefits in light of technological, political, and financial constraints.

The last part of the chapter departs from pricing theory to discuss real-world pricing applications. Pricing strategies to achieve fuller cost recovery include congestion pricing, highway cost allocation methods, and VMT fees. The chapter also examines equity issues that can arise from transportation pricing (and other) policies, such as impacts on specific socio-economic groups and/or people with special mobility needs.

2.6 References

American Transportation Research Institute (2008) An Analysis of the Operational Cost of Trucking. Arlington, Virginia.

American Trucking Association (2008) American Trucking Trends: 2007–2008. Arlington, VA.

Boos, M.A. & Moruza, A.K. (2008) Vehicle Miles Traveled (VMT) Tax: An Alternative to the Gas Tax for Generating Highway Revenue. Virginia Department of Transportation, Research Synthesis Bibliography No. 19. Available at: http://vtrc.virginiadot.org/rsb/RSB19.pdf

Boyer, K. D. (1998) Principles of Transportation Economics. New York: Addison Wesley Longman.

Button, K. (2010) Transport Economics. Northampton, MA: Edward Elgar Publishing.

Caitlin, P. (2009) Acting Director, Colorado Tolling Enterprise. Email communication. March 17.

DeCorla-Souza, P. (2000) FAIR Lanes: A New Approach to Managing Traffic Congestion *ITS Quarterly*, 3(2).

De Corla-Souza, P. (2009) Value Pricing Pilot Program Manager USDOT. Phone conversation. March 16.

Eliasson, J. & Mattsson, L. (2006) Equity Effects of Congestion Pricing: Quantitative Methodology and a Case Study for Stockholm. *Transportation Research*, A(40), 602–620.

Fagnant, D., Kockelman, K., & Xie, C. (2010) Anticipating Roadway Expansion and Tolling Impacts: A Toolkit for Abstracted Networks. Available online at: http://www.ce.utexas.edu/prof/kockelman/public_html/TRB11Toolkit.pdf.

Holguín-Veras, J., et al. (2010) Integrative Freight Demand Management in the New York City Metropolitan Area. Center for Infrastructure, Transportation and the Environment, Troy, New York. Available online at http://transp.rpi.edu/~usdotp/index.shtml.

Jara-Diaz, S. (2007) *Transport Economic Theory*. Amsterdam: Elsevier.

Kalmanje, S. & Kockelman, K. (2004) Credit-Based Congestion Pricing: Travel, Land Value and Welfare Impacts. *Transportation Research Record,* 1864, 45-53.

Levinson, D. & Smalkowski, B. (2003) *Value of Time for Commercial Vehicle Operators in Minnesota.* University of Minnesota, TRB International Symposium on Road Pricing.

Lindsey, R. & Verhoef, E.T. (2001) Traffic Congestion and Congestion Pricing In Button and Henshers (Eds.), *Handbook of Transport Systems and Traffic Control*. New York: Pergamon.

Litman, T. (2011a) Evaluating Transportation Equity: Guidance for Incorporating Distributional Impacts in Transportation Planning. Victoria Policy Institute. Available online at http://www.vtpi.org/equity.pdf.

Litman, T. (2011b) Socially Optimal Transport Prices and Markets: Principles, Strategies, and Impacts. Victoria Policy Institute. Available online at http://www.vtpi.org/sotpm.pdf.

Luskin, D., Garcia-Diaz, A., Walton, C. M., & Zhang, Z. (2002) Highway Cost Allocation in Texas: Executive Summary. Center for Transportation Research at University of Texas at Austin, Research Report 1810-3. Available online at http://www.utexas.edu/research/ctr/pdf_reports/1810_3.pdf.

Mohring, H. & Harwitz, M. (1962) *Highway Benefits: An Analytical Framework.* Evanston, IL. Northwestern University Press.

Orange County Transportation Authority (OCTA) (2009) 91 Express Lanes, Toll Schedules. Accessed July 2011 from http://www.91expresslanes.com/tollschedules.asp.

Oregon Department of Transportation (ODOT) (2011) Road User Fee Task Force. Accessed at http://www.oregon.gov/ODOT/HWY/RUFPP/ruftf.shtml.

Pigou, A. C. (1920) The Economics of Welfare. London: Macmillan.

Podgorski, K. & Kockelman, K. (2006) Public Perceptions of Toll Roads: A Survey of the Texas Perspective. *Transportation Research,* 40A, 888–902.

Rosenthal, E. (2011, August 10) In Auto Test in Europe, Meter Ticks Off Miles, and Fee to Driver. *The New York Times*.

Roth, G. (2009) The Independent Institute. Email communication. March 16.

Safirova, E., Houde, S., & Harrington, W. (2007) Marginal Social Cost Pricing on a Transportation Network: A Comparison of Second-Best Policies. *Resources for the Future,* DP 07–52.

Samuel, P. (2009) Editor TOLLROADnews. Phone conversation. March 12

Schweitzer, L. & Taylor, B. (2008) Just Pricing: The Distributional Effects of Congestion Pricing and Sales Taxes. *Transportation,* 35(6), 797–812.

Small, K. (1999) NCHRP Report 43: Valuation of Travel-Time Savings and Predictability in Congested Conditions for Highway User-Cost Estimation. Transportation Research Board, Washington, DC.

Small, K. A. & Verhoef, E. T. (2007) The Economics of Urban Transportation. New York, Routledge.

Taylor, B. D. (2004) The Geography of Urban Transportation Finance. In S. Hanson and G. Giuliano (Ed.), *The Geography of Urban Transportation.* New York, Guilford Press.

Texas A&M Transportation Institute (2007) 2007 Annual Urban Mobility Report. College Station, Texas. http://www.commutercars.com/downloads/UrbanMobility07.pdf

The Tioga Group (2011) New Dedicated Revenue Mechanisms for Freight Transportation Investment. Presentation P11-0307 at Transportation Research Boards 90th Annual Meeting. Washington, DC.

US Department of Transportation, Bureau of Transportation Statistics (BTS) (2006) North American Freight Transportation. Washington, DC. http://www.bts.gov/publications/north_american_freight_transportation/pdf/entire.pdf.

US Department of Transportation, Federal Highway Administration (FHWA) (2000) Comprehensive Truck Size and Weight Study. Available online at http://www.fhwa.dot.gov/reports/tswstudy/.

US Department of Transportation, Federal Highway Administration (FHWA) (2007) FAF^2 Freight Traffic Analysis. Available online at http://ops.fhwa.dot.gov/freight/freight_analysis/faf/index.htm.

US Department of Transportation, Federal Highway Administration (FHWA) (2009) Freight Facts and Figures 2009. Accessible online at http://www.ops.fhwa.dot.gov/freight/freight_analysis/nat_freight_stats/.

Verhoef, E. (2000) The Implementation of Marginal External Cost Pricing in Road Transport: Long Run vs. Short Run and First-Best vs. Second Best. *Papers in Regional Science,* 79, 307–332.

Verhoef, E. & Small, K. (2004) Product Differentiation on Roads: Constrained Congestion Pricing with Heterogeneous Users. *Journal of Transport Economics and Policy,* 38, 127–156.

Vickrey, W. S. (1963) Pricing in Urban and Suburban Transport. *American Economic Review, Papers and Proceedings,* 53, 452–65.

Whitty, J. M. (2007) Oregon's Mileage Fee Concept and Road User Fee Pilot Program: Final Report. Oregon Department of Transportation. Available online at: http://www.oregon.gov/ODOT/HWY/RUFPP/docs/RUFPP_finalreport.pdf.

Wiseman, J. (1957) The Theory of Public Utility Price—An Empty Box. *Oxford Economic Papers,* 9, 56–74.

Chapter 3. Regulation and Competition

3.1 Introduction

As the U.S. transportation system has evolved over the decades, the policies and regulations surrounding its use have developed accordingly. Regulatory policy affects competition, which is addressed as a major concept in this chapter. The variation of competition by market (air, trucking, and public transit) and the effects on final price are discussed and presented as examples of how regulations can impact markets and affect users. Landmark cases involving regulation (and deregulation) are discussed in the context of their impacts on the public and private sectors. Concepts of competitive behavior are also applied to the construction industry, where guidelines are presented for improving bid selection and reducing costs.

Key Terms

- ✓ **NEPA:** National Environmental Policy Act

3.2 Regulations

To govern and control procedures and behaviors, governments set regulations that fall into two general categories: economic and social. **Economic regulation** traditionally has been designed to prevent monopolistic behavior in private-sector firms by controlling:

Regulation is defined as "controlling human or societal behavior by rules or restrictions" (Koops et al. 2006)

- Maximum prices
- Rates-of-return on investment (profits)
- Conditions of service provision
- Market entry and exit
- Mergers and acquisitions, and
- Accounting practices

In transportation, independent regulatory agencies—such as the Interstate Commerce Commission and the Civil Aeronautics Board—were established by the U.S. Congress to first investigate and then render administrative decisions on the above-mentioned aspects of railroad, motor carrier, and airline behavior.

The federal government also exercises some control over private-sector activities relating to health, product and worker safety, and the environment. Such **social regulatory** tools include:

- Promulgation of standards
- Financial penalties
- Outright prohibition of harmful activities, and
- Requirements to monitor and measure adverse impacts

The extent of social regulation has grown over the years as the general public has become increasingly alarmed by any number of issues related to the pollution of the environment, defective products, worker safety, hazardous materials, and highway crashes.

The many regulations in the transportation sector address a wide range of issues such as the environment, safety, and workers' wages, which are discussed in more detail below.

Environmental

The transportation sector accounted for nearly 30% of total U.S. energy consumption and greenhouse gas emissions (GHG), as Figures 3.1 and 3.2 show.

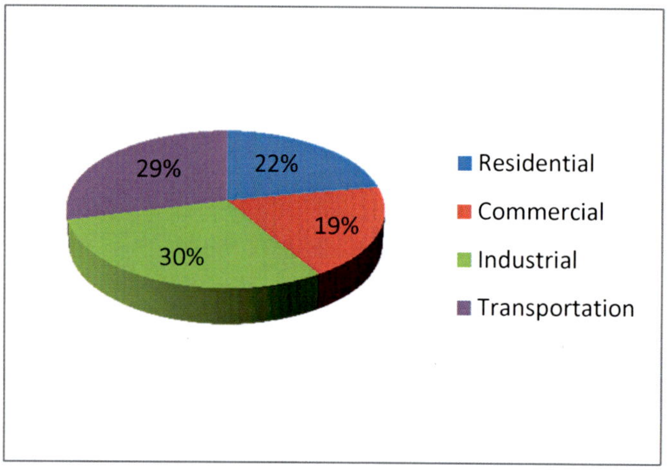

Figure 3.1: End-Use Sector Shares of Total Energy Consumption, 2009 (Source: EIA 2009)

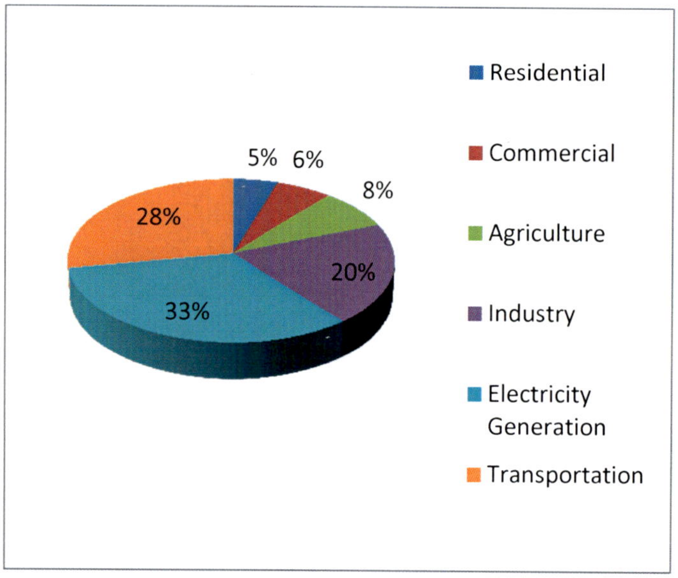

Figure 3.2: Percentage of U.S. GHG Emissions from Different Sectors, 2006 (Source: EPA 2008)

Regulating Emissions and MPG

Many regulations govern vehicle emissions and fuel consumption, as well as air quality, which is impacted by transportation systems. Following the 1973 oil crisis, fuel economy was first regulated in the form of Corporate Average Fuel Economy (CAFE) standards. As Figure 3.3 shows, the average new passenger vehicle fuel economy has improved from 13.5 miles per gallon

(MPG) in 1975 to 25.8 MPG in 2010. Manufacturers are under continuing pressure to improve fuel economies as standards increase over time; light-duty vehicles must achieve over 35 MPG by 2016 (USDOT 2010) and proposed standards of nearly 55 MPG by 2025 (NHTSA 2011d). Despite this increase, average heavy and medium duty truck fuel economies (as well as those of motorcycles and recreational vehicles) have remained relatively constant over the same period. Federal regulators (the Environmental Protection Agency [EPA], USDOT, and National Highway Traffic Safety Administration [NHTSA]) addressed part of this issue in 2008 by setting CO_2 emissions and fuel consumption standards for new medium- and heavy-duty vehicles.

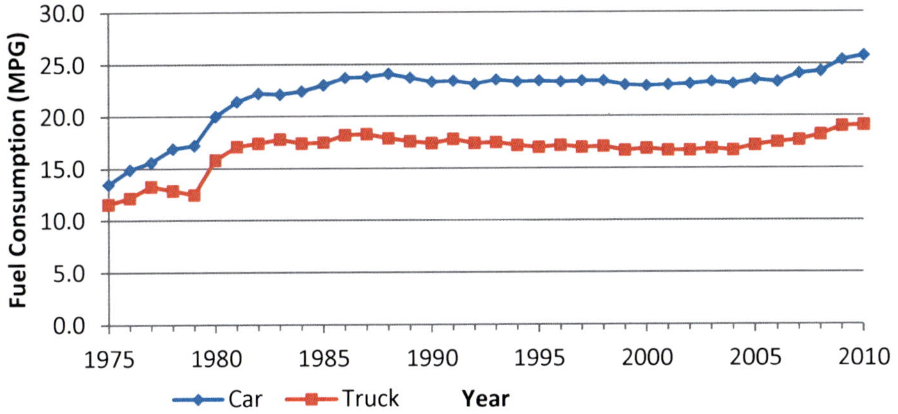

Figure 3.3: Vehicle Fuel Consumption Trend from 1975 to 2010 (Source: EPA 2011b)

Despite ambitious plans to improve fuel economy in the future, the U.S. currently lags well behind the rest of the world in fuel economy standards, and will likely continue to do so, even with the newly approved U.S. standards of 35 MPG by 2016. Figure 3.4 compares global light-duty vehicle fuel economies, converted to the U.S. CAFE test cycle standards.

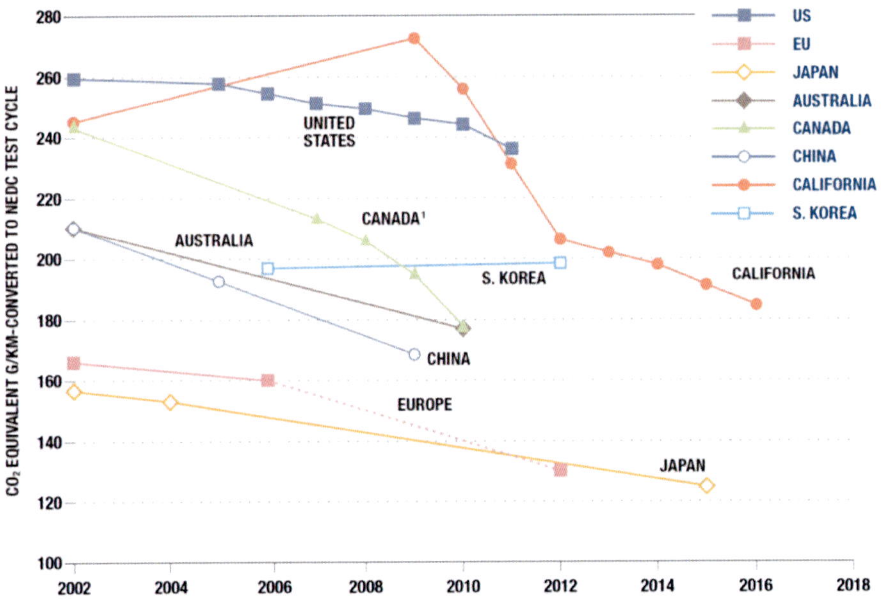

Figure 3.4: Global Fuel Economy Standards (Source: An et al. 2007)

The EPA limits new-vehicle emissions of non-methane hydrocarbons (NMHC), carbon monoxide (CO), oxides of nitrogen (NO_x), particulates, and formaldehyde. If a manufacturer's products do not meet emission standards, the EPA can mandate a recall under Section 207 of the Clean Air Act. In addition to new vehicle regulations, conformity-based policies can also govern current fleet emissions. As vehicles age, their emissions control devices become less effective, so inspection and maintenance (I&M) programs exist in many U.S. non-attainment regions. Manufacturers must ensure emissions meet standards for the "useful life" of the vehicle, which the EPA has set at 120,000 miles for cars and light trucks (EPA 2006). The EPA's emissions regulations have significantly improved air quality across the U.S. Since 1970, on-road CO and PM emissions have decreased by 50% and hydrocarbon emissions by 30% (EPA 2012). NO_x emissions have increased, but are projected to be less than 1970 levels by 2020 (EPA 2012).

Infrastructure's Role in Emissions

Transportation infrastructure also influences emissions by changing travel patterns, shaping land use, and affecting speeds and acceleration profiles. Billions of dollars are invested in transportation infrastructure each year to improve connections between housing, jobs, and other destinations. The location and timing of constructing new and expanding existing roads, airports, railroads, and other transportation facilities can change travel modes and distances, and, as a result, impact energy consumption and conformity with National Ambient Air Quality Standards (NAAQS). The Clean Air Act and National Environmental Policy Act processes require evaluation of all significant environmental impacts before FHWA funds can be used.

Transportation infrastructure also influences vehicle emissions by changing travel patterns and shaping land use.

National Environmental Policy Act (NEPA)

NEPA created a framework for environmental policy in the U.S. by requiring impact assessments for projects involving federal agencies. NEPA requirements apply to state and local transportation projects involving federal funding or approval. NEPA requires agencies to analyze social, economic, and environmental impacts, consider alternatives, inform and involve the public, and implement measures to avoid, minimize, or mitigate environmental impacts. Sensitive areas and resources addressed by NEPA include floodplains, historic and archeological sites, wetlands, endangered species, parklands, and wildlife habitats. Under NEPA, projects must provide an environmental impact statement (most extensive), an environmental assessment (less stringent), or prove the project will have minimal impacts and thus qualifies for categorical exclusion (CE).

An environmental impact statement (EIS) is required for any project or action that may significantly impact the environment, typically for projects such as new freeways or new separated high-occupancy vehicle (HOV) or bus lanes, as well as rapid, light, or commuter rail facilities. By first notifying the public of project intent and receiving input in return, as well as coordinating with other affected agencies, transportation agencies can determine whether the project will have significant environmental impacts. If such impacts are expected, a draft EIS is prepared. Otherwise an environmental assessment or categorical exclusion is required. An EIS must be approved by the FHWA before a project can begin.

Pursuing projects eligible for CE or a Finding of No Significant Impact (FONSI) avoids extensive analysis costs and expedites implementation. Examples of projects eligible for categorical exclusion are as follows:

- Highway resurfacing, restoration, rehabilitation, reconstruction.
- Adding shoulders and auxiliary lanes for parking, weaving, turning, or hill climbing.
- Highway safety or traffic operations improvements, including ramp metering devices and lighting.
- Bridge rehabilitation, reconstruction, or replacement and grade separations to replace existing at-grade railroad crossings.

Safety

As shown in Figures 3.5 and 3.6, almost 38,000 people are killed and over 2 million injured in motor vehicle crashes each year in the U.S. While these numbers appear to be trending downward (in part due to an economic recession), traffic crashes deprive the nation of more person-years and young lives than almost any other activity or disease.

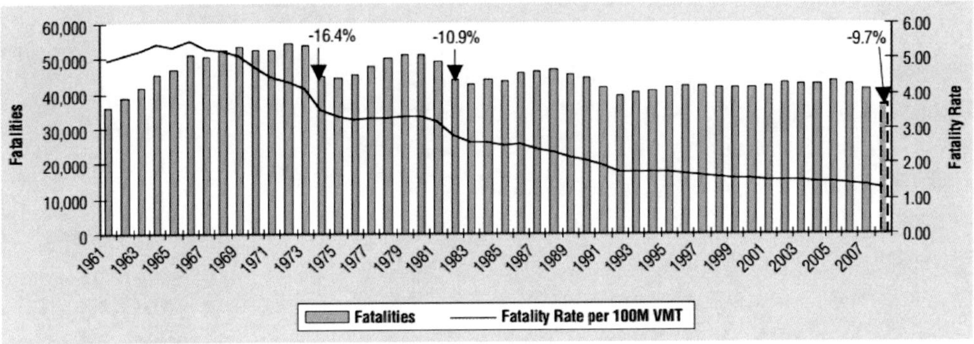

Figure 3.5: Fatalities and Fatality Rates per 100 Million VMT from 1961 to 2008 (Source: NHTSA 2009)

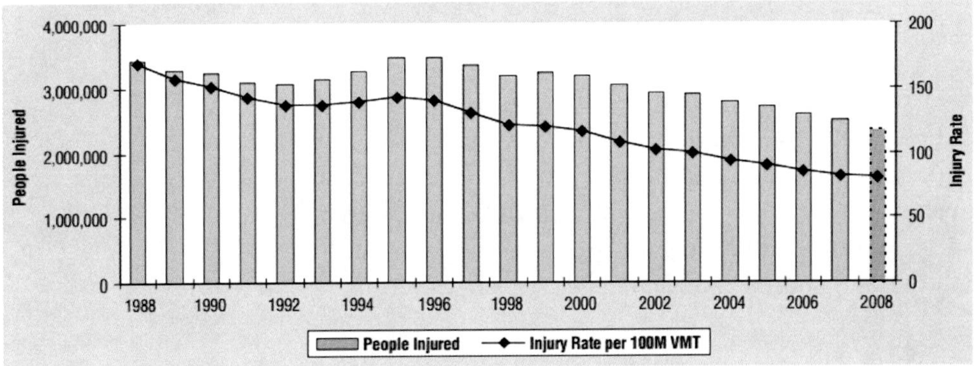

Figure 3.6: People Injured and Injury Rate per 100 Million VMT from 1988 to 2008 (Source: NHTSA 2009)

In 2000, U.S. motor vehicle crashes cost $230 billion (NHTSA 2008). Approximately half of these costs come from direct market productivity losses and property damage, as shown in Figure 3.7. Medical care costs and emergency services account for another 14% of all costs, while travel delays caused by crashes are estimated to account for 11% (NHTSA 2002).

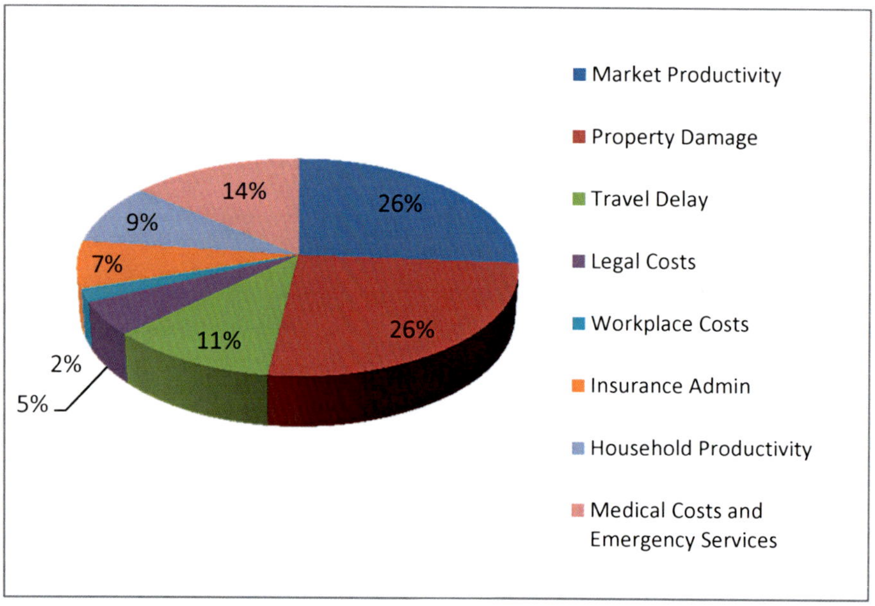

Figure 3.7: Distribution of Motor Vehicle Crash Cost (Source: NHTSA 2002)

Reducing Fatalities

The NHTSA regulates vehicle design, including air bags, brakes, car seats, seat belts, and tires. In September 1997, the NHTSA required that all passenger cars be equipped with air bags. Then, in January 1999, the NHTSA required all multipurpose passenger vehicles, trucks, and buses heavier than 10,000 pounds to be equipped with anti-lock brake systems. If new vehicles do not meet all safety standards, the NHTSA can require a recall. In addition to safety regulations for new vehicles, current fleet safety components are inspected during annual registration renewal to reduce existing or potential safety deficiencies on the road.

In an effort to reduce vehicle fatalities and injuries, several types of regulations exist to govern vehicle and highway design safety regulations.

In another step to increase traveler safety, transportation engineers identify and survey crash sites to make those locations safer. The Safe, Accountable, Flexible, Efficient Transportation Equity Act: A Legacy for Users (SAFETEA-LU) established the Highway Safety Improvement Program (HSIP) to allow states to use funds to correct or improve hazardous road locations and address other highway safety problems. In 2009, the funds available for the HSIP program totaled about $1.3 billion (FHWA 2011). The HSIP requires each state to develop and implement a Strategic Highway Safety Plan (SHSP) to improve highway design, construction, and maintenance so that the number of traffic crashes and costs will fall.

Designing Safer Work Zones

In addition to vehicle and highway safety regulations, designing safer work zones is a priority for improving transportation safety. In 2008, the U.S. had 716 work zone fatalities (The National Work Zone Safety Information Clearinghouse 2011). To decrease the number of fatalities in work zones, the National Cooperative Highway Research Program (NCHRP) released a guide to address work zone safety with these recommended strategies:

1. Implement improved methods to reduce the number and duration of activities.
2. Adopt improved procedures to ensure more effective practices including traffic control devices for managing work zone operations.
3. Enhance and extend training for the planning, implementation, and maintenance of work zones to maximize safety.
4. Enhance safe work zone driving through education and enforcement actions.

Workers' Wages

While the Fair Labor Standards Act (FLSA) covers most U.S. workers, those who work for contractors performing federally funded construction, alteration, or repair projects in excess of $2,000 are covered by the Davis-Bacon and Related Acts (DBRA). The FLSA requires that nonsupervisory private sector employees be paid at least $7.25 per hour and DBRA minimum wages vary by job type, as illustrated in Table 3.1.

Table 3.1: Highway Construction Workers' Minimum Wage for Federally Funded Projects (USDOL 2011)

Activity	Minimum Wage
Asphalt Distributor Operator	$11.45/hr
Asphalt Raker	$9.30/hr
Bulldozer Operator	$11.80/hr
Laborer, Common	$8.69/hr
Scraper Operator	$10.29/hr

3.3 Deregulation

Perceived economic regulatory failure can become a catalyst for regulatory reform.

Deregulation is the removal or simplification of government rules and regulations to facilitate a more efficient operation of markets. Deregulation acts seek to relax or remove excessive government control. Economic deregulation has not been total and varies by mode. However, its net result has been positive for freight and long-distance passenger transport growth. Most importantly, regulatory reforms have enabled an increased interaction and cooperation between modes (intermodalism). The following sections discuss the effects of deregulation on the operation of these industries.

Railroad and Motor Carrier Deregulation

The U.S. freight network links businesses with suppliers and markets by moving an incredible volume of goods each year. In 2009, over 16 billion tons of goods, worth about $14.6 trillion, were moved (Figure 3.8) (Center for Transportation Analysis 2011). The truck and rail modes carry the largest portion of freight shipments. Bulk goods—such as grain, coal, and ores—have a large share of the tonnage moved on the U.S. freight network. High-value, high-velocity goods such as electronics, machinery, textiles, and leather comprise a large share of the value moved.

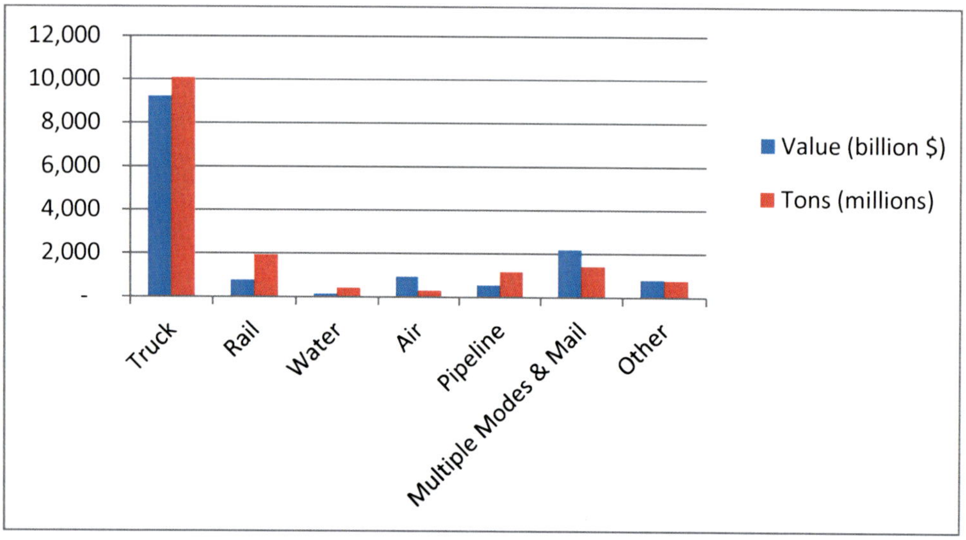

Figure 3.8: U.S. Freight Shipments by Mode in 2009 (Source: Center for Transportation Analysis 2011)

Before 1980, heavy rail and trucks were subject to rate, entry, and exit regulations, and freight rates were based on the value of the commodities shipped (Winston et al. 1990). Regulations increased costs for potential new carriers, which were already substantial, due to the cost of vehicles, along with ROW, rail yards, and track, in the case of railroads. In addition, a railroad's exit from the market was costly and time consuming, as the Interstate Commerce Commission would not approve an abandonment request if shippers or local governments opposed the decision (Winston et al. 1990). As a result, one-third of the U.S. rail industry was bankrupt or close to failure by 1975 (Rodrigue et al. 2009).

Moreover, trucking firms could enter the market only if they could justify their entry as necessary and convenient to the public. If existing carriers could make a case proving that new carrier entry would hurt them financially, new entry applications could be denied. In addition, truckers were required to have route- and commodity-specific operating authority. For example, if a carrier with the authority to go from Austin to Houston and Houston to Denver wanted to carry goods from Austin to Denver, it had to carry goods through Houston, even though the direct route was shorter. Such regulations raised costs, and shippers often had to deal with multiple carriers because each carrier had limited location authority.

To overcome such inefficiencies, the Staggers Rail Act of 1980 allowed direct negotiation between railroads and shippers, enabling them to set rates and facilitate entry to and exit from the railroad industry. The Motor Carrier Act of 1980 loosened restrictions on trucking companies' shipping rates and removed geographical constraints on their service regions.

Trucking deregulation resulted in lower trucking rates, as shown in Figure 3.9, and an increase in productivity, thanks to better use of labor and equipment, as shown in Figure 3.10. In addition, freight costs decreased from 16.2% of gross domestic product (GDP) in 1981 to 7.7% in 2009 (ICF & HLB 2002). The number of interstate trucking firms increased from 18,000 in 1975 to over 500,000 in 2000 (BTS 2001). Increases in truck VMT on U.S. highways have raised concerns about safety. Nevertheless, the crash rate has been steady since deregulation while fatality and injury rates have fallen.

Motor carrier industry deregulation resulted in no adverse safety effects.

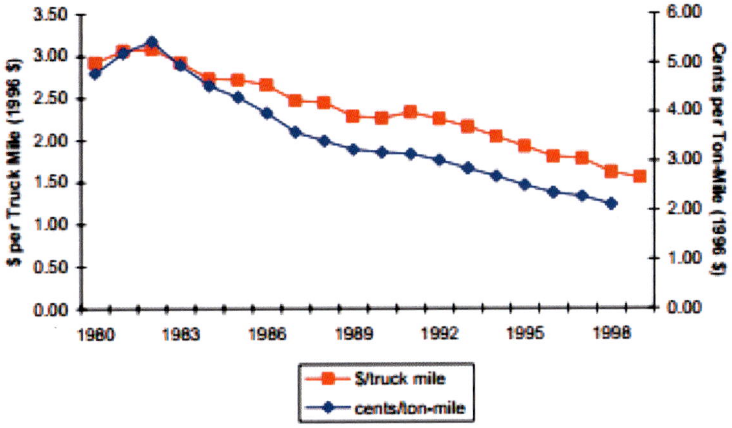

Figure 3.9: Truck and Rail Revenue Rates (Source: ICF & HLB 2002)

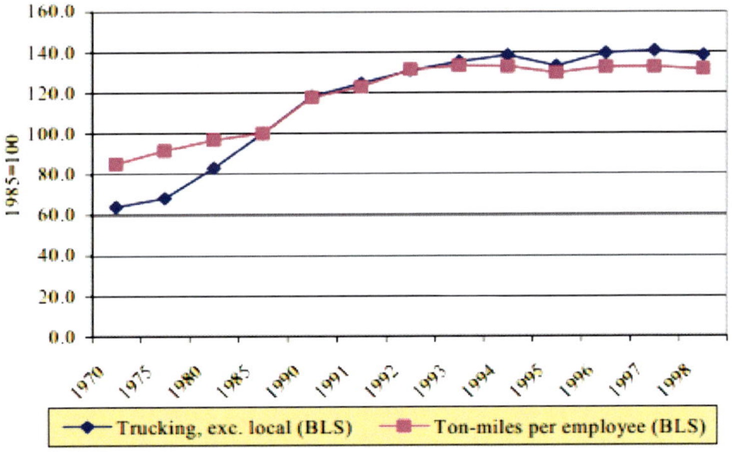

Figure 3.10: Productivity Trends in the Trucking Industry (Source: ICF & HLB 2002)

Airline Deregulation

Prior to 1978, the Civil Aeronautics Board (CAB) regulated airfares, the number of flights, and which cities airlines could fly between. Also, airlines were subject to lengthy delays for CAB permission to establish new routes or eliminate services, making the industry less productive. To address such issues, President Jimmy Carter signed the Air Transportation Regulatory Reform Act in 1978 to deregulate the airline industry. Following are some provisions of this act:

- Facilitated new entry to the air transportation industry.
- Eliminated the CAB's authority to set fares, routes, schedules, and market entry.
- Authorized international carriers to offer domestic service.

Deregulation Effects in the Airline Industry

The first consequence of this deregulation was a reduction in the average fare level, as Figure 3.11 shows, due to changes in the fare structure. By 1986, about 90% of passenger-miles were flown on discounted tickets, with average costs 61% below the standard coach fares—translating into $11 billion in total savings for passengers in 1986 alone (Kahn 1988).

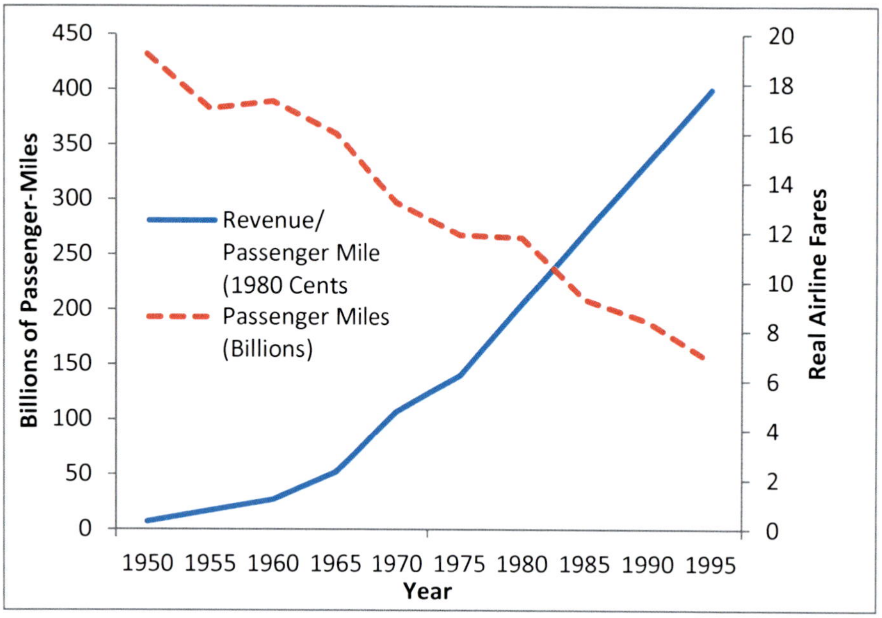

Figure 3.11: Trends in U.S. Airline Fares and Traffic (Source: Boyer 1997)

The second consequence of deregulation was a surge in the number of air passengers. Table 3.2 shows the percentage changes in passengers for different U.S. markets between 1950 and 1995. To accommodate demand increases, airlines used larger planes, put more seats in existing planes, and increased flight frequency. As a result, the average number of seats for long-haul flights increased by 21% (Moore 1986). Smaller planes were used for short- and medium-haul flights.

Type of Markets	Number of Passengers	Passengers per Departure
Long-Haul (over 800 miles)—Major Cities (over 1.2 million population)	+ 63%	+ 20%
Medium-Haul (200–700 miles)—Large Cities (over 200,000 population)	+ 29%	+ 20%
Short-Haul (under 200 miles)—Large Cities	- 16%	- 30%
Medium-Haul—Small Cities (under 200,000 population)	- 12%	- 2%
Short-Haul—Small Cities	- 60%	- 77%

Table 3.2: Passengers and Departures—Percentage Change (Source: Moore 1986)

Deregulation has increased productivity with more seats per flight filled via sale of discounted tickets. In 1982, the average percentage of occupied seats, also called the *industry load factor*, was 59% (prior to deregulation in 1976, it was 55%). Note, however, that these increased load factors have resulted in decreased service quality, amenities, comfort, and on-time service.

As a result of airline deregulation, passengers paid less but also got fewer amenities.

Another consequence of airline deregulation was that the number of passenger and freight carriers more than doubled. In 1976, the numbers of certified carriers and certified passenger carriers were 33 and 28, respectively. By 1983, after deregulation, these numbers had increased to 98 and 61 (Moore 1986). As new non-union airlines offering lower fares with higher load factors entered the market, the industry saw less restrictive work rules and lower wages. By the late 1980s and mid-1990s, almost all of these airlines went bankrupt or merged because of mismanagement and crashes of under-maintained aircrafts (Boyer 1997).

3.4 Competition

In general, as the number of companies providing services increases, prices fall, thanks to competitive forces. Urban transit systems used to be privately owned and relatively competitive, but many such systems faced serious economic issues in the 1950s. As the U.S. recovered from the war, incomes rose and automobile ownership became quite common.

Competition in Public Transit Systems

In order to help public transit systems, the 1964 Urban Mass Transportation Act (UMTA) authorized $2.23 billion (in 2003 dollars) of federal initiatives (Hess and Lombardi 2005) and many cities purchased private transit operations. Federal funding for public transit has continued through the 1956 Federal-Aid Highway Act, the 1982 Surface Transportation Assistance Act, and others. The most recent is the 2005 SAFETEA-LU, which authorized approximately $2.3 billion to address transit programs in fiscal year 2009 (APWA 2005).

As U.S. public transit systems are largely funded by government sources, some feel that they have become overly subsidized and inefficient monopolies. Subsidies have a negative effect on performance and productivity by reducing incentives for innovation and initiative, and financial mismanagement of transit properties may occur. With direct competition, transit firms may face less pressure to reduce costs and operate more efficiently. For example, Anderson (1983) estimated that the average operating cost per bus-hour of public firms is 28% more than private firms.

In contrast to public transit systems, privately owned transit companies grow in response to competition.

Although urban or intra-city busses are mostly operated publicly, *inter*-city busses are operated privately. In 2010, inter-city bus service was the fastest growing mode of intercity transportation for the third year in row as a result of rising travel demand and fuel prices, investment in new routes, and the emergence and expansion of low-cost operators. For example, the Chicago-based Megabus handled 500,000 passengers in its first 15 months of operation beginning in 2006, and has since expanded hub locations to include Atlanta, New York, Toronto, Pittsburgh, Philadelphia, and Washington D.C. As a result, conventional operators have upgraded their product by offering free WiFi connections, more spacious cabins, television monitors, and other amenities.

Competition between Bidders

Construction project bidding represents another source of competition. In general, an increase in competition among bidders reduces construction costs as more bidders compete to win a project. Factors that impact the number of bidders include bid timing and project type, duration, and size. For example, fewer companies bid on long-duration projects because they are subject to more price fluctuations (Sanderson 2006). While many highly qualified large contractors do not bid for small projects, small or medium-sized firms may not have the resources to bid for large projects. According to AASHTO survey results, bundling smaller projects together and splitting large projects into smaller projects are two top strategies to achieve lower bid prices.

3.5 Summary

The transportation sector is responsible for a large share of urban and rural air quality issues, and tens of thousands of premature deaths each year in the U.S., along with hundreds of billions of dollars in pollution, crash, and delay costs. Transportation is also a key component of the U.S. economy, responsible for nearly 20% of the U.S. GDP. Environmental and safety regulations help reduce external and other costs.

Competition also affects costs and benefits. Regulatory policies affect the level and nature of competition in different transportation sectors. Airline carriers and public transit operators work in very different contexts. Competition adds bidders and reduces project costs. Regulations are not uniform and vary by industry, function, need, and beneficiary. The broad application of competition to various sectors reflects the importance in striking a balance in regulatory policy that promotes healthy growth in the private sector while moderating negative impacts on the public.

3.6 An In-Depth Look

This section provides supplementary information on the history of transportation regulation and deregulation.

Regulatory Evolution

Shifts in user demands, mode choice, technological development, market forces, and international trade have prompted the federal government to pass laws aimed to optimize the transportation system for both public and private sector use. Government regulations generally aim to improve welfare of the public in terms of health, safety, and efficient use of public funds. Federal regulations like the Clean Air Act and automotive safety standards set by the NHTSA are examples of proactive legislation aimed to benefit public interests. Over time, the federal government recognized the benefit of relaxing certain regulations on transportation industries and repealing many laws that governed certain business and trade requirements, generally resulting in more efficient businesses with lower fares, ticket prices, and shipping costs.

Most social regulatory agencies belong to the executive branch of the federal government. Examples are the Food and Drug Administration (FDA), Occupational Safety and Health Administration (OSHA), Consumer Product Safety Commission (CPSC), Environmental Protection Agency (EPA), National Highway Traffic Safety Administration (NHTSA), and the Federal Railroad Administration (FRA). Many of these agencies have their counterparts in individual states.

Prior to 1980, many regulations governed the operations of railroads, motor carriers, and airlines. Governments had control of fares, routes, and market entry and exit through the ICC and the CAB. Beginning in the late 1970s, perceived economic regulatory failure in the electric utility, telecommunications, banking, and transportation sectors of the economy became a catalyst for regulatory reform. In transportation, Congress passed several deregulation acts that relaxed market entry and exit, increased freedom to set rates, permitted horizontal and vertical mergers, and increased competition within and between modes. Some of the more important acts include the following:

- Air-Cargo Deregulation Act of 1977
- Airline Deregulation Act of 1978
- Motor Carrier Act of 1980
- Staggers Rail Act of 1980

- Bus Regulatory Reform Act of 1982
- Shipping Act of 1984
- Freight Forwarder Deregulation Act of 1986
- Interstate Commerce Commission Termination Act of 1995

Regulation and deregulation has significant impacts on competition (both within and across modes), as discussed in this chapter.

3.7 References

An, F. (2007) Passenger Vehicle Greenhouse Gas and Fuel Economy Standards: A Global Update. Prepared for the International Council on Clean Transportation Available online at http://www.theicct.org/sites/default/files/publications/PV_standards_2007.pdf.

Alexander, D. L. (1992) Motor Carrier Deregulation and Highway Safety: An Empirical Analysis. Southern Economic Journal, 59(1), 28–38.

American Association of State Highway and Transportation Office (2004) AASHTO Strategic Highway Safety Plan. Available online at: http://safety.transportation.org/doc/Safety-StrategicHighwaySafetyPlan.pdf.

American Public Works Association (APWA) (2005) SAFETEA-LU: A Guide to Provisions Related to Local Governments. Available online at: http://www2.apwa.net/documents/Advocacy/SAFETEA/APWA-SAFTEA-LU.pdf.

Anderson, S. H. (1983) The Effect of Government Ownership and Subsidy on Performance: Evidence from the Bus Transit Industry. Transportation Research A 17A(3), 191–200.

Boyer, K. (1997) Principles of Transportation Economics. Addison Wesley Longman, Inc. United States.

Center for Transportation Analysis, Freight Analysis Framework Data Extraction Tool. Available at: http://cta-gis.ornl.gov/faf/Extraction1.aspx. Last access on 3/6/2011.

Federal Register (2011) Greenhouse Gas Emissions Standards and Fuel Efficiency Standards for Medium- and Heavy-Duty Engines and Vehicles. U.S. Environmental Protection Agency, U.S. Department of Transportation, and National Highway Traffic Safety Administration. 76 179).

Hajek, J. J. & Billing, J. (2007) Trucking Trends and Changes that Affect Pavements. *Transportation Research Record* 96-103.

Hess, D. B. & Lombardi, P. A. (2005) Governmental Subsidies for Public Transit History, Current Issues, and Recent Evidence. *Public Works Management & Policy*, 10(2), 138-156.

ICF Consulting & HLB Decision-Economics (ICF & HLB) (2002) Economic Effects of Transportation: The Freight Story. Available online at http://www.ops.fhwa.dot.gov/freight/documents/freightstory_12902.pdf.

Jacobsen, M. (2012) Evaluating U.S. Fuel Economy Standards in a Model with Producer and Household Heterogeneity. University of California at San Diego. Available at http://econ.ucsd.edu/~m3jacobs/Jacobsen_CAFE.pdf.

Kahn, A. E. (1988) Surprises of Airline Deregulation. The American Economic Review, 78(2), 316–322.

Koops B. J., Lips, M., Prins, C., & Schellekens, M. (2006) Starting Points for ICT Regulations, Deconstructing Prevalent Policy One-liners. Cambridge University Press, Cambridge.

Kralaftis, M. G. & McCarthy, P. S. (1999) Subsidy and Public Transit Performance: A Factor Analytic Approach. *Transportation,* 24, 253–270.

Lewis, D. & Williams, F. L. (1999) Policy and Planning as Public Choice Mass Transit in the United States. Ashgate Publishing Company, USA.

Luther, L. (2005) The National Environmental Policy Act: Background and Implementation. Congressional Research Service Report for Congress. The Library of Congress. Available at http://www.fta.dot.gov/documents/Unit1_01CRSReport.pdf.

Moore, T. G. (1986) U. S. Airline Deregulation: Its Effects on Passengers, Capital, and Labor. *Journal of Law and Economics,* 29(1), 1–28.

National Highway Traffic Safety Administration (NHTSA) (2011a) Uniform Guidelines for State Highway Safety Programs, Title No. 1. Available online at http://www.nhtsa.gov/nhtsa/whatsup/tea21/tea21programs/402guide.html#g1. Last access on 3/12/2011.

National Highway Traffic Safety Administration (NHTSA) (2011b) Uniform Guidelines for State Highway Safety Programs, Title No. 21. Available online at http://www.nhtsa.gov/nhtsa/whatsup/tea21/tea21programs/402guide.html#g21. Last access on: 3/12/2011.

National Highway Traffic Safety Administration (NHTSA) (2011c) Federal Motor Vehicle Safety Standards and Regulations. Available online at http://www.nhtsa.gov/cars/rules/import/fmvss/index.html#SN212. Last access on: 3/12/2011.

National Highway Traffic Safety Administration (NHTSA) (2011d) President Obama Announces Historic 54.5 mpg Fuel Efficiency Standard. July 29, 2011. Available at http://www.nhtsa.gov/About+NHTSA/Press+Releases/2011/President+Obama+Announces+Historic+54.5+mpg+Fuel+Efficiency+Standard.

National Highway Traffic Safety Administration (NHTSA) (2009) Traffic Safety Facts. Available online at: http://www-nrd.nhtsa.dot.gov/pubs/811172.pdf. Last access on: 3/12/2011.

National Highway Traffic Safety Administration (NHTSA) (2008) Traffic Safety Facts. Available online at: http://www-nrd.nhtsa.dot.gov/Pubs/811162.PDF. Last access on: 3/12/2011.

National Highway Traffic Safety Administration (NHTSA) (2002) The Economic Impact of Motor Vehicle Crashes 2000. Available online at: http://www.cita-vehicleinspection.org/Portals/cita/autofore_study/LinkedDocuments/literature/NHTSA%20the%20economic%20impact%20of%20motor%20vehicle%20crashes%202000%20USA%202002.pdf.

National Research Council (NRC) (2010) Technologies and Approaches to Reducing the Fuel Consumption of Medium- and Heavy-Duty Vehicles. Washington, DC: The National Academies Press.

National Work Zone Safety Information Clearinghouse (2011) Fatalities in Motor Vehicle Traffic Crashes by State and Construction/Maintenance Zone (2008). Available at: < http://www.workzonesafety.org/crash_data/workzone_fatalities/2008>.

Rodrigue, J.P., Comtois, C., & Slack, B. (2009) The Geography of Transportation System. Routledge, Oxon, UK.

Sanderson, V. (2006) Current Strategies to Address Increasing Highway Construction Costs and Reduced Competition. Technical Agenda, AASHTO Subcommittee on Construction, Washington, D.C.

Schwieterman, J. P. & Fischer, L. (2010) The Intercity Bus: America's Fastest Growing Transportation Mode 2010 Update on Scheduled Bus Service. Chaddick Institute for Metropolitan Development, DePaul University.

Sullivan, A. & Sheffrin, S. M. (2002) Economics: Principles in Action. New Jersey: Pearson Prentice Hall.

Texas Department of Transportation (TxDOT). Texas DOT's Cost Control Task Force Recommendations Highway Construction and Maintenance Cost Control Ideas. Available at : < http://www.fhwa.dot.gov/programadmin/contracts/costcont.cfm>. Last Access on 1/3/2011.

Texas Department of Transportation (TxDOT) (2004) Manual Notice 2004-2. Available at: http://onlinemanuals.txdot.gov/txdotmanuals/env/env_mns.htm. Last Access on 12/12/2012.

Texas Department of Transportation (TxDOT) (2008) Manual Notice: 2008-1. Available at :< http://onlinemanuals.txdot.gov/txdotmanuals/hsi/manual_notice.htm>. Last Access on4/22/2011.

Texas Department of Transportation (TxDOT) (2011) RM 1431 Improvement Project Archeological Review Meeting Notes. November 1, 2011, TxDOT Austin District.

U.S. Department of Labor (DOL) (2011) Minimum Wage Laws in the States. Available at: < http://www.dol.gov/whd/minwage/america.htm#Texas>.

U.S. Department of Transportation (USDOT) (2010) Transportation's Role in Reducing U.S. Greenhouse Gas Emissions, Volume 1: Synthesis Report. Available online at: http://ntl.bts.gov/lib/32000/32700/32779/DOT_Climate_Change_Report_-_April_2010_-_Volume_1_and_2.pdf.

U.S. Department of Transportation, Bureau of Transportation Statistics (BTS) (2001) National Transportation Statistics 2000. BTS01-01. Washington, DC.

U.S. Department of Transportation, Federal Highway Administration (FHWA) (1992, August 21) NEPA and Transportation Decisionmaking. Environmental Review Toolkit. Available online at http://environment.fhwa.dot.gov/projdev/tdmpdo.asp.

U.S. Department of Transportation, Federal Highway Administration (FHWA) (2002) The Freight Story: A National Perspective on Enhancing Freight Transportation. Available online at http://ops.fhwa.dot.gov/freight/publications/fhwaop03004/index.htm.

U.S. Department of Transportation, Federal Highway Administration (FHWA) (2004) Programmatic Agreement for the Review and Approval of NEPA categorically Excluded Transportation Projects between the Federal Highway Administration Texas Division and the Texas Department of Transportation. Accessed April 10, 2012. Available at: http://environment.transportation.org/pal_database/agreement_details.aspx?pal_id=132.

U.S. Department of Transportation, Federal Highway Administration (FHWA) (2008) 2008 Status of the Nation's Highways, Bridges, and Transit: Conditions & Performance. Available online at http://www.fhwa.dot.gov/policy/2008cpr/pdfs/cp2008.pdf.

U.S. Department of Transportation, Federal Highway Administration (FHWA) (2011) Highway Safety Improvement Program (HSIP). Available online at http://safety.fhwa.dot.gov/safetealu/fact_sheets/ftsht1401.cfm. Last access on 4/22/2011.

U.S. Energy Information Administration (EIA) (2009) 2009 Annual Energy Review. Available at: http://www.eia.doe.gov/aer/pdf/pages/sec2_4.pdf.

U.S. Environmental Protection Agency (EPA) (2006) Emission Durability Procedures for New Light-Duty Vehicles, Light-Duty trucks and Heavy-Duty Vehicles. 40 CFR Part 86. Available online at http://www.epa.gov/EPA-AIR/2006/January/Day-17/a074.htm.

U.S. Environmental Protection Agency (EPA) (2008) 2008 Inventory of U.S. Greenhouse Gas Emissions and Sinks: 1990–2006. Available online at http://www.epa.gov/climatechange/emissions/downloads/08_CR.pdf.

U.S. Environmental Protection Agency (EPA) (2011a) Cars and Light Trucks. Available at:< http://www.epa.gov/oms/recall.htm>. Last access on 3/12/2011.

U.S. Environmental Protection Agency (EPA) (2011b) Light-Duty Automotive Technology, Carbon Dioxide Emissions, and Fuel Economy Trends: 1975 Through 2010 Available at:< http://www.epa.gov/otaq/fetrends.htm>. Last access on 3/12/2011.

U.S. Environmental Protection Agency (EPA) (2012) Mobil Source Emissions—Past, Present, and Future. Updated on January 3, 2012. Available online at http://www.epa.gov/oms/invntory/overview/results/onroad.htm.

U.S. Code (2011) *Title 23, United States Code.* House Transportation and Infrastructure Committee. Available online at http://www.fhwa.dot.gov/legsregs/title23.pdf.

U.S. Senate (1984) Oversight of the Motor Carrier Act of 1980: Hearing Before the Committee on Commerce, Science, and Transportation. September 18, 1984.

Winston, C., Corsi, T. M., Grimm, C. M., & Evans, C. A. (1990) The Economic Effects of Surface Freight Transportation. Washington, D.C., The Brookings Institution.

Chapter 4. Movement, Transportation, and Location

4.1 Introduction

Travel is an important facet of most people's lives, and the permanent nature of transportation infrastructure directly shapes urban form. This form, in turn, impacts land use, land values, and wages. Travel is a byproduct of the need to work, shop, run errands—essentially, produce and consume. The need for travel is a **derived demand**, as opposed to a **direct demand** (which consumers get direct satisfaction from), such as the need for food, clothes, and other consumer goods. When transportation infrastructure improvement is the trigger that changes accessibility and mobility, what are the subsequent impacts on business and residential location choice? When businesses and households change their choice of locations, what happens to the land values and wages? This chapter explores the relationship between transportation investment and all of these issues.

Key Terms

- ✓ **CBD:** central business district

4.2 Accessibility and Mobility

Two key concepts describe the relationship between transportation and location: mobility and accessibility. **Mobility** conveys the efficiency and amount of movement, usually in terms of travel speeds and distances. Increased mobility generally means more ability to move from one point to another. **Accessibility** conveys the ease of reaching quality destinations, reflecting both the attractiveness of potential destinations and the ease of reaching them. Those residing in highly accessible locations can more easily reach attractive or desirable activity sites (i.e., within a certain distance or travel time) than those dwelling in less accessible places.

Although greater mobility can provide more accessibility for travelers by reducing travel times, accessibility does not depend solely on mobility. Accessibility, especially walk, bike, and transit accessibility, tends to rise with higher densities. In contrast, enhanced mobility can contribute to increased separation of land uses as people are able to travel farther given the same travel times or budgets.

4.3 Transportation and Location Choice

Do business and household locations determine the design expansion of the transportation system? Or does the design and extent of the transportation system determine locations choices? Button (2010) likens the continual cycles of cause and effect between transportation and land use to the age-old "chicken or egg" dilemma. Given the longevity of transport infrastructure, system changes often have long-term effects on economic activity. Subsequent changes to residential and employment location patterns will, in turn, influence future transportation demand.

The practical decision of whether to treat transportation as a cause or an effect of land use depends upon the research question at hand. Urban and regional planners treat transportation as an influential variable that impacts development decisions, whereas transportation planners traditionally model transportation decisions using a four-step process based on trips generated according to land use. The NCHRP's *Land Use Impacts of Transportation Guidebook* (1999, p.12) identifies the land use/transportation relationship as "an interaction of supply and demand for accessibility that is further affected by public policies," as illustrated in Figure 4.1. While land use and transportation influence the supply of accessible residential and commercial properties, demand is affected by the preferences of individuals and businesses (which interrelate with public policies).

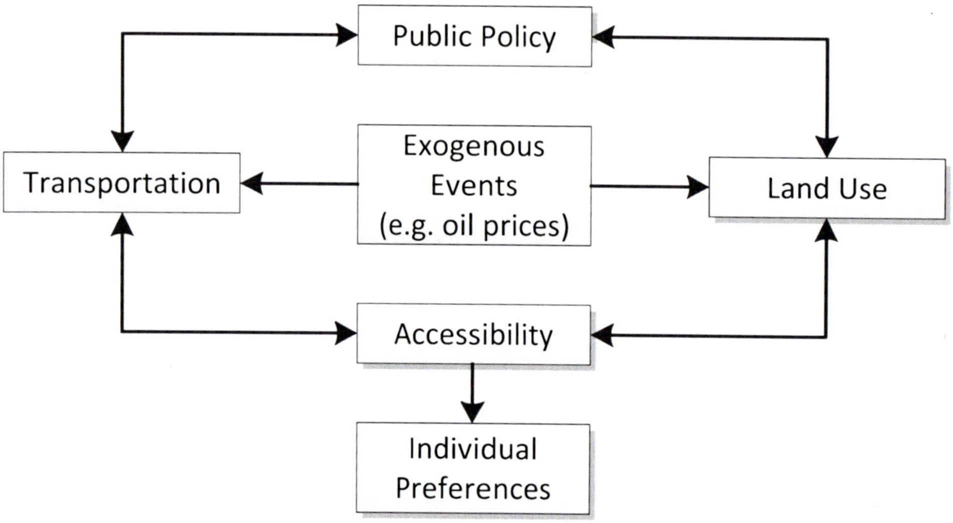

Figure 4.1: Relationship between Transportation, Land Use, and Accessibility (Source: NCHRP 1999, p.13)

The relative accessibility of locations has important impacts on their land values. Existing land use conditions determine near-term travel behaviors and regional mobility, which influence transportation investment decisions, both long term and near term.

Long-term traffic forecasts benefit greatly from long-term land use forecasts, which are affected by near-term (and longer-term) transportation system changes. The spatial distributions of a region's jobs and households are essential inputs to transportation planning (travel demand) models. To this end, land-use forecasting models, such as the Gravity Land Use Model (G-LUM) developed by researchers at The University of Texas at Austin, help analysts forecast future travel times across a region by forecasting job and household locations.

Resource

G-LUM is open source software, freely available at http://www.ce.utexas.edu/prof/kockelman/G-LUM_Website/homepage.htm.

Theories of Business Location

Various business location theories address the different industry types. Many economic activities, such as retail sales and services, depend on access to consumers, while others do not. **Central place theory** (Christaller 1966) describes a distribution of market centers based on *consumer range* (the distance that consumers are willing to travel for a certain type of good), while meeting a business's *market threshold* (the minimum sales volume or customer base required to meet profit goals).

When transportation costs fall, central place theory predicts larger and more dispersed market centers as workers and consumers are willing to travel greater distances to access jobs and goods. For example, the introduction of an access-controlled facility in place of an existing arterial allows consumers to travel greater distances for the same amount of travel time. As a result, in many cities, developments near freeway interchanges have become secondary market centers to central business districts (CBD). Likewise, when transportation costs rise, the theory predicts smaller and more concentrated market centers.

Per central place theory, the optimal location of a market-sensitive business is one that minimizes total consumer transportation costs while meeting the minimum market threshold requirements.

In contrast, **industrial location theory** suggests that non-market-sensitive businesses choose locations based on transportation cost considerations such as the distance and weight of goods carried (Weber 1929). Unlike the central place theory, industrial location theory focuses on goods movement rather than employee and customer access.

On a local level, passenger facilities may be more influential on business location selection than freight transportation, particularly in retail where customer access is key.

Research suggests that industrial location models are no longer adequate for predicting U.S. business location choices due to a shift over the last 30 years away from basic industries to manufacturing and services. Siting decisions are now less sensitive to transportation investments, particularly at the inter-regional level. With the nation's transportation network well established, the cost of moving goods has declined more rapidly than commuting costs, making employee and customer access more relevant for business location. In fact, since the 1960s, jobs have been steadily decentralizing from urban city centers to the suburbs. Allen and Robertson (1983) studied the factors that influence location choices of high-technology businesses in Pennsylvania found that proximity to the market and desired workers ranked higher than proximity to regional surface transport and airports. In a survey of Texas businesses in Lubbock, Houston, San Antonio, and Dallas, accessibility and convenience were most often cited for reasons for business location (Buffington et al. 1997).

The NCHRP's Land Use Impacts of Transportation (1999) looks at the relative importance of factors influencing business location primarily from the perspective of access (to workers, customers, and suppliers), as summarized in Table 4.1.

Relative Importance	Factor	Comments
High	Cost and availability of space	Businesses weigh the advantages of various locations with the costs of leasing or owning. The availability of lower-cost space outside the CBD leads to suburbanization of businesses.
	Access to labor	Businesses have different labor needs. Some locate in the CBD to have the best access to a more diverse pool of highly skilled labor. Some locate near residential areas, which may be preferred by key technical and managerial staff.
	Access to customers	Customer base is critical to retail and customer service industries. Customer access is also important to manufacturing firms, although to a lesser degree because manufacturers also consider the locations of suppliers.
	Access to highways	Highways receive higher importance as the dominant form of transportation. Highway interchanges give some suburban locations accessibility that rivals the CBD. Highways also support the movement of large manufacturing facilities to suburban and rural sites where land is less expensive.
Moderate	Near like businesses	Colocation (agglomeration) of similar businesses improves access to workers and customers while facilitating information. Examples include auto dealerships and retailers of the same product locating together to facilitate comparison shopping.
	Near suppliers and support services	Proximity to suppliers is the most important for manufacturers but also relevant for office location choices.
Low	Amenities, quality of life, prestige	These factors are most important for firms with many professional and technical workers.
	Quality of public services	Public services are most important to manufacturing firms requiring large amounts of water and sewer services for production.
	Property tax rates	Manufacturing industries are more sensitive to taxes due to the land-intensive nature of their facilities.
	Access to airports	The rise of business travel increases the importance of airport access for headquarters of national and global businesses.
	Economic development incentives	Incentives influence the amount and location of redevelopment by reducing costs of development in a specific community or area.
	Location of competitors	Retailers want access to large customer bases in areas with multiple stores, but generally do not want to be too close to competitors.

Table 4.1: Factors Influencing the Location Decision of Businesses (Adapted from NCHRP 423A 1999)

Theories of Residential Location

Alonso (1964) developed theories of residential location choice based on agricultural land rents and usage. These theories view household location choice as a *utility maximization problem* constrained by resources such as housing cost, commuting cost, and costs of all other goods and services. Residential location models typically take the following form:

Chapter 7 addresses utility functions.

$$D = f(P_h, P_t, P_g) \tag{4.1}$$

where housing demand D is a function of housing price P_h, transportation cost P_t, and price of all other goods P_g. Households located near employment centers experience lower travel costs, and so can allocate more to housing. In the traditional model with a single CBD, the highest land values are at the CBD, and population density and land values both fall with increased distance from the CBD as accessibility to employment decreases (illustrated in Figure 4.2).

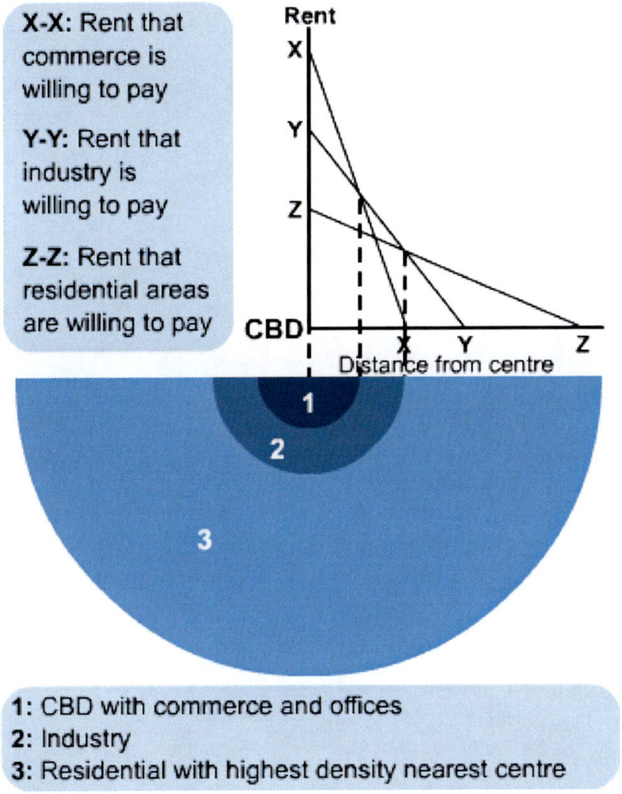

Figure 4.2: Bid-Rent Curve (Source: S-cool.co.uk, undated)

A survey of Texas households revealed that proximity to work and schools, freeway access, neighborhood type, and school quality were the top considerations for residential location (Buffington et al. 1997).

However, the models assume identical relative preferences for location and saving across households, which is far from reality. Proximity to public assets such as parks and schools, ethnic and family loyalty to specific neighborhoods, and preference for architectural styles and housing type or size all influence residential location choice (Giuliano 2004). The basic models also do not account for the growing number of multi-worker households that must accommodate more than one commuter's work trip. Further, not all jobs are in a CBD. Lastly, the trend towards higher rates of job turnover and greater moving costs suggests that households may locate to increase accessibility to future employment opportunities instead of reducing commute costs to current jobs. Table 4.2 summarizes the relative importance of factors influencing household location.

Table 4.2: Factors Influencing the Location Decision of Households (Adapted from NCHRP 423A 1999)

Relative Importance	Factor	Comments
High	Housing costs	Household budgets must accommodate cost of housing and other goods.
Moderate	Access to jobs	High access to jobs is especially important in large metropolitan areas, where travel distances can be long and transport systems congested.
	Access to goods and services	Preferences vary with household type. Singles prefer to live near entertainment, families with school age children prefer access to good schools and parks, and empty nesters seek leisure and culture.
	Community residents	Most people prefer to live near others who are like them.
Low	Quality of non-school public services	Sometimes households consider police protection and/or other public services when selecting a neighborhood.
	Amenities and quality of life	Residences with appealing views, attractive design, and low crime rates are preferred over those located directly adjacent to industry or heavy traffic.
	Property tax rates	Study results are mixed on whether or not property taxes influence household location choice.

As out-of-pocket travel costs have declined more rapidly than housing costs, jobs have decentralized and commute distances have increased. The average work trip distance rose from 8.55 miles in 1983 to 12.08 miles in 2001, and then to 12.20 miles in 2009 (USDOT 2004, USDOT 2011). Though suburbanization has served as the primary pattern of residential growth in urban areas, note that low-income and minority groups have not been able to decentralize to the same degree as other groups. As jobs decentralize, disadvantaged groups face decreasing access to employment opportunities. The term "spatial mismatch" was coined to describe this geographic mismatch between inner-city workers' home locations and suburban employment locations (Kain 1968).

Policy Impacts

The business and residential location theories discussed here assume a perfectly competitive land market. But municipal policies such as zoning regulations and infrastructure provisions contribute to a highly regulated land market—far from perfect competition. Most communities have long-range land use plans that drive zoning ordinances and growth management policies.

Because urban development cannot occur without public services and utilities, policy decisions regarding infrastructure largely determine the available supply of developable land.

Concerned with tax revenue generation, local governments tend to encourage land uses that contribute positively to the tax base and discourage land uses that incur public costs. The appetite for tax revenue often drives neighboring municipalities to compete for economically beneficial developments, often with incentives like property tax abatements and subsidized loans (Giuliano 2004). Due to low property taxes, California cities are increasingly relying on sales, hotel, and other consumption taxes to generate revenue. To boost consumption taxes, municipalities encourage the development of high-volume big box retailers. At the same time, low-income housing developments are discouraged because they increase public expenditures service requirements such as police departments and schools, which often cannot be covered by the meager property taxes generated by these housing developments (Altshuler et al. 1993).

4.4 Transportation and Land Values

Transportation investments are often viewed as growth generators. Transit investments, particularly rail, are frequently perceived as triggers for economic revitalization in central cities, while highway facilities encourage growth by increasing access to inexpensive land farther from the city center (Ryan 1999). Such perceptions have made transportation investment decisions hot topics of debate among policy makers.

Theoretical Expectations

Many theories seek to describe the land value/accessibility interaction. Per the theory of residential location choice, households with lower commute costs can allocate more of their budget to housing, thereby bidding up property values. Based on this bid-rent curve, when commuting costs fall, city center rents will decrease due to a relative decrease in location advantage. Consequently, consumers taking advantage of decreased commute costs will increase commute distances, extending the city's boundaries. When commute costs rise, the theory predicts the reverse: households will reduce travel by locating closer to the city center and thus bid up land values (Giuliano 2004).

When new transportation infrastructure is introduced to a network, destinations served by the new facilities experience decreased travel times or travel cost, thus increasing these locations' relative accessibility. Households and businesses served by the new transportation infrastructure should experience a rise in their land value (Ryan 1999).

Rail Transit

Unlike highway investment, rail transit investment generally changes accessibility only in the immediate vicinity of the rail corridor, specifically near access points, or rail stations. When activity increases with rail station access, a corresponding increase in land values should reflect the uptick in activity. A traditional approach to assessing the impact of rail transit investments on land use is to examine how property values vary with distance to a station. Results of these price-vs.-distance case studies have been inconsistent, as indicated by the following examples:

- A 1992 analysis of home sale prices within 1.25 miles of the Metropolitan Atlanta Rapid Transit Authority's (MARTA) New East Line found that the line had positive effects on property values to the south, a neighborhood with lower-middle-class households, but negative effects on property values to the north, in a neighborhood of primarily middle class households with some wealthy households. To the south, values increased approximately $1,045 for every 100 feet a property was closer to the line. To the north, property values fell approximately $965 for every 100 feet a property was closer to the line (Nelson 1992).
- Gatzlaff and Smith (1993) examined the sale prices of specific properties in successive sales before and after the 1984 opening of Miami's Metrorail, which was planned to revive development in economically depressed areas of the city. In higher priced neighborhoods experiencing growth, the Metrorail weakly increased existing property values. For neighborhoods already in economic decline, the Metrorail provided no benefit to property values.
- A 1997 study of the impact of the Northern California Bay Area Rapid Transit (BART) system on property values also yielded contradictory results. While apartment units in booming suburban Contra Costa County near the Pleasant Hill BART station rented for 15¢ more per square foot than apartments farther than a quarter-mile from the station, apartment rents did not differ with proximity to the Richmond BART station in northern Alameda County, where a poor local economy and high crime rates deterred station-area development (Cervero & Landis 1997).

In the Atlanta case, low-income households are more likely to use transit and benefit from improved transit service, so their property values rise. Higher income households, in this case, are less likely to use the transit service and experience insignificant time savings. Property value benefits arise only in areas that value the service. In Miami's case, with low Metrorail ridership, property values saw no impact from the investment. Systems that have the highest ridership rates and reach more locations experience the greatest gain in property values from rail transit investment. Furthermore, rail transit's impact on property values is felt within very limited distances from the stations. The highest impacts are localized and experienced in fast-growing

These case studies support the idea that property values increase when travel times fall—but only in areas that value the service.

core areas that are already economically robust. For areas in economic decline, rail investment alone is not sufficient to induce development.

While the previous three case studies used only distance to measure the effect of rail transit on nearby property values, Bowes and Ihlandfeldt (2001) developed a price model to measure impacts of four potential variables: 1) reduction in commute costs, 2) attraction of retail activity to a neighborhood, 3) increase in noise and emissions, and 4) potential rise in crime. Using Atlanta homes sales data, they estimated that properties within a quarter mile of MARTA rail stations are sold for 19% less than properties more than 3 miles away, while properties between 1 and 3 miles from a station enjoyed significantly higher values. These results suggest the following:

- Properties adjacent to rail stations may be negatively impacted by externalities such as noise and traffic, but those at an intermediate distance (beyond the externality effects) still benefit from transportation access.
- Additionally, houses within a half-mile of a rail station with parking lots experienced higher densities of crime while houses between half a mile and 3 miles of the rail station experienced decreased crime, suggesting that the presence of a parking lot at the rail station changes the crime distribution.
- Property values rose more for intermediate distances in high-income neighborhoods than in low-income neighborhoods.

Highway Investment

Though the greatest land value increases tend to occur near infrastructure investments, relative accessibility changes occur throughout the roadway network. Furthermore, investments in already extensive urban networks will have less pronounced impacts than those in rural areas where access is relatively limited to begin with.

Because highways exist as a component of a larger network, their impacts on land value are not as directly measurable as those of rail transit.

Early highway studies following the highway building boom of the 1950s and 1960s showed significant increases in land values alongside new interstates (see Adkins 1959 and Mohring 1961). Almost all studies, however, focus on nearby parcels and do not examine whether land values fall elsewhere. More recent studies yielded the following insights:

- Gamble et al. (1974) examined residential sales prices within 4000 feet of Virginia's IH 495 corridor and found average property value increases of about $3,000 in 1972 dollars. However, for houses abutting the highway (within 400 feet), the new freeway reduced land values, presumably due to noise and boundary effects, and perhaps emissions impacts.
- A 2000 study of home sales in Toronto found that the presence of a highway within 2 kilometers had negative effects on property values

(Haider & Miller 2000).

- Boarnet and Chalermpong (2001) examined residential sales price impacts of the new Orange County tollway system in the Foothill and San Joaquin Hill corridors between 1988 and 1999 (before and after the tollways' construction) as compared to sales prices along a control corridor. The nearest highway interchange was found to be a significant and positive factor.

- Lewis et al. (1997) examined the impact of elevated, at-grade, and depressed highway sections on Texas property values, and found that cities with strong land use regulations more frequently experienced land value increases following highway construction. The type of highway improvement and property also played a role: the highest residential land values occurred near depressed freeway sections (which reduce the nuisance of noise, emissions, and other negative effects), whereas community properties benefitted most near at-grade freeway sections, thanks to added visibility. Both types of land use were estimated to exhibit their lowest values near elevated freeway sections.

Like rail transit, the impact of highway investment on adjacent property values is context-specific. In general, economically thriving communities experiencing population and employment growth with available land for development tend to benefit the most from investments, at least from a *local* perspective. When growth is not present, investment impacts are not significant. However, from a *regional* perspective, there is little evidence that highway investment generates net economic development.

Local economic growth is often a result of activity redistribution from adjacent locations.

Just as added transportation capacity can improve the accessibility and property value of specific locations, the temporary delay caused by major highway construction can hamper access and economic activity and value locations. Luskin and Chandrasekaran (2005) surveyed Dallas office tenants who had experienced traffic delays during the construction of the nearby High Five Interchange (US 75 at IH 635). The researchers asked for tenants' stated preferences for office rents based on temporary commute time increases and estimated an average rent decrease of $22 per person-hour of delay higher than the average hourly wage of private-sector U.S. employees at that time ($15.71 in July 2004).

4.5 Transportation and Wages

In addition to transportation costs affecting land values, the traditional model of the city implies that transportation costs affect wages. All else equal, a worker residing in the outlying areas of a city would be willing to give up a higher-paying job in the CBD for a lower-paying job close to home, as long as the commute cost savings makes up the wage difference. Then, as is the case with land values, the highest wages should exist at the city core, with wages declining as jobs migrate outward. The concept is described in the term **urban wage gradient**, which refers to the theoretical decrease in wages with

A recent study of the urban wage gradient in two metro areas found that up to 15% of wage differences could be attributed to commute time difference (Timothy and Wheaton 2001).

increasing distance to the urban core.

When a secondary employment center is introduced, some workers residing near the CBD may reverse commute to this secondary center. Because reverse commuting costs are typically less than those of the traditional commute (due to decreased congestion), the slope of the *reverse-commute wage gradient* is theoretically less steep than that of the *commuting wage gradient* (Button 2010).

The nature of transportation's influence on wage rates is far more complex than the basic theory holds.

Although the studies are not completely consistent, the evidence supports the idea that urban wage gradients exist in some form, particularly for cities with significant CBDs and for workers commuting in from the suburbs. However, imperfections in the transportation market (e.g., public transportation subsidies and employer-subsidized parking) and job benefits not reflected in wages (e.g., flexible work hours and free meals) complicate the simple nature of the urban wage gradient, along with the specifics of different occupations and uniqueness of each job and worker.

4.6 Transportation and Economic Development

Transportation infrastructure enjoys a special relationship with regional economic development. Investment in transportation systems typically facilitates economic growth, but it does not necessarily guarantee direct economic gains. Clearly, efficient transportation systems and economic development depend heavily upon each other, but in a way that is difficult to quantify or state explicitly.

In general, transportation infrastructure investments are most beneficial to regional development in the early stages of economic growth. Opening up new markets and creating a more mobile network for workers and consumers is vital to initiate growth. Transportation improvements bring not only direct injections of cash in a local economy throughout the construction process (thanks to having local workers involved in planning, design, and construction) but provide multiplier effects that reverberate on a larger spatial and temporal scale. Transportation economists have identified distinct types and levels of economic benefits from infrastructure investments:

- **Primary** – economic gains as a direct result of infrastructure construction. These benefits are muted by outside bidders and out-of-area planning and design work.
- **Secondary** – benefits derived from operation and maintenance of the facilities. Upon construction, more steady local employment is generated through tasks such as road maintenance, toll collection, airport facilities operation, etc.
- **Tertiary** – typically emphasized as a critical source of benefits, this level includes the economic development that is drawn by the newly available infrastructure. For example, a developer may decide to bring a

shopping center to the area after the completion of a major highway that brings large volumes of traffic through a previously underdeveloped region. In turn, the shopping center generates revenue and creates jobs.

- **Perpetuity** – more abstract concept of large-scale shifts in economic structure. For example, through the extensive interstate connections, airports, and warehousing investments in Memphis, Tennessee, the area has developed into an economy supported greatly by the shipping industry. This stage is a long-term result of major investments.

Figure 4.3 illustrates how these economic multiplier benefits are related to long-term and widespread economic growth.

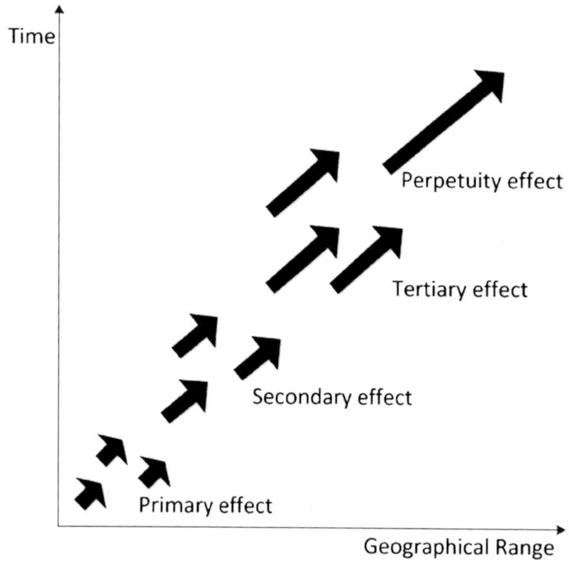

Figure 4.3: Types and Levels of Economic Benefits from Transportation Investment (Source: Button 2010)

While providing a certain level of mobility and accessibility is key to stimulating economic growth, a law of diminishing returns tends to apply to infrastructure development and economic growth. Once basic levels of mobility and access are provided, further system improvements do not bring the same magnitude of economic gains (Button 2010). In established areas, relief routes (such as loops and bypasses) are typically built to relieve traffic pressures along the main roads through the downtown. This relief can increase the function of a previously established system, acting as a catalyst for economic growth.

Most studies of highway impacts report great variability in the results of infrastructure improvements.

The economic benefits of infrastructure improvements are often moderated economic losses from business relocations, sale slumps (and reduced tax revenues) during construction, and shifts in employment and land values. Buffington et al.'s (1997) 4-year study of economic impacts from Texas freeway improvements indicated a variety of positive and negative impacts on rural and urban areas. Among the conclusions were indications that construction processes negatively impact sales and tax revenues, while user costs and employment effects depend on freeway type. Depressed freeways tended to be less productive for business, while elevated sections were less desirable for residents. Land value effects varied widely, based mostly on "factors of location and accessibility, overall economic health of the locale, growth rates, and subsequent demand for various types of property" (Buffington et al. 1997, p. 34).

Economic Impact of Relief Routes

Relief route construction is a common practice for growing suburban and rural communities in Texas. While their construction relieves congestion in the town center arterials, a community's CBD can be significantly impacted by the reduced traffic. Studies of relief routes in Texas indicate a positive public reaction to the reduced traffic, a negative reaction from downtown business owners due to a noticeable decline in sales, and overall changes in the economic structure (Handy et al. 2000). The short-term business loss in CBD is often counteracted by business relocations, resulting in overall economic rebounds. In Wisconsin, Kansas, Iowa, Texas, and North Carolina, the general consensus was mixed, in that bypasses neither spurred substantial economic growth nor hindered it overall (Leong & Weisbrod 1999). Outside of economic considerations, the decline of an historic and aesthetically pleasing CBD is frequently cited as a loss. Of course, counter arguments claim that new investment potentially captured by the relief route, in term of generated taxes, could be used to revitalize the CBD, creating a safer, quieter community more separated from the busy flow of commerce on the periphery (Handy et al. 2000).

The economic impact of relief routes is discussed in detail in Chapter 10's Case Study 2.

Economic Impact of Access Management

Access management controls traffic movement between roadways and adjacent land uses with medians, driveway design, turn lanes, traffic signals, and similar facilities. The goal is to improve safety and facilitate travel while maintaining access to abutting properties. Roadway designs that provide a structured pattern of flow with fewer conflict points expose travelers to fewer risks, allow a greater response time to potential hazards, and reduce delay (TxDOT 2011). Driveway density and median type are features that can significantly impact performance and safety. As compared to two-way left turn lanes and undivided roadways, raised medians are most effective at reducing crashes (Gluck 1999) and improving flow (TRB 2000). A nontraversable median removes left turns from through travel lanes, clearly defines upcoming conflict points, and creates pedestrian refuges. Controlling driveway density is another effective method of increasing safety and flow. Most studies find a direct rise in crash rates with increasing driveway frequency; for example, Gluck (1999) estimated a 4% increase in crashes for each additional access point per mile. Driveway density has a very similar relationship with flow as well; the Highway Capacity Manual indicates a 2.5 mph drop in free flow speed for every 10 access points per mile, up to 40 or more access points (TRB 2000).

A nontraversable median can reduce pedestrian crash counts by up to 50% (ODOT 1996).

Although access management techniques may improve safety and flow along a corridor, adjacent businesses fear a reduction in sales, mainly from left-turn restrictions and driveway access changes that reduce customer access.

Local economies may be largely unaffected by access management projects, but individual businesses can experience varied results. For example, Eisele (2000) found that after a median project was completed in Texas, overall corridor land values increased by nearly 7% while a property owner in another study (Weisbrod & Neuwirth 1998) indicated rent rates on a property fell from $6.50 to $5.00 per square foot after a left-turn restriction was implemented.

Studies have found that business owners typically overestimate the negative impacts of access management (Eisele 2000) and generally experience minimal negative impacts (TxDOT 2011).

Individual cases of these negative impacts may be explained by considering the store type and customer base: establishments that rely on passing traffic tend to be more negatively impacted by access management than destination-based stores with a more permanent set of customers (Rose et al. 2005). Specifically, gas stations, convenience stores, and motels are more likely to be negatively impacted than restaurants and grocery stores, but the former tend to be more easily relocated, resulting in economic shifts rather than losses (Weisbrod & Neuwirth 1998). Negative effects on businesses are most likely to occur during construction phase of a project (Eisele & Frawley 2000). While sales may drop during this time, recovery typically occurs within a few months (Weisbrod & Neuwirth 1998). Some businesses report that increased advertising is necessary to maintain competitiveness during and immediately after construction.

A study of customer opinion in Texas indicated that site accessibility was less important than customer service, product quality, and product price, which are all factors under the control of the business owners (Eisele & Frawley 2000). Surveys indicate that most customers are unaffected by or unaware of left-turn restrictions. Furthermore, 80% of patrons surveyed reported that left-turn restrictions did not affect the frequency of their visits to the business, and 84% reported making a U-turn or multiple-turn maneuver to access the business, which indicates a willingness to travel out of the way (Weisbrod & Neuwirth 1998).

Vu (2002) used probit and logit models with willingness-to-pay (WTP) considerations to further interpret the relationships between access management and public perception of impacts. The higher a business's WTP, the more pessimistically the owner viewed access management's effect on business. The results suggest that "economic thresholds exist for any particular business, and at a certain limit the loss of patronage due to access will force it to consider relocating to maintain economic viability" (Vu 2002, p. 29). Vu's 2002 Washington State study also noted that retail businesses (e.g., salons, banks, and clinics) perceive less negative impacts due to their more established customer base. Interestingly, convenience stores tend to view access management in a positive light, perhaps because most are already located at strategic corner locations, rather than mid-block.

Chapter 8 details probit and logit model techniques.

4.7 Summary

To fully appreciate the "chicken and egg" relationship between transportation infrastructure and location choice of households and businesses, the distinction between access and mobility must be understood. Relative accessibility is the foundation for location choice theories, providing a basis for land values and wage gradients. The complex relationship between transportation and urban form will likely grow more complicated as we move towards an information-based economy that allows more flexibility in household and business location. When job turnover rates increase, households tend to value access to overall job opportunities over access to any specific job. When flexible work relationships (e.g., telecommuting) are introduced, the existing location choice models become even more inadequate, given the complex relationship between transportation and land use in modern metropolitan areas.

4.8 References

Adkins, W. (1959) Land Value Impacts of Expressways in Dallas, Houston, and San Antonio, Texas. *Highway Research Bulletin* 227, 50–65.

Allen, D. N. & Robertson, G. E. (1983) Silicon, Sensors and Software: Listening to Advanced Technology Enterprises in Pennsylvania. Institute of Public Administration, Pennsylvania State University.

Alonso, W. (1964) Location and Land Use. Cambridge, MA: Harvard University Press.

Altshuler, A., Gomez-Ibanez, J., & Howitt, A. (1993) Regulations for Revenue: The Political Economy of Land Use Exactions. Washington, DC: Brookings Institution.

Boarnet, M. & Chalermpong, S. (2001) Travel by Design: The Influence of Urban Form on Travel. New York: Oxford University Press.

Bowes, D. R. & Ihlanfeldt, K.R. (2001) Identifying the Impacts of Rail Transit Stations on Residential Property Values. *Journal of Urban Economics* 50, 1–25.

Buffington, J., Vadali, S., Womack, K., Zimmer, R., McCully, W., Nikolaou, M., & Lewis, C. (1997) Social, Economic, and Environmental Effects of Elevated, Depressed, and At-Grade Level Freeways in Texas. Texas Transportation Institute, Texas A&M University. Available online at http://tti.tamu.edu/documents/1327-6F.pdf.

Buffington, J. L., Womack, K. N., & Lewis, C. (1997). Social and Economic Effects of Elevated, Depressed, and At-Grade Freeways in Texas, TxDOT Research Report 1327-1.

Button, K. (2010) Transport Economics. Northampton, MA: Edward Elgar Publishing.

Cervero, R. & Landis, J. (1997) Twenty Years of the Bay Area Rapid Transit System: Land Use and Development Impacts. *Transportation Research* , 31A(4), 309–333.

Christaller, W. (1966) Central Places in Southern Germany. Translated by C. Bushin. Englewood Cliffs, NJ: Prentice Hall.

Cohen, J. P. & Paul, C. M. (2007) The Impacts of Transportation Infrastructure on Property Values: A Higher-Order Spatial Econometrics Approach. *Journal of Regional Science*, 47(3), 457–478.

Crane, R. (1996) The Influence of Uncertain Job Location on the Urban Form and the Journey to Work. *Journal of Urban Economics,* 39(3), 342–358.

Eisele, W. & Frawley, W. (2000) A Methodology for Determining Economic Impacts of Raised Medians: Final Project Results, Texas Department of Transportation, Texas Transportation Institute, Texas A&M University. Available online at http://tti.tamu.edu/documents/3904-4.pdf.

Gamble, H., Sauerlender, O., & Langley, J. (1974) Adverse and Beneficial Effects of Highways on Residential Property Values. *Transportation Research Record* 508, 37–48.

Gatzlaff, D. & Smith, M. (1993) The Impact of the Miami Metrorail on the Value of Residences Near Station Locations. Land Economics, 69(1), 54–66.

Gluck, J., Levinson, H. S., & Stover, V. G. (1999) NCHRP Report 420: Impacts of Access Management Techniques, National Cooperative Highway Research Program, Transportation Research Board, Washington, D.C., National Academy Press. Available online at http://www.accessmanagement.info/pdf/420NCHRP.pdf.

Guiliano, G. (2004) Land Use Impacts of Transportation Investments: Highway and Transit. Chapter 9 in *The Geography of Urban Transportation.* New York: The Guilford Press.

Haider, M. & Miller, E. (2000) Effects of Transportation Infrastructure and Location on Residential Real Estate Values: Application of Spatial Autoregressive Technique. *Transportation Research Record* 1772, 1–8.

Handy, S. L., Kubly, S., Jarrett, J., & Srinivasan, S. (2000) Economic Effects of Highway Relief Routes on Small and Medium-Size Communities Center for Transportation Research, The University of Texas at Austin. Available online at http://www.utexas.edu/research/ctr/pdf_reports/1843_1.pdf.

Handy, S. L. & Niemier, D. A. (1997) Measuring Accessibility: An Exploration of Issues and Alternatives. *Environment and Planning* 29, 1175–1194.

Kain, J. F. (1968) Housing Segregation, Negro Employment, and Metropolitan Decentralization, *Quarterly Journal of Economics* 82(2),175–197.

Leong, D. & Weisbrod, G. (1999) Summary of Highway Bypass Studies. Wisconsin Department of Transportation and Economic Development Research Group. Available online at http://www.edrgroup.com/pdf/Town-Bypass-Case-Studies.pdf.

Lewis, C.A., Buffington, J. L., Vadali, S. R., & Goodwin, R. E. (1997) Land Value and Land Use Effects of Elevated, Depressed, and At-Grade Level Freeways in Texas. TxDOT Research Report 1327-2.

Luskin, D. & Chandrasekaran, A. (2005) Employer Valuations of Employee Commuting Time: Case Study of Office Tenants in North Dallas, Texas. *Transportation Research Record* 1902, 10–17.

Mohring, H. (1961) Land Values and the Measure of Highway Benefits. *Journal of Political Economy* 72, 236–249.

National Cooperative Highway Research Program (1999) NCHRP Report 423A: Land Use Impacts of Transportation: A Guidebook. National Cooperative Highway Research Program, Washington, DC: National Academy Press.

Nelson, A. C. (1992) Effects of Elevated Heavy-Rail Transit Stations on House Prices with Respect to Neighborhood Income. *Transportation Research Record* 1359, 127–132.

Oregon Department of Transportation (ODOT) (1996) Discussion Paper No. 4: Medians. Oregon Department of Transportation. Available online at http://www.oregon.gov/ODOT/HWY/ACCESSMGT/docs/Medians.pdf?ga=t.

Rose, D., Gluck, J., Williams, K., & Kramer J. (2005) NCHRP Report 548: A Guidebook for Including Access Management in Transportation Planning, National Cooperative Highway Research Program, Transportation Research Board, Washington, D.C., National Academy Press. Available online at http://onlinepubs.trb.org/onlinepubs/nchrp/nchrp_rpt_548.pdf.

Ryan, S. (1999) Property Values and Transportation Facilities: Finding the Transportation-Land Use Connection. *Journal of Planning Literature* 13(4), 412–427.

S-cool, the Revision Website (undated) Central place and Bid-rent theories. Available at http://www.s-cool.co.uk/a-level/geography/urban-profiles/revise-it/central-place-and-bid-rent-theories.

Srour, I., Kockelman, K. M., & Dunn, T. (2002) Accessibility. *Transportation Research Record* No. 1805, 25–34.

Timonthy, D. & Wheaton, W. (2001) Intra-Urban Wage Variation, Employment Location, and Commuting Times. *Journal of Urban Economics* 50, 338–366.

Transportation Research Board (TRB) (2000) Highway Capacity Manual Transportation Research Board.

Texas Department of Transportation (TxDOT) (2011) Access Management Manual. TxDOT Design Division. Austin, Texas. Available at http://onlinemanuals.txdot.gov/txdotmanuals/acm/acm.pdf.

US Census Bureau (2000) Working at Home: 2000. PHC-T-35. Available online at http://www.census.gov/population/www/cen2000/briefs/phc-t35/index.html.

US Department of Transportation (USDOT) (2011) Summary of Travel Trends: 2009 National Household Travel Survey. Available online at http://nhts.ornl.gov/2009/pub/stt.pdf

Vu, P. (2002) Economic Impacts of Access Management. Washington State Department of Transportation, Washington State Transportation Center, University of Washington. Available online at http://www.wsdot.wa.gov/research/reports/fullreports/554.1.pdf.

Wang, Y., Kockelman, K. M., & Wang, X. (2011) Anticipating Land Use Change Using Geographically Weighted Regression Models for Discrete Response. Forthcoming in *Transportation Research Record.*

Weber, A. (1929) Theory of the Location of Industry. Chicago: Chicago University Press.

Weisbrod, G. & Neuwirth, R. (1998) Economic Effects of Restricting Left Turns. National Cooperative Highway Research Program. Available online at http://www.edrgroup.com/pdf/left-turns-digest.pdf.

Williams, K. (2000) Economic Impacts of Access Management. Center for Urban Transportation Research, University of South Florida. Available online at http://www.cutr.usf.edu/research/access_m/pdf/Econeffects.pdf.

Zax, J. (1991) Compensation for Commutes in Labor and Housing Markets. *Journal of Urban Economics* 30(2), 192–207.

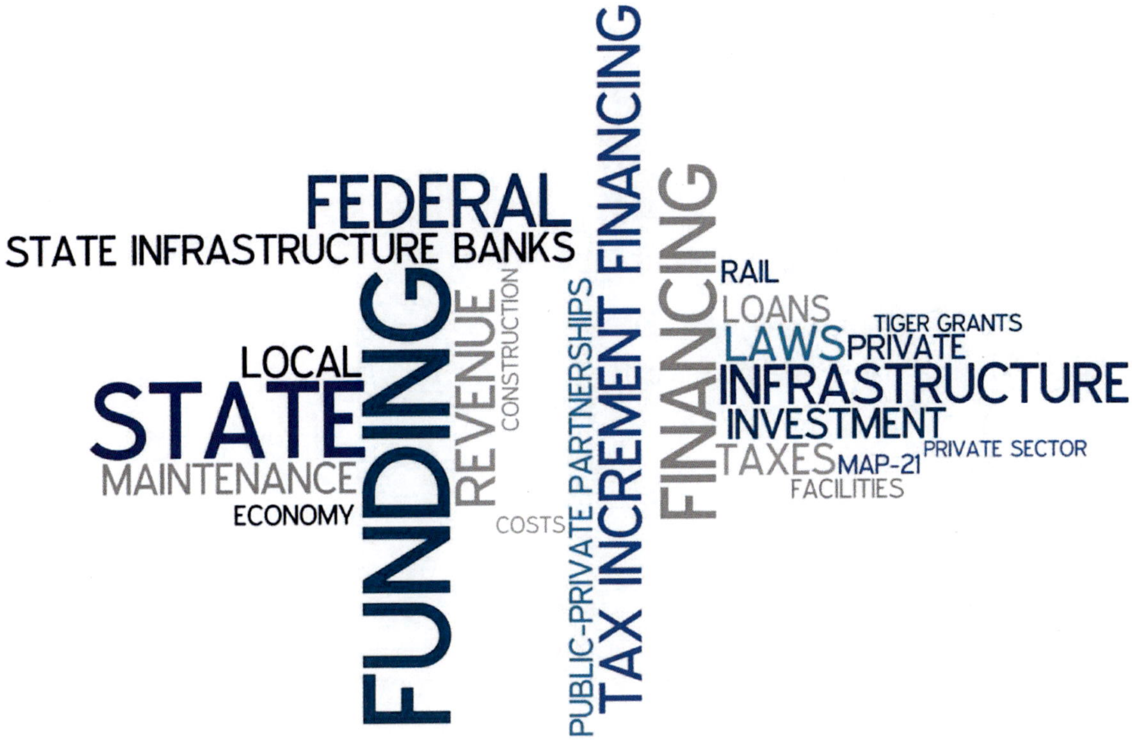

Chapter 5. Investment and Financing

5.1 Introduction

Transportation agencies face municipal, state, and federal budget constraints, so awareness of funding priorities based on the physical condition of transportation systems is key. The decisions to rehabilitate, expand, and construct new systems depend on the conditions of existing systems and competing modes. For example, expanding rail capacity may decrease the need for building new roadways.

Furthermore, the scarcity of traditional transportation funding is contributing to a growing gap between the funds required for improvements and the funds available to do so. This chapter discusses the importance of infrastructure improvements and the innovative financing methods designed to fill the funding gap, such as Section 129 loans, state infrastructure banks (SIBs), the Transportation Infrastructure Finance and Innovation Act (TIFIA), tax increment financing (TIF), private activity bonds (PABs), and public-private partnerships.

Key Terms

- ✓ **VMT:** vehicle miles traveled
- ✓ **SIB:** state infrastructure banks
- ✓ **PAB:** private activity bond
- ✓ **TIFIA:** Transportation Infrastructure Finance and Innovation Act
- ✓ **TIGER:** Transportation Investment Generating Economic Recovery
- ✓ **TIF:** tax increment financing
- ✓ **TIRZ:** Tax Increment Reinvestment Zone
- ✓ **PPP:** public-private partnership

5.2 U.S. Railroad, Road, and Bridge Investment Needs

Importance of Transportation Infrastructure

A well-functioning and efficient transportation system depends on both its capacity and infrastructure condition. The impacts of investment in transportation infrastructure fall into four categories:

1. **User Impacts:** monetary cost of travel, safety, travel time, comfort, and reliability.
2. **Economic Impacts:** change in employment, personal income, property values, business sales volume, and business profit.
3. **Government Fiscal Impacts:** changes in public revenues and expenditures.
4. **Other Social Impacts:** effects on social indicators such as air quality and other environmental conditions.

One way to calculate the impact of transportation investment is to use an **economic multiplier**. The economic multiplier approach assesses the direct and indirect impacts of transportation projects on business attraction, expansion, retention, or tourism. For example, Weisbrod and Weisbrod (1997) reported that the economic output multiplier for the national value for most transportation investments is 2.5 to 3.5, while the value for state and local impacts is 2.0 to 2.5 and 1.5 to 2.0, respectively.

See Chapter 7 for more details on the economic multiplier approach.

Concept Example: Economic Multiplier

If $100 million is invested in building a highway, economic activity at the state level would be expected to receive a net increase of $200–$250 million.

In addition, the USDOT reports that every $1 billion invested in transportation infrastructure creates 42,000 jobs and generates more than $2 billion in economic activity (Sinha & Labi 2007).

U.S. Railroad Infrastructure

The U.S. railroad system is divided into two parts: freight and passenger rail. Railroad tracks are privately owned (as opposed to highways, which are predominantly publicly owned). Although railroads are vastly more efficient than roadways, railroad networks in the U.S. are shrinking.

Federal financing programs are available to refurbish and expand the nation's valuable rail resources.

Freight Rail

- Approximately 42% of all intercity freight on a ton-mile basis travels via rail (Government Accountability Office 2006).
- Freight transportation demand is forecasted to increase from 19.3 billion tons in 2007 to 37.2 billion in 2035 (Cambridge Systematics 2007).
- Approximately $148 billion is needed to improve the railroad system to accommodate the rising increase in freight demand (Cambridge Systematics 2007).

A number of federal loans and grants are available to increase investment in railroad systems. The Federal Railroad Administration (FRA) offers various loan options through the Railroad Rehabilitation and Improvement Financing Program (RRIF) for acquiring, improving, rehabilitating, and refinancing intermodal or rail equipment or facilities. RRIF can also be used for developing and establishing new intermodal or railroad facilities. **The FRA is authorized via SAFETEA-LU to provide direct loan and guarantees up to $35 billion** (FRA 2011b).

Passenger Rail

- The Passenger Rail Working Group (PRWG) anticipates a required investment of $7.4 billion from 2007 to 2016 to address the total capital cost of an intercity rail network (with an additional total of $290.9 billion required between 2016 and 2050.
- Per PRWG estimates, approximately $4 billion in fuel will be saved annually by diverting passengers to rail if the proposed investments are made.
- Investment in passenger rail may decrease required investment in other modes of transportation.

Assisting investment in passenger rail, the FRA provides Amtrak grants for both operation and capital improvement. **The Passenger Rail Investment and Improvement Act of 2008 (PRIIA) authorized funds for the USDOT to issue grants for three new federal intercity rail capital assistance programs:** Intercity Passenger Rail Service Corridor Capital Assistance, High Speed Rail Corridor Development, and Congestion Relief.

U.S. Roads and Bridges

Texas roads and bridges require billions of dollars in maintenance and repair; state funding alone will not meet the demand.

Current Condition

The Unites States has over 4 million miles of public roads and nearly 3.7 billion square feet of bridges as of 2008—and on those roads, 3 trillion vehicle miles traveled (VMT)(USDOT 2008b). As Figure 5.1 shows, transportation infrastructure in Texas has not met the increase in VMT in the past 15 years. As a result, a gap is growing between VMT growth and new lane miles.

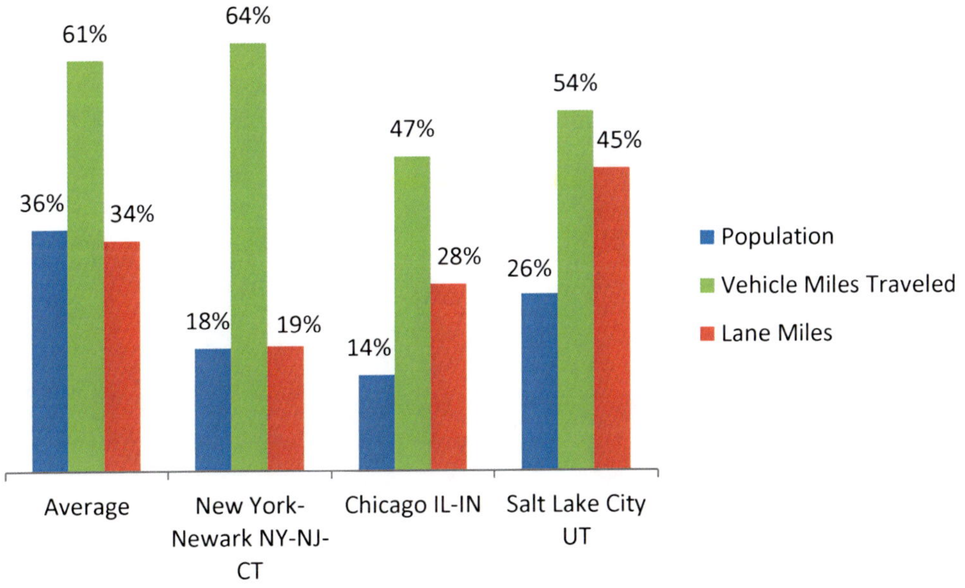

Figure 5.1: Growth, 1990–2010 (Source: Schrank et al. 2012)

Because of the disparity between highway growth and VMT rise, roadway congestion is increasing. Americans spend 4.2 billion hours each year in traffic, which amounts to an annual cost of $78.2 billion in wasted time and fuel (TRIP 2010a). Greater roadway congestion has increased the quantity of fuel wasted from 1.7 billion gallons in 1999 to 2.9 billion gallons in 2005 (TTI 2007). Inadequate roadway conditions play a significant role in approximately one-third of traffic fatalities and cost American motorists $67 billion a year in extra vehicle repairs and operating costs (TRIP 2010b). **From 1990 to 2008, new road mileage increased by 4%, while vehicle travel increased by 39%** (TRIP 2010a). A National Surface Transportation Policy and Revenue Commission study showed that for the 15-year period from 2005 to 2020, $130 billion to $240 billion should be invested in highways for adequate capacity and maintenance, yet the current spending level falls well short of that number at $70.3 billion (ASCE 2009).

A key element in roadway infrastructure is bridges. **Across the U.S., about 1 in 9 of all bridges are considered structurally deficient, while the average bridge age is 42 years (ASCE 2013).** A *structurally deficient* rating reflects the bridge's integrity with regard to the condition of the bridge deck, superstructure, and substructure, while *functionally obsolete* refers to inadequate geometry, clearance, or alignment to the roadway approaching the bridge. Though recent trends do suggest modest improvements in reducing deficient and obsolete bridges, there are still a large number of structures that do not meet standards of safe use.

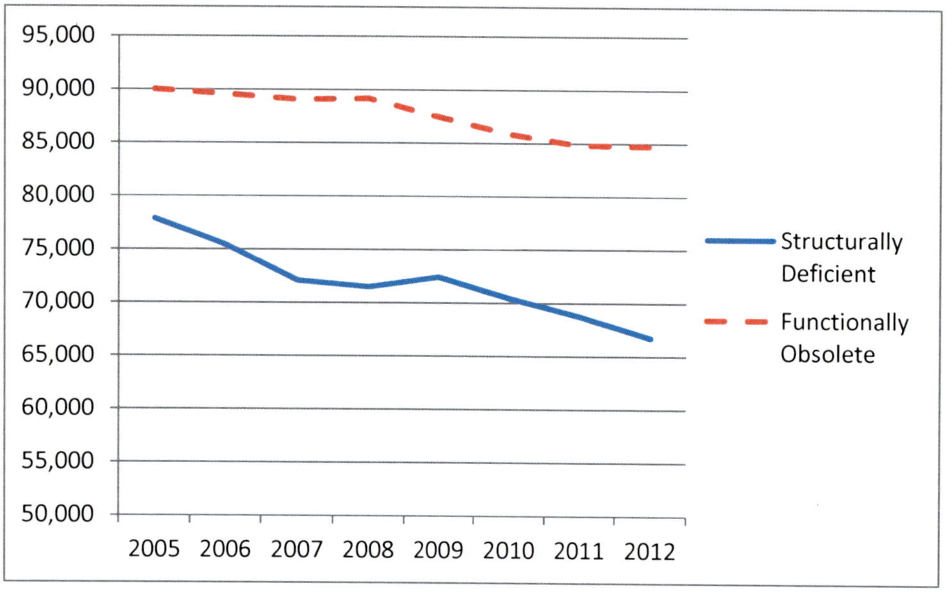

Figure 5.2: Number of Structurally Deficient or Functionally Obsolete Bridges in the U.S. (Source: ASCE 2013)

Pavement Maintenance

Because 32% of America's major roads are in "poor" or "mediocre" condition, U.S. motorists spend $67 billion on extra vehicle repairs and operating costs (ASCE 2013). Improving these road conditions will likely require an estimated $85 billion between 2008 and 2028, not counting the $170 billion required to maintain the current state of roadways during the same period (ASCE 2013).

Aside from federal funding, states must find solutions to address poor road conditions. Figure 5.2 compares pavement conditions between two different 21-year scenarios evaluated in Texas. In one scenario, Texas spends $1.2 billion per year for maintenance and rehabilitation (M&R), compared to an average spending of only $325 million per year (2030 Committee 2009). This graph shows that at $325 million a year, the percentage of roads in good or better condition would fall to 50% in just 7 years and 0% in 20 years. In this scenario, more than $1.2 billion annually would need to be spent to keep pavement in good condition.

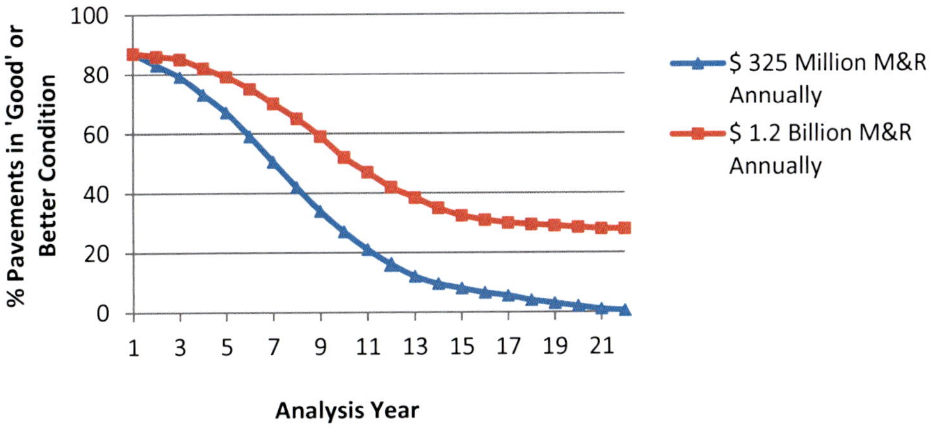

Figure 5.3: Change in Percent of Pavements in "Good" or Better Condition for Two Scenarios in Texas (Source: 2030 Committee 2009)

Bridge Maintenance

Figure 5.4 shows the distribution of construction age for all highway bridges in the United States (FHWA 2011). It indicates that most bridges were built in the 1960s, 1970s, and 1990s. The life span of a bridge is usually considered 50 years (ASCE 2009), meaning that a high percentage of bridges constructed in the 1960s and 1970s will need to be replaced over the next 20 years. ASCE (2013) estimates that replacing all deficient bridges by 2028 would require an annual investment of $20.5 billion As of 2013, little over half of that is being spent to replace and repair bridges nationwide.

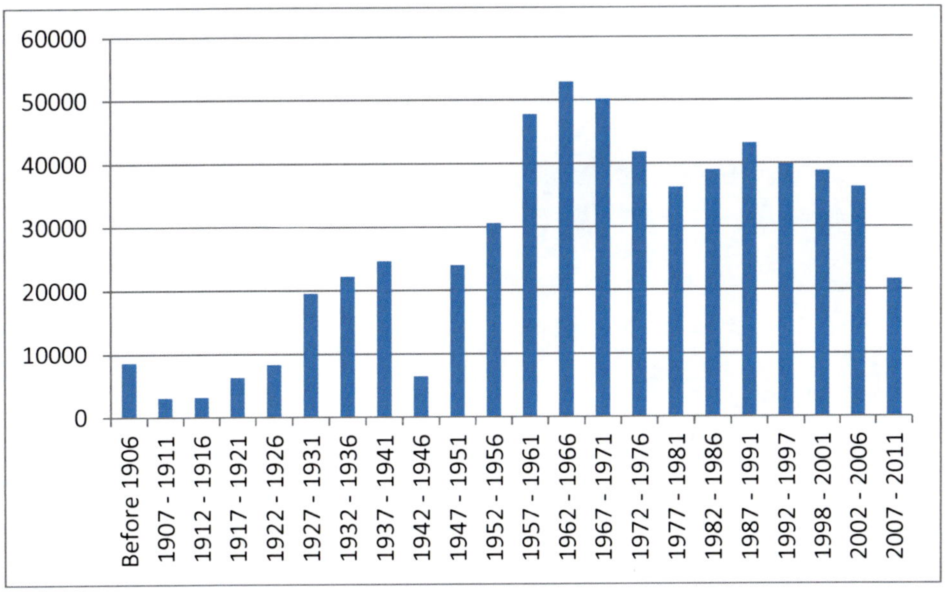

Figure 5.4: Distribution of U.S. Bridges by Construction Period (Source: FHWA 2011)

5.3 Financing

Revenue Sources

Traditionally, highway infrastructure has been built whenever state and federal financing sources are available. Project funding may come from a number of sources, including the following:

1. **Federal Funding**: The U.S. government allocates funding from federal motor fuel taxes, truck tire excise, truck and trailer sales, and heavy vehicle use taxes to different states through the Federal Highway Trust Fund.
2. **Bond Proceeds**: State and local government entities issue bonds to raise money for transportation projects. A bond is a written promise to pay back borrowed money and interest over the life of a bond.
3. **Tolls:** Because transportation agencies face constrained budgets, tolls are used to support transportation investment. Tolls can be considered payment against bonds issued to construct, operate, maintain, and expand facilities, and as a resource to attract private capital to invest in transportation infrastructure.
4. **State Motor Fuel Tax**: Each state sets a motor fuel tax rate on gasoline, diesel, and other special fuels. Fuel tax revenues depend on the amount of fuel consumed; fuel consumption, in turn, depends on vehicle type. Some increasingly common technologies reduce traditional gasoline use (such as electric vehicles).
5. **State Motor Vehicle Tax**: All states levy motor vehicle registration fees based on vehicle characteristics.
6. **Other State Funding:** Other sources of state transportation revenue include property taxes and motor vehicle operator license fees.
7. **Local Funding:** Local government revenue from different sources includes local motor fuel taxes, local motor vehicle registration fees, local option sales taxes, value capture, property taxes, and tolls.

Revenue to fund infrastructure projects primarily stems from seven types of sources.

Resource

CTR provides a model (Vcost) to estimate revenue from Texas fuel taxes; see Welter 2009.

Figure 5.5 shows this revenue source distribution for Texas in 2008.

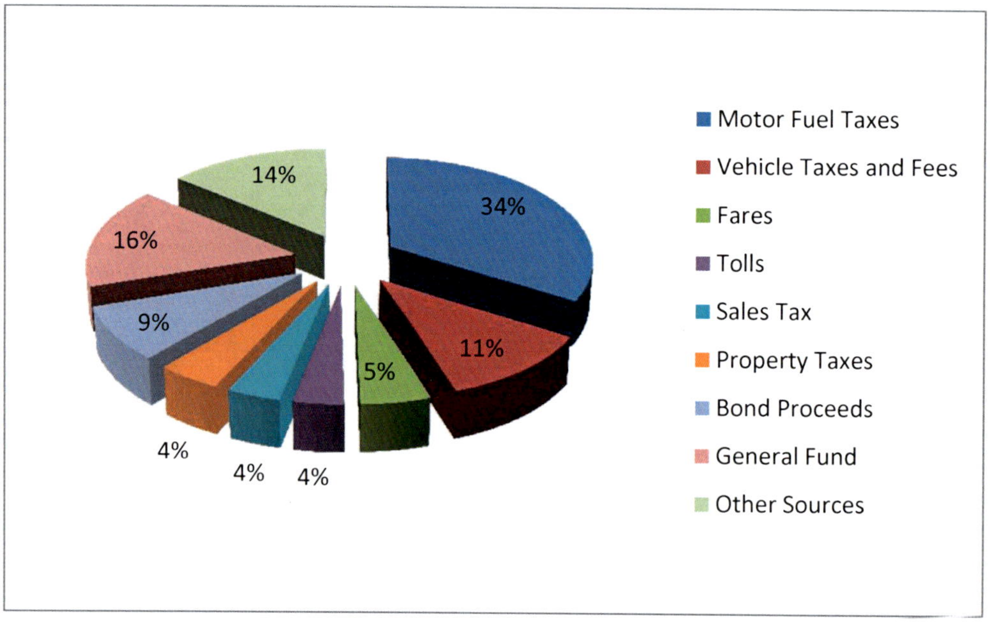

Figure 5.5: Surface Transportation Revenue Sources (FHWA 2004)

Expenditure Sources

Transportation agencies are responsible for using revenue to provide a safe, reliable, and efficient transportation system for the movement of people and goods. To achieve this goal, many state agencies focus on projects that reduce congestion, enhance safety, expand economic opportunity, improve air quality, and preserve the value of transportation assets. Similar projects may frequently compete under the same funding category and must be compared in economic terms, to ensure the project that best meets specific an agency's responsibilities and goals is selected.

State DOT projects may be divided into different funding categories. Texas's DOT (TxDOT) divides projects as follows:

- **Category 1 – Preventive Maintenance and Rehabilitation**: To perform preventive maintenance and rehabilitation of the existing State Highway System.
- **Category 2 – Transportation Metropolitan Area (TMA) Corridor Projects**: To address the mobility needs in all major metropolitan areas with populations greater than 200,000.
- **Category 3 – Urban Area (Non-TMA) Corridor Projects:** To address the mobility needs in metropolitan areas with populations greater than 50,000 and less than 200,000.
- **Category 4 – Statewide Connectivity Corridor Projects:** To address mobility and added capacity needs of major highway system corridors.
- **Category 5 – Congestion Mitigation and Air Quality Improvement:** To improve the air quality in the non-attainment urban areas.
- **Category 6 – Structures Replacement and Rehabilitation:** To replace and rehabilitate deficient existing bridges and deficient railroad underpasses on the state highway system and construct grade separations at existing highway-railroad crossings.
- **Category 7 – Metropolitan Mobility and Rehabilitation**: To address transportation needs within the boundaries of urban MPOs with populations of 200,000 or greater.
- **Category 8 – Safety**: To reduce highway fatalities and major injuries, and provide an appealing environment for primary and middle school children to walk and bicycle to schools. Also, to eliminate hazards at highway-railroad crossings.
- **Category 9 – Transportation Enhancements**: To enhance the transportation system in less conventional areas such as historic preservation or tourism programs.
- **Category 10 – Supplemental Transportation Projects:** To address projects that are not qualified for funding in other categories such as state park roads, curb ramp programs, railroad grade crossing re-

TxDOT expenditures fall into 12 equally important funding categories.

planking programs, and truck weight stations.

- **Category 11 – District Discretionary**: To address projects selected at the district engineer's discretion.
- **Category 12 – Strategic Priority:** To fund projects with specific importance to the state based on an "as-needed" basis.

Across these categories, TxDOT's total disbursements were over $4.9 billion in 2008 (Saenz 2008) and distributed as shown in Figure 5.6.

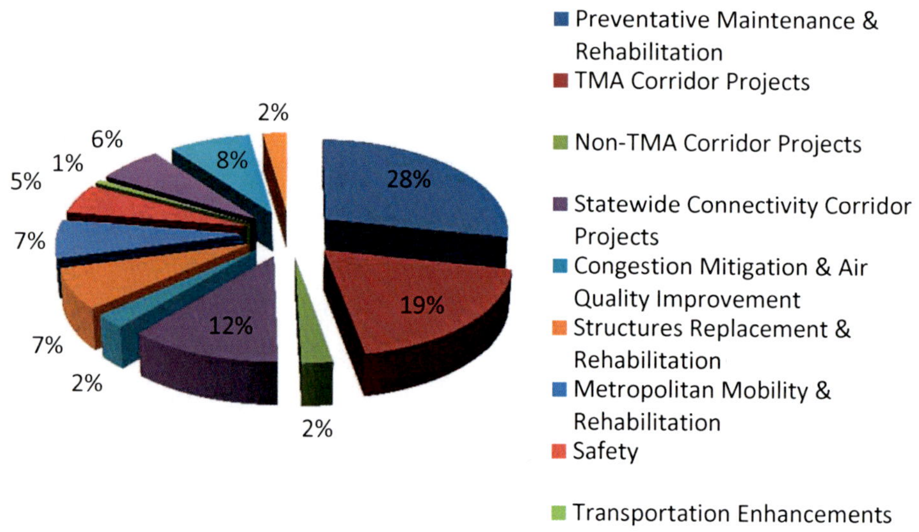

Figure 5.6: TxDOT 2008 Disbursements

Traditional Project Delivery Methods

Many state DOT's have traditionally used, and were once legally limited, to using a design-bid-build (DBB) project delivery method. DBB deliberately separates design and construction by contractors and by sequence (i.e., design must be complete before separate construction firms begin bidding). This deliberate separation was originally favored because competitive bidding was encouraged and public funds were protected against "graft and favoritism" (Walewski et al. 2001). However, in the 2000s, DOTs began adopting project delivery methods like design-build (D-B) to expedite project completion and attract private investment for large projects like the IH 635 Managed Lanes project in Dallas, the Denver Union Station, and the I-15 Corridor Reconstruction in Salt Lake City, among many others . Details on public-private partnerships and their funding advantages and disadvantages are presented later in the following section.

Innovative Financing

Beginning in the mid-1990s, the FHWA began developing a series of flexible financing methods to fill the gap between state infrastructure needs and federal funding. A series of new federal loans and credit enhancement programs were provided under the National Highway System Designation Act in 1995, the Transportation Equity Act in 1998, and most recently SAFETEA-

LU in 2005. Finance methods developed through these acts allowed for greater public and private investment at lower interest rates through credit enhancement and provided states greater access to capital through loans. Types of financing options developed in this period include Section 129 loans, state infrastructure banks, and public-private partnerships. Many of these innovative methods have now become typical finance procedures, and are discussed in detail here.

Flexible financing approaches help DOTs meet their infrastructure needs.

Section 129 Loans

Section 129 loans use federal highway apportionments to fund loans for toll and non-toll projects alike. These loans are extended to public or private entities for projects with a dedicated revenue source (user fees or taxes). Section 129 loans help borrowers (i.e., states) establish credit on the loaned funds, which allows reinvestment and "recycling" in other projects. Up to 80% of the maximum federal share may be lent through Section 129 loans, and the borrower must begin return payments within 5 years of project completion, with all borrowed funds repaid within 30 years of loan authorization (USDOT 2010a). This allotted time in repayment allows flexibility in cases of start-up delays and extended project lengths. Regardless of these benefits, the popularity of Section 129 loans has waned after the introduction of the more recent TIFIA credit program (described later in this section).

The federally funded Section 129 loans have flexible repayment terms.

Example

Section 129 Loan Financing Example

The first project to receive funding though a Section 129 loan was the President George Bush Turnpike (Highway 190) in Dallas. The four-to-eight lane urban toll facility required funding beyond what is typically available at a state level. TxDOT and the Texas Turnpike Authority (TTA) received a $135 million Section 129 loan, which was disbursed in five different payments over a 4-year period (USDOT 2010a). These separated disbursements allowed TxDOT to avoid one single repayment of $135 million and allowed for advanced construction. Use of the Section 129 loan in this manner, alongside toll revenue bonds, reportedly accelerated the project's completion date by over a decade.

State Infrastructure Banks (SIBs)

State infrastructure banks (SIBs) are revolving funds that provide loans and credit assistance for surface transportation projects established and administered by states. SIB assistance attracts non-federal public and private investments to increase the efficiency of states' transportation investments and leverage federal resources. Highway projects under Title 23 of the U.S. Code, federally aided projects, or Title 49 transit capital projects can use SIB funds. If a community deems a local project vital, SIB funds may be requested to

For highway projects, state-administered SIBs offer loans at or below market rates.

expedite the project's completion.

SIBs act much like private banks and offer assistance either by loans or credit enhancement, which can accelerate project completion; incentivize state, local, and private investment; and recycle funds for future projects. Historically SIBs were available only to a limited number of states, but have since become an option for all states as part of SAFETEA-LU.

SIB assistance includes "loans (at or below market rates), loan guarantees, bond insurance, and other forms on non-grant assistance" (USDOT 2010a). Loans through SIBs have a maximum term of 30 years with no more than 5 years allowed between commencement of repayment and project completion. SIBs are considered "revolving" because, although initially capitalized, federal funds are to be repaid directly to the state after they are lent out. These returned funds are then eligible for Title 23 project investment in the state.

SIB-funded projects may benefit from flexible project financing and increased private investment, and can aid the state by accelerating certain projects and recouping investments for future works.

Private Activity Bonds (PABs)

A section of SAFETEA-LU created private activity bonds (PABs), which are tax-exempt bonds eligible for disbursement on privately developed highway and freight transfer facilities. To be eligible for PAB support, projects must be receiving Title 23 federal assistance or be an international bridge or freight transfer facility initiative. No more than $15 billion can be provided in the form of PABs (USDOT 2005).

Example

PAB Financing in the U.S.

As of 2010, PABs have been approved for seven projects, totaling $6.3 billion in financing (US DOT 2011).

	PAB Allocation
Capital Beltway HOT Lanes, Virginia	$589 million
North Tarrant Express, Texas	$400 million
IH635 (LBJ Freeway), Texas	$2,650 million
RidgePort Logistics Center, Will County, Illinois	$554 million
CenterPoint Intermodal Center, Joliet, Illinois	$1,340 million
Knik Arm Crossing, Alaska	$600 million
Mississippi DOT Jackson Airport Parkway	$200 million

PABs are U.S. bonds issued for up to $15 billion.

The Transportation Infrastructure Finance and Innovation Act (TIFIA)

The Transportation Infrastructure Finance and Innovation Act (TIFIA), approved in YEAR?, fills market gaps and leverages substantial private co-investment. The TIFIA provides supplemental and subordinate capital amounting to 33% of eligible project costs in the form of secured (direct) loans, loan guarantees, or standby lines of credit (AASHTO 2010c). Highway and transit projects already eligible for federal funding and with a revenue source such as tolls or local sales taxes are eligible for TIFIA credit assistance. As of 2009, projects are given an investment rating, based on the following criteria (Table 5.1), with the associated relative weights of importance.

Federal assistance from TIFIA is designed to expedite large projects.

Private Participation	20%
Environmental Impact (Sustainability and Repair)	20%
National or Regional Significance (Livability, Economic Competitiveness , and Safety)	20%
Project Acceleration	12.5%
Creditworthiness	12.5%
Use of New Technologies	5%
Reduced Federal Grant Assistance	5%
Consumption of Budget Authority	5%

Table 5.1: Criteria for TIFIA Project Investment Rating and Relative Weights (Source: adapted from USDOT 2010a)

TIFIA assistance is intended to expedite large projects by providing more flexible repayment terms and reduced interest rates. Thus, restrictions state that TIFIA-supported projects must be no less than $50 million for typical surface transportation projects; Intelligent Transportation Systems (ITS) projects may be eligible at a size of $30 million and above (USDOT 2010a). As of 2010, TIFIA assistance has drawn $29 billion in project investment and provided $7.7 billion for 21 projects (USDOT 2010a).

Example

TIFIA Financing in the U.S. (USDOT 2011)

Project	Project Cost	TIFIA Direct Loan
Central Texas Turnpike System (SH 45 North, Loop 1, SH 130)	$3,277.8 million	$900 million
Triangle Expressway Raleigh-Durham, NC	$1,171.7 million	$386.7 million
Crenshaw/LAX Transit Corridor Los Angeles, CA	$1,749 million	$545.9 million
Capital Beltway HOT Lanes Fairfax County, VA	$2.068 billion	$589 million
SR 520 Floating Bridge Seattle, WA	$2,845 million	$300 million
Port of Miami Tunnel	$1,113 million	$341.5 million

TIGER Grants and MAP-21

Definition

The Transportation Investment Generating Economic Recovery (TIGER) grant program began with the passage of the American Recovery and Reinvestment Act of 2009 (ARRA). Title 12 of the act appropriated funds for supplementary discretionary grants "awarded on a competitive basis for capital investments in surface transportation projects that have a significant impact on the nation, a metropolitan area, or a region." Eligible applicants included state and local governments, transit agencies, port authorities, MPOs, and multi-state groups. Eligible projects included highways or bridges, public transportation, passenger/freight rail infrastructure, and port infrastructure. With the passing of the latest transportation bill, Moving Ahead for Progress in the 21st Century (MAP-21), large regional projects will be funded under the Projects of National and Regional Significance (PNRS) program (with a year-2013 budget of $500 million), through a competitive process similar to an infrastructure bank or TIGER.

The federally funded TIGER grants are available to many types of jurisdictions and agencies.

Process

Evaluation criteria for TIGER grants fell under four categories: job creation and economic stimulus; innovation and partnership; project-specific benefits; and long-term outcomes relating to economic competiveness and improvements to the condition of transport facilities and systems.

Applicants for TIGER grants were required to identify, quantify, and compare expected project benefits and costs compared to a base case. Monetized costs include construction, design, right-of-way (ROW) acquisition, operations and maintenance, life-cycle costs, noise, congestion, emissions, and anticipated user costs during construction.

The potential project benefits fell under these categories:

- **Livability**: Improved access to jobs, amenities, and for disadvantaged communities, land use changes, transit, pedestrian and bicycle improvements, and transportation and housing affordability.
- **Economic competitiveness**: Changes in operating costs, travel times, user out-of-pocket costs, and reliability. Job creation benefits should only be measured using productivity increases while avoiding multiplier effects and double counting.
- **Safety**: Savings in fatality, injury, and crash costs.
- **State of good repair**: Reductions in long-term maintenance costs and closure time (travel time savings).
- **Sustainability**: Emissions savings valuation for carbon dioxide, sulfur dioxide, nitrogen oxide, and particulate matter.

The program required discounting monetized estimates of yearly benefits and costs to represent present value. From 2009 to 2012 the TIGER grant program awarded $2.5 billion to passenger and freight transportation projects. Awards tended to favor public transit projects that encourage environmental sustainability.

Tax Increment Financing (TIF)

Definition

A special provision in most state laws allow cities to designate areas as Tax Increment Finance (TIF) districts—also called Tax Increment Reinvestment Zones (TIRZs)—to promote development or redevelopment of the area if development is not projected to occur solely through private investment. After it is designated a TIF district, a portion of the tax increase from a district's investment-related increases in property values can be used to pay off capital bonds for public investments. A state agency has no legal authority to create a TIF; TIFs can be used only by local jurisdictions (such as cities) that can levy and spend the related taxes. Counties, school districts, and other special distracts can participate in the TIF agreement after a city has started to establish a TIF zone.

TIF districts represent creative funding at the level of local jurisdictions.

See the In-Depth Look section at this chapter's end for details on TIF and TRIZ legislation and valuation.

Process

Local residents may petition for their neighborhood to be designated a TIRZ, and a city council may initiate the TIF process for an area if it meets at least one of the following criteria:

- The area's present condition substantially impairs the city's growth, provision of housing, or constitutes an economic or social liability to the public health, safety, morals, or welfare. This condition must exist because of the presence of infrastructure problems, unsafe conditions, taxes or assessments that exceed fair market value of the land, or title problems.

- The area is predominately open, and because of obsolete platting, deteriorating structures, or other factors, the open area substantially impairs the growth of the city.
- The area is in or adjacent to a "federally assisted new community" as defined under Tax Code Section 311.005(b).

Example

TIF Financing

The Flats East Bank neighborhood in Cleveland, Ohio was revitalized with a TIF agreement beginning in 2005. The $400 million redevelopment created a mixed-use district and required significant infrastructure redevelopment. Over $100 million of the total project costs involved road and utility improvements, which were funding through infrastructure bonds, partnerships with city and regional utilities, and grants from a number of state and federal sources. This development also ensured that the Cleveland Municipal School District received 50% of newly generated property taxes, ensuring that the original funding was not transferred unfairly from the local school district (CDFA 2007)

Public-Private Partnerships (PPP)

Public-private partnership (PPP) agreements are another way to overcome budgetary constraints and achieve higher standards and operation efficiencies for infrastructure services. PPPs are agreements between a public agency (federal, state, or local) and the private sector. Through public-private agreements, the private sector can contribute to the delivery of the highway infrastructure by participating in the financing, development, and operation of such projects for a specific period of time under a concession agreement. The private investor intends to recover their investment through a guaranteed revenue stream such as fees, tolls, and tax increment financing. These agreements are not always straightforward, and governments have concerns with issues such as transparency and competitiveness of the bids and appropriate allocation of risks. Table 5.2 presents the various PPP structures based on level of responsibility taken by private sector.

PPPs are project-specific solutions to budgetary shortfalls.

Type of Facilities	Public-Private Partnership Options		
	Public ⟷		Private
New Build Facilities	Design-Build (DB)	Design-Build-Operate-Maintain (DBOM)	Design-Build-Finance-Operate (DBFO)
Existing Facilities	Operation and Maintenance Concession		Long-Term Lease
Hybrid			Lease-Develop-Operate (LDO)

Table 5.2: Various PPP Structures (Source: USDOT 2010a)

New Build Facilities (DB Partnerships)

PPP is a project delivery method in which owners may have a single fixed-fee contract with the private sector for its services, such as designing, constructing, maintaining, operating, and/or financing. The private sector representative can be a single firm, joint venture, consortium, or other organization assembled for a particular project. To build new facilities, three different DB partnership options are available with varying levels of private sector responsibilities: Design-Build, Design-Build-Operate-Maintain, and Design-Build-Finance-Operate. This type of project delivery method has both advantages and disadvantages.

Advantages of DB partnerships

- High level of communication between the design and construction teams often allows projects to be easily fast-tracked and decreases unforeseen problems (Gould & Joyce 2003).
- Risk-sharing makes it easier for both parties to build large, costly infrastructure projects (Dutzik & Schneider 2011).
- These projects demonstrably accelerate project completion time and provide a more efficient and quicker method of project delivery (FHWA 2010b).
- Overlapping design and construction tasks result in both construction and design influencing each other in a continuous and dynamic manner (FHWA 2010b).

One example of DB risk occurred in July 2011, when a Las Vegas contractor abruptly went bankrupt and quit construction of two flyover bridges in Austin, causing months of delay (Austin Business Journal 2011).

Disadvantages of DB partnerships

- They lack the checks and balances of traditional partnerships—because the designer and the contractor are co-workers in a DB setup, the designer might not oversee the contractor's work or identify potential deficiencies work (Gould & Joyce 2003).

- Risk-sharing—a potential benefit—become less beneficial when the project's outcome is utterly dependent upon the success (or failure) of a private company (Dutzik & Schneider 2011).
- Poorly written contracts may allow private companies excessive power to slow work or alter terms of delivery in their favor; the public agency could lose oversight control of PPP projects that have become "too big to fail."

Responsibilities

The private sector's responsibility is based on the type of the contract. In a Design-Build setup, the private sector is responsible for the majority of the design work and all construction activities, while in the Design-Build-Operate-Maintain setup, the private sector is responsible for designing, building, and providing long-term operation and maintenance services. In the Design-Build-Finance-Operate setup, the private sector has the responsibility for almost all activities in a project and retains the operation revenue risk as well as any surplus operating revenue, while the owner retains only ownership over the contract. Figures 5.7, 5.8, and 5.9 illustrate the contract setup for the three different build options.

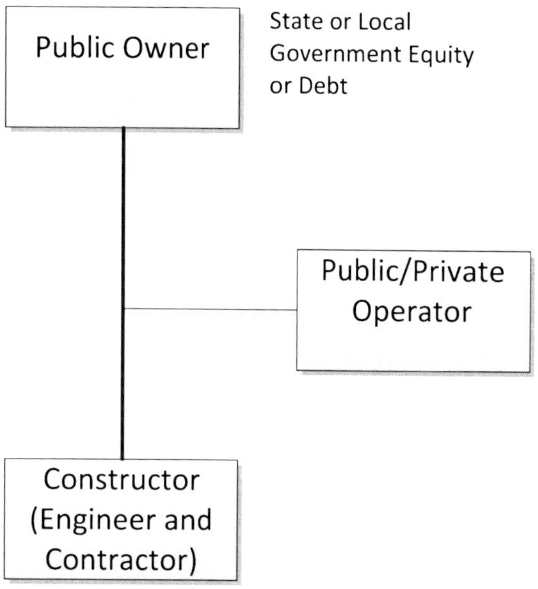

Figure 5.7: Design-Build Chart (Source: USDOT 2010a)

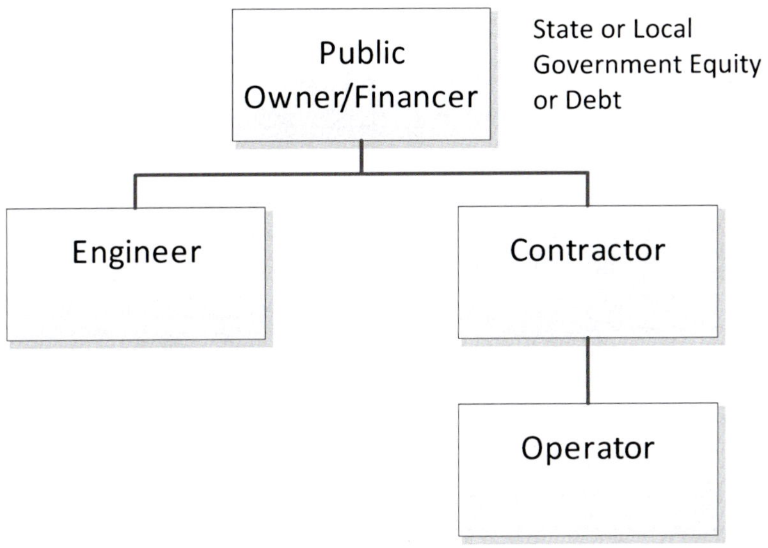

Figure 5.8: Design-Build-Operate-Maintain Chart (Source: USDOT 2010a)

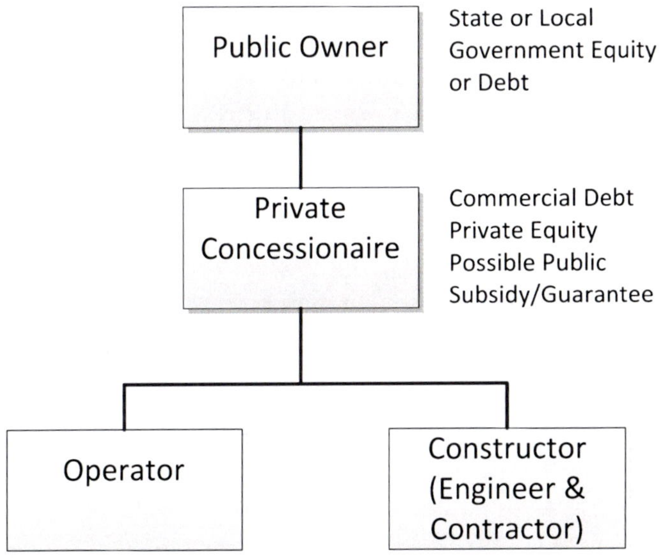

Figure 5.9: Design-Build-Finance-Operate Chart (Source: USDOT 2010a)

Example

New Build PPP

To build 14 miles of the Capital Beltway High Occupancy Toll Lanes, the Virginia Department of Transportation made an agreement with a joint venture of Fluor and Transurban, two private transportation firms. The project delivery system was Design-Build-Finance-Operate, with a lump sum contract and a fixed contract time. The total length of the concession is 5 years of construction and 80 years of operation, totaling 85 years (USDOT 2011).

Existing Facilities (Concession or Lease)

For existing facilities, public operating agencies can transfer operations and maintenance concession of existing facilities to the private sector. This transfer can be done in two ways: as an Operation and Maintenance Concession, or as a Long-Term Lease. In the Operation and Maintenance Concession scenario, the private sector can be paid on either a fixed fee or on an incentive basis, including premiums for meeting specified service levels or performance targets. In the Long-Term Lease option, the private sector has the right to collect tolls on the facility (USDOT 2010a).

Advantages

The owner can more easily take advantage of life cycle cost and asset management practices, emphasizing cost effective planning and resource allocation for the preservation, upgrade, and timely replacement of highway assets. Also, the public sector can benefit from private sector operational and maintenance efficiencies to reduce ongoing operating and maintenance costs.

Disadvantages

Long-term operating contracts can lead to inefficiencies due to lack of competition. The private entity with the exclusive rights to the infrastructure facility is essentially operating a monopoly.

Responsibilities

In the Operation and Maintenance Concession scenario, the owner retains ownership and overall management of the public facility. The contractors are responsible for ongoing activities such as snow removal, mowing, maintenance, and major repairs. In the Long-Term Lease option, the owner retains only project ownership and the contractors may be asked to make improvements to the facility. Figures 5.10 and 5.11 illustrate these contracts.

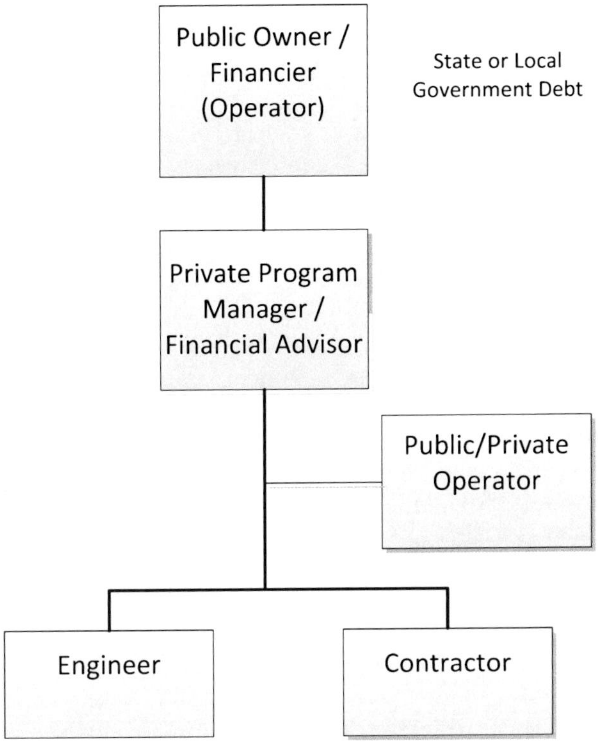

Figure 5.10: Operation and Maintenance Concession Chart (Source: USDOT 2010a)

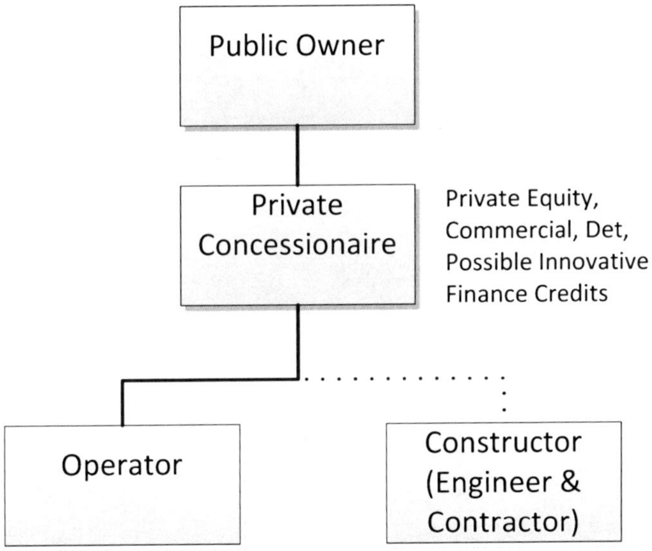

Figure 5.11: Long-Term Lease Chart (Source: USDOT 2010a)

Example

Long-Term Lease PPP

In October 2004, Cintra/Macquarie bid $1.83 billion to operate and maintain the Chicago Skyway for a period of 99 years. The private company acquired the right to collect all tolls and concession revenue. This project was the first long-term lease of an existing facility in the U.S. This agreement funded a $500 million long-term and $375 million medium-term reserve for the city of Chicago and gave the city a $1.83 billion cash infusion (Skyway Concession Company, LLC 2005).

Hybrid Partnerships for Existing Facilities

In this type of agreement, a private firm leases an existing public facility and invests its own capital to improve and expand the facility under a revenue-sharing contract for a fixed term. The private sector can recover its investment plus a reasonable return from the facility's revenue. Legal ownership is still maintained by the public in Lease-Develop-Operate agreements, but in this arrangement private investment can bring alternative solutions to projects that are losing money for the public agency. **This leasing flexibility allows private firms to share profits of a project without having to fully purchase the project**, which may be outside of their purchasing ability (Finnerty 2007). This arrangement is charted in Figure 5.12.

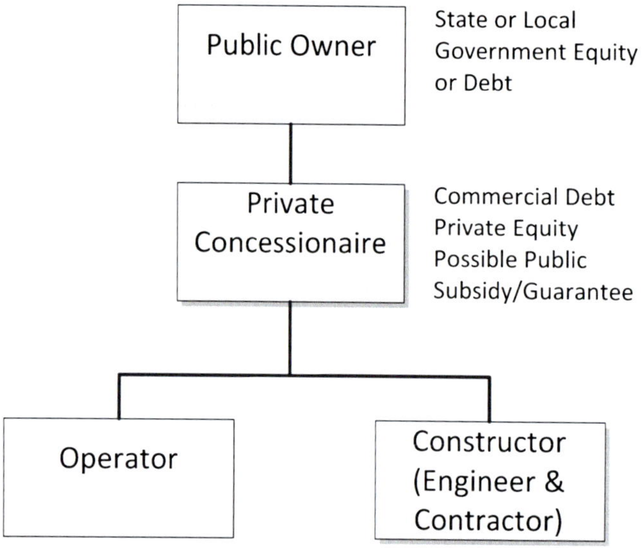

Figure 5.12: Lease-Develop-Operate Chart (Source: USDOT 2010a)

Example

Lease-Develop-Operate PPP

The private company Transurban has been given the rights to enhance, manage, operate, maintain, and collect tolls on the Pocahontas Parkway/Richmond Airport Connector for 99 years. Also, Transurban was made responsible for Pocahontas Parkway Association's underlying debt and was obliged to construct a 1.58-mile extension of the toll road.

The financing techniques described in this section are meant to fill the gap between traditional government funding sources and transportation needs in order to improve the nation's transportation infrastructure systems. Many government organizations across the United States have used these tools.

5.4 Summary

This chapter examined the impact of transportation infrastructure on society by looking at economic outputs, the condition of U.S. rail, roads, and bridges, and the gap between available and needed funds. Innovative methods such as Section 129 loans, SIBs, and TIFIA assistance can provide alternative financing options when conventional funding sources are inadequate. To encourage the private sector to participate in designing, building, operating, and maintaining transportation infrastructure, various types of public-private partnership agreements are included as part of modern transportation policy.

5.5 References

Akintoye, A., Beck, M., & Hardcastle, C. (2003) Public Private Partnerships Managing Risks and Opportunities. Oxford, UK. Blackwell Science Ltd.

American Society of Civil Engineers (ASCE) (2009) Report Card for America's Infrastructure. Available online at http://www.infrastructurereportcard.org/.

American Society of Civil Engineers (ASCE) (2013) Report Card for America's Infrastructure. Available online at http://www.infrastructurereportcard.org/.

Austin Business Journal (2011) Mopac flyover work stopped after contractor files bankruptcy. Available online at http://www.bizjournals.com/austin/news/2011/07/18/mopac-flyover-construction-stopped.html.

Burchell, R. W. (1978) Fiscal Impact Handbook. The Center for Urban Policy Research, New Brunswick, NJ.

Cambridge Systematics, Inc. (2007) National Rail Freight Infrastructure Capacity & Investment Study. Available online at: http://www.camsys.com/pubs/AAR_Nat_%20Rail_Cap_Study.pdf.

Council of Development Finance Agencies (CDFA) (2007) Tax Increment Finance Best Practices Reference Guide. http://www.icsc.org/government/CDFA.pdf.

Dutzik, T. & Schneider, J. (2011) High Speed Rail: Public, Private or Both? The Frontier Group. US PIRG Education Fund. Available online at http://cdn.publicinterestnetwork.org/assets/85a40b6572e20834e07b0da3e66e98bf/HSR-PPP-USPIRG-July-19-2011.pdf.

Federal Highway Administration (FHWA) (2004) Revenue Sources for Surface Transportation Funding. FHWA Office of Innovative Financing. http://www.transportation-finance.org/funding_financing/funding/.

Federal Highway Administration (2011) National Bridge Inventory. U.S. Department of Transportation. http://www.fhwa.dot.gov/bridge/nbi/yrblt11.cfm.

Finnerty, J. D. (2007) Project Financing: Asset-Based Financial Engineering. John Wiley & Sons, Inc., Hoboken, NJ.

Gould, F. E. & Joyce, N. E. (2003) Construction Project Management. Upper Saddle River, NJ: Pearson Education, Inc.

Government Accountability Office (GAO) (2006) Freight Railroads: Industry Health Has Improved, but Concerns about Competition & Capacity Should Be Addressed. Available online at http://www.gao.gov/new.items/d0794.pdf.

Malhorta, A. K. (1997) Private Participation in Infrastructure: Lessons from Asian's Power Sector. *Finance and Development*, December, 33–35.

National Council for Public-Private Partnerships (2010) How PPPs Work. Available online at http://www.ncppp.org/howpart/index.shtml#define.

Passenger Rail Working Group (PRWG) (2007) Vision for the Future: U.S. Intercity Passenger Rail Networking through 2050. Available online at http://www.dot.state.wi.us/projects/state/docs/prwg-report.pdf.

Persad, K., Walton, C. M., & Franco, P. (2008) Financing Tools and Partnerships for Rural and Semi-Rural Transportation Projects. Center for Transportation Research, University of Texas at Austin. Available online at http://www.utexas.edu/research/ctr/pdf_reports/0_6034_P1.pdf.

Saenz, A. (2008) Transportation Program Expenditures – Fiscal Year 2008. Texas Department of Transportation. Available online at http://ftp.dot.state.tx.us/pub/txdot-info/library/reports/expenditures/fy2008.pdf.

Shrank, D., B. Eisele, & T. Lomax (2012) TTI's 2012 Urban Mobility Report. Texas A&M Transportation Institute. The Texas A&M University System. Available online at http://d2dtl5nnlpfr0r.cloudfront.net/tti.tamu.edu/documents/mobility-report-2012.pdf.

Sinha, K. C. & Labi, S. (2007) Transportation Decision Making—Principles of Project Evaluation and Programming. Hoboken, NJ: John Wiley & Sons, Inc.

Skyway Concession Company, LLC (2005) Chicago Skyway. Available online at http://www.chicagoskyway.org/.

Texas Department of Transportation (TxDOT) (2001) Transportation Planning Manual. Available online at http://onlinemanuals.txdot.gov/txdotmanuals/pln/pln.pdf.

Texas Department of Transportation (TxDOT) (2011d) Pass-Through Financing Program. Available online at http://www.txdot.gov/business/governments/pass_finance.htm.

Texas Department of Transportation (TxDOT) (2011e) TxDOT Tracker: Semi-Annual Results – FY 2011. Available online at http://www.txdot.gov/txdot_tracker/.

Texas Legislative Budget Board Fiscal Note (2011, May 12) Available online at http://www.legis.state.tx.us/tlodocs/82R/fiscalnotes/html/HB00563F.htm.

Texas Transportation Code Chapter 222 (2011). Available online at http://www.statutes.legis.state.tx.us/Docs/TN/htm/TN.222.htm.

Texas Transportation Institute (TTI) (2007) The 2007 Urban Mobility Report. Available online at http://www.commutercars.com/downloads/UrbanMobility07.pdf.

TRIP, A National Transportation Research Group (2010a) Key Facts about America's Surface Transportation System and Federal Funding. Available online at http://www.tripnet.org/Fact_Sheet_National.pdf.

TRIP, A National Transportation Research Group (2010b) Key Facts about Texas's Surface Transportation System and Federal Funding. Available online at http://www.tripnet.org/Fact_Sheet_TX.pdf.

US Department of Transportation (USDOT), AASHTO Center for Excellence in Project Finance (2010a) State Infrastructure Banks. Available at: http://www.transportation-finance.org/funding_financing/financing/credit_assistance/state_infrastructure_banks.aspx.

US Department of Transportation (USDOT), AASHTO Center for Excellence in Project Finance (2010b) TIFIA. Available at: http://www.fhwa.dot.gov/ipd/tifia/index.htm.

US Department of Transportation (USDOT), AASHTO Center for Excellence in Project Finance (2010c) Project Finance State by State. Available online at http://www.transportation-finance.org/tools/state_by_state/revenue_sources.aspx?state=tx.

US Department of Transportation (USDOT), Federal Highway Administration (2002) Innovative Finance Primer. Available online at http://www.fhwa.dot.gov/innovativefinance/ifp/index.htm.

US Department of Transportation (USDOT), Federal Highway Administration (2005) A Summary of Highway Provisions in SAFETEA-LU. Available at: http://www.fhwa.dot.gov/safetealu/summary.htm.

US Department of Transportation (USDOT), Federal Highway Administration (2008b) Highway Statistics 2008. Available at: http://www.fhwa.dot.gov/policyinformation/statistics/2008/.

US Department of Transportation (USDOT), Federal Highway Administration (2010a) Innovative Finance Primer. Available online at http://www.fhwa.dot.gov/innovativefinance/ifp/index.htm.

US Department of Transportation (USDOT), Federal Highway Administration (2010b) U.S. Department of Transportation, Public-Private Partnerships. Available at: http://www.fhwa.dot.gov/ipd/p3/index.htm.

US Department of Transportation (USDOT), Federal Highway Administration (2010c) Case Studies. Available at: http://www.fhwa.dot.gov/ipd/case_studies/tx_sh130.htm.

US Department of Transportation (USDOT), Federal Highway Administration (2011) Innovative Program Delivery. Available online at http://www.fhwa.dot.gov/ipd/tifia/projects_project_profiles/index.htm.

US Department of Transportation (USDOT), Federal Highway Administration (2011b) Federal Debt Financing Tools: Private Activity Bonds (PABs). Available online at http://www.fhwa.dot.gov/ipd/finance/tools_programs/federal_debt_financing/private_activity_bonds/index.htm.

US Department of Transportation (USDOT), Federal Railroad Administration (2011a) Financial Assistance: Passenger Rail. Available at: http://www.fra.dot.gov/rpd/passenger/34.shtml.

US Department of Transportation (USDOT), Federal Railroad Administration (2011b) Railroad Rehabilitation & Improvement Financing (RRIF) Program. Available at: http://www.fra.dot.gov/rpd/freight/1770.shtml.

Walewski, J., Gibson Jr., G. E., & Jasper, J. (2001) Project Delivery Methods and Contracting Approaches Available for Implementation by the Texas Department of Transportation. Center for Transportation Research, the University of Texas at Austin. Prepared for Texas Department of Transportation.

Weisbrod, G. & Weisbrod, B. (1997) Assessing the Economic Impact of Transportation Projects: How to Choose the Appropriate Technique for Your Project. *Transportation Research Circular*, 447. Washington, DC: Transportation Research Board.

Welter, D., Ates, M., Loftus-Otway, L., Matthews, R., & Harrison, R. (2009) Estimate Texas Motor Vehicle Operating Cost. Center for Transportation Research, Austin, TX. Available at http://www.utexas.edu/research/ctr/pdf_reports/0_5974_1.pdf.

Window on State Government (2004) Title 3. Local Taxation, Subtitle B. Special Property Tax Provisions, Chapter 311. Tax Increment Financing Act. Available at: http://www.window.state.tx.us/taxinfo/proptax/tc04/ch311.htm.

Window on State Government (2004a) Tax Increment Financing Registry. Available at http://www.window.state.tx.us/taxinfo/proptax/registry04/zone.html.

Window on State Government (2004b) Tax Increment Financing Registry – Bexar County. Available at http://www.window.state.tx.us/taxinfo/proptax/registry04/bexarco.pdf.

Chapter 6. Project Evaluation

6.1 Introduction

While private companies or individuals generally evaluate how new projects will impact their own financial accounts, transportation agencies must consider **how potential projects and policies will impact a much larger community's well-being**—including economic, environmental, equity, and other impacts. In either setting, the decision to pursue a project or policy requires structured processes to ensure choice of the most beneficial alternatives.

Fortunately, a variety of valuable tools exist for project evaluation and decision-making, such as discounting of future costs and generating benefit-cost ratios. Such accounting can provide a priceless transparency for an agency and its many stakeholders.

These concepts and metrics presented are important in quantifying the net benefits of different alternatives, particularly important in budget-constrained situations, which all state DOTs face.

Traditional engineering techniques alone generally cannot address a variety of less tangible agency concerns regarding important economic consequences that are difficult to measure in monetary terms alone, such as environmental justice and community preferences. Thus, this chapter presents multi-criteria analysis methods to assess these non-economic elements.

Moreover, investment in transportation infrastructure is risky, as uncertainty surrounds construction costs, future demand, and maintenance costs. Example of sensitivity analysis and breakeven analysis are therefore also included in this chapter.

Key Terms

- ✓ **NPV:** net present value
- ✓ **IRR:** internal rate of return
- ✓ **MARR:** minimum accepted rate of return
- ✓ **ΔROR:** incremental rate of return
- ✓ **CBA:** cost-benefit analysis
- ✓ **B/C ratio:** benefit/cost ratio
- ✓ **LCCA:** life cycle cost analysis
- ✓ **MCA:** multicriteria analysis
- ✓ **SAW:** simple additive weighting
- ✓ **DEA:** data envelope analysis
- ✓ **DMU:** decision-making unit

6.2 Engineering Economic Analysis

This section presents an overview of the traditional engineering economic concepts of discount rate, internal rate of return, payback period, and several other essential economic calculations used to generate the necessary economic analysis data.

Discount Rate and Time Value of Money

The purchasing power of money normally decreases over any given period of time due to inflation and uncertainty. **A discount rate adjusts the value of money for time,** expressing expected future monetary quantities in terms of their worth today. Following are the two different kinds of interest rates:

1. Real interest: rate exclusive of inflation
2. Nominal interest: rate inclusive of inflation

Net Present Value (NPV)

A net present value (NPV) calculation is used to state a project's worth or cost for its entire life cycle in today's dollars or at some specific point in time. NPV is calculated as follows:

$$NPV = -C_i + SV(\frac{1}{1+DR})^{proj\ life} - IPC(\frac{1}{1+DR})^{year} + \sum_{y=1}^{proj\ life} (B_y - C_y)(\frac{1}{1+DR})^{y}$$

where C_i is the initial project cost, SV is the salvage value, IPC are interim project costs (generally involving highway capacity additions and upgrades), DR is the discount rate (as a proportion, rather than as a percentage, e.g., 0.05 instead of 5%), B_y denotes the monetized benefits realized in year y, and C_y denotes costs realized in year y.

To calculate the present value of money, the following formulas can be used, where ***F*** is future value, ***P*** is present value, ***i*** is the discount rate per period, and ***N*** is the number of compounding periods.

Note these variable definitions for the equations in this section.

Single Payment

To calculate the present value of a single payment in the future:

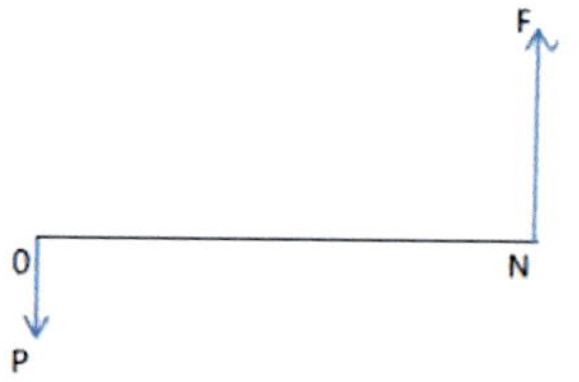

$$To\ find\ P, given\ F\ (P|F,i,N): P = F \times (1+i)^{-N}$$

Example of Simple Compound Interest

If you have $100 today and invest it at 10% simple annual compound interest rate per year for 2 years, you will have the following:

Interest earned in year 1: $100 x %10 = $10

Interest earned in Year 2: $110 x %10 = $21

Total interest earned in 2 Years = $31

Present value = $100

Future value at the end of Year 2 = $131

Example of Interest Calculation

Suppose a planned project is suddenly delayed for 2 years. Construction, labor, and materials costs are expected to increase 2% annually during the delay, but the unused funds could meanwhile accrue 1.75% interest in other investments.

Current construction cost: $10,000,000

Present value of funds from interest:

10,000,000=F ×(1+0.0175)$\$10{,}000{,}000 = F \times (1 + 0.0175)^{-2}$ => F_i = $10,353,063

Therefore, in 2 years, the money not spent on construction could earn $353,063 from interest.

Present value of inflation costs:

10,000,000=F ×(1+0.02)$\$10{,}000{,}000 = F \times (1 + 0.02)^{-2}$ => F_f= $10,404,000

Rising materials and labor costs would increase construction costs $404,000 in two years. Considering both interest and inflation, a 2-year delay would cost overall $50,937.

Equal Payment Series

To calculate the present value of constant cash flow (one payment "*A*" per period):

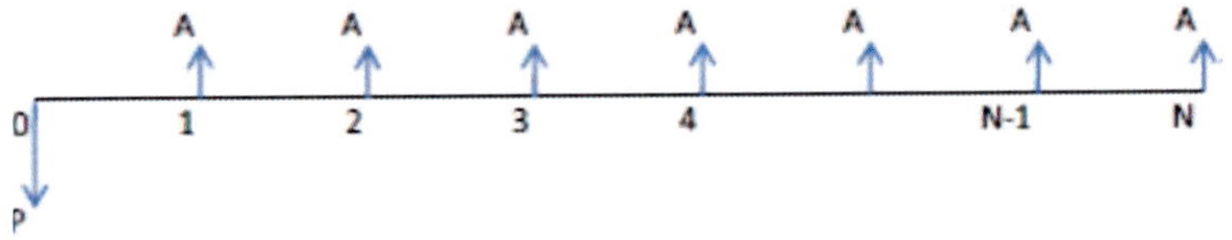

$$To\ find\ P, given\ A\ (P|A, i, N){:}\ P = \frac{A}{i}[1 - (1 + i)^{-N}]$$

$$P = \frac{A}{i}\ (if\ N = \infty)$$

Example of Equal Payment Series

$$P = \frac{1{,}000{,}000}{0.05}[1 - 1.05^{20}] = \$33{,}065{,}954$$

Linear Gradient Series

To calculate the present value of a cash flow series with a gradient (G) component that either increases or decreases by a constant rate over N time periods:

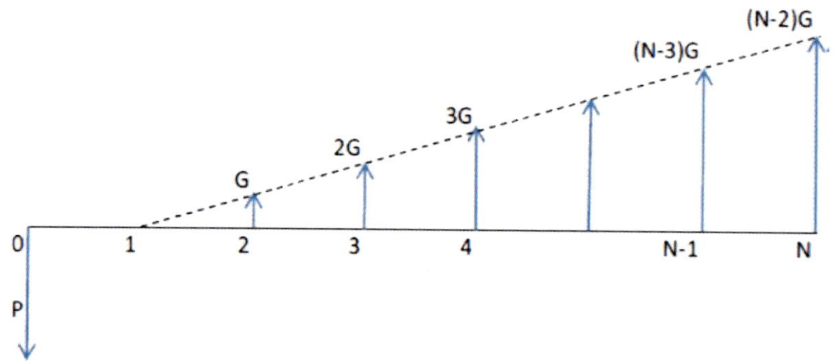

$$To\ find\ P, given\ G\ (P|G,i,N): P = G\left[\frac{(1+i)^N - iN - 1}{i^2(1+i)^N}\right]$$

Example of Linear Gradient Series

A transportation department considers benefits of a project as shown in this table. Toll revenue is assumed to increase by a rate of $80,000 each year. What is the NPV of the project? Assume the discount rate is 10%.

Year	End-of-Year Payment
1	$1,000,000
2	$1,080,000
3	$1,160,000
4	$1,240,000
5	$1,320,000

The total project cash flow consists of two cash flows:

1. Annual Cash Flow A = $1,000,000/period
2. Linear Gradient Cash Flow G = $80,000/period

So, the NPV of this series of payments at a 10% discount rate is

$$NPV = \frac{1{,}000{,}000}{0.10}[1 - 1.1^{-5}] + 80\left[\frac{(1+0.1)^5 - 0.1 \times 5 - 1}{0.1^2 \times (1+0.1)^5}\right]$$
$$= \$4{,}339{,}731$$

Geometric Gradient Series

To calculate the present value of a cash flow that changes by a fixed percentage (g) each time period:

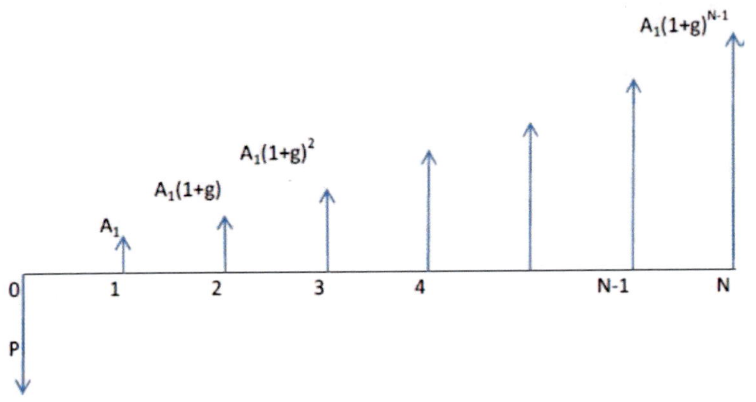

$$To\ find\ P, given\ A_1\ and\ g{:}\ (P|A, g, i, N){:}\ P = A_1\left[\frac{1-(1+g)^N(1+i)^{-N}}{i-g}\right]$$

$$(if\ i \neq g)$$

$$P = A_1\left(\frac{N}{1+i}\right)\ (if\ i = g)$$

Example of Geometric Gradient Series #1

The first-year maintenance cost of a dump truck is estimated to be \$1,000, and is expected to increase at a uniform rate of 2% per year. Using a 10% discount rate, calculate the NPV of the cost of the first 20 years of maintenance.

$$P = A_1\left[\frac{1-(1+g)^N(1+i)^{-N}}{i-g}\right] = 100.00$$

$$\left[\frac{1-(1+0.10)^{20}(1+0.08)^{-20}}{0.10-0.02}\right] = \$9739$$

Thus, the present worth of the cost of maintenance for the first 20 years is \$9,739.

Example of Geometric Gradient Series #2

Planners have determined that a continuous flow intersection (CFI) alignment at a busy intersection will significantly reduce delay. Engineers report the following details for the intersection:

Average no-build intersection delay: 425.0 $\frac{sec}{veh}$

Average CFI delay: 40.0 $\frac{sec}{veh}$

Current ADT: 36,000

Traffic growth rate: 2.5% (exponential)

Assuming for the users a value of travel time (VOTT) of $25 per hour, and an interest rate of 2%, estimate the total cost savings benefit of the continuous flow intersection over 10 years using the following formula and a geometric gradient series.

$$Delay\ Cost = VOTT \times Delay \times ADT \times Days \times \frac{1\ hr}{3600\ sec}$$

Annual no-build delay costs:

$$\$25 \times 425.0 \times 36{,}000 \times 365 \times \frac{1}{3600} = \$38{,}781{,}250$$

Annual CFI delay costs:

$$\$25 \times 40.0 \times 36{,}000 \times 365 \times \frac{1}{3600} = \$3{,}650{,}000$$

Annual delay cost savings:

$$\$38{,}781{,}250 - \$3{,}650{,}000 = \mathbf{\$35{,}131{,}250}$$

Because traffic is increasing exponentially at 2.5% every year, a geometric gradient series can be used to determine present value delay cost savings over a 10-year interval.

$$P = \$35{,}131{,}250 \left[\frac{1 - (1 + 0.025)^{10}(1 + 0.02)^{-10}}{0.02 - 0.025}\right] = \mathbf{\$352{,}121{,}780}$$

So, **$352.122 million** in user delays will be eliminated over 10 years by constructing the CFI.

Microsoft Excel's NPV function calculates NPV with the input NPV(rate, value1, value 2,...), where *rate* is discount factor, *value1* is the cash flow input for the end of the first period, *value2* is the cash flow input for the end of the second period, and so on. The returned NPV refers to the value at the end of the initial year. See Figure 6.1.

Use of Excel's NPV function

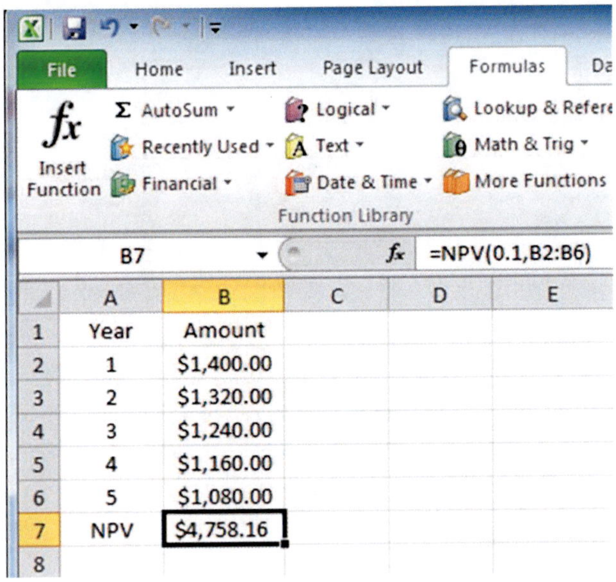

Figure 6.1: Excel's NPV Function Screen Shot

In some cases, NPV is not the best criterion for selecting between alternatives.

Example of Non-NPV-Based Selection

Suppose a state transportation agency considers building a new highway with alternatives A and B with respective NPVs of $2,000,000 and $3,500,000. The required investment for alternatives A and B are $10,000,000, and $30,000,000, respectively. Although alternative B's NPV is greater than that of alternative A, alternative B requires significantly more investment. To have a better understanding of alternatives evaluation, internal rate of return and payback period can also be used.

Internal Rate of Return (IRR)

As described in Kockelman et al. (2010), **"the project's Internal Rate of Return (IRR) determines the discount rate at which the sum of discounted costs equals the sum of discounted benefits (at their present-year worth):"**

$$C_i + \sum_{y=1}^{proj\,life} (C_y)\left(\frac{1}{1+IRR}\right)^y - SV\left(\frac{1}{1+IRR}\right)^{proj\,life} + IPC\left(\frac{1}{1+DR}\right)^{year}$$

$$= \sum_{y=1}^{proj\,life} (B_y)\left(\frac{1}{1+IRR}\right)^y$$

Essentially, IRR is the effective (equivalent) interest rate used to measure the value of an investment. IRR can be used only when the project will generate income. To evaluate alternatives using IRR, the alternatives' IRRs should be greater than the minimum accepted rate of return (MARR), which is also known as the *hurdle rate*. **MARR is the lowest interest rate that investors would accept,** given the risk of the investment and the opportunity cost of foregoing other projects.

Example of IRR

A transportation department considers building a new toll or managed lane highway with the following cash flow for first 5 years. What is the IRR for this period?

Year	Cash Flow	Year	Cash Flow
0	-$10,000,000	3	$2,345,000
1	$1,340,000	4	$2,680,000
2	$2,010,000	5	$2,847,500

$$10{,}000{,}000 = \frac{1{,}340{,}000}{(1+IRR)} + \frac{2{,}010{,}000}{(1+IRR)^2} + \ldots + \frac{2{,}847{,}500}{(1+IRR)^6}$$

IRR = 3.6%

Use of Excel's IRR function

To calculate IRR in Microsoft Excel, the function IRR(value) can be used, where *value* is a reference to the range of cells for which the user would like to calculate the IRR (Figure 6.2).

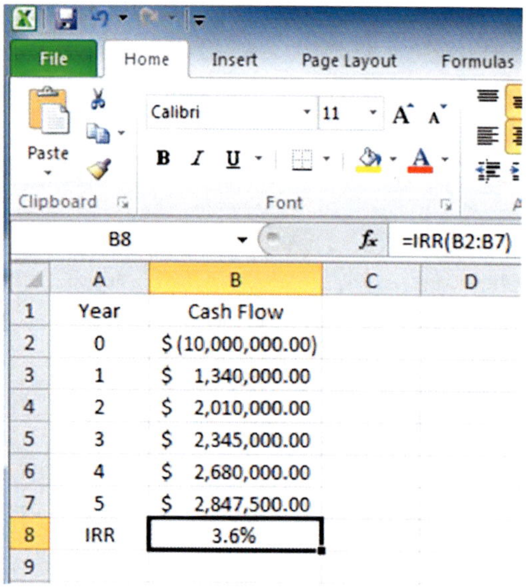

Figure 6.2: Excel's IRR Function Screen Shot

Incremental Rate of Return (ΔROR)

As mentioned, the NPV of one alternative can be greater than its competing alternative but require greater investment. In this situation, incremental rate of return (ΔROR) can be used. **ΔROR is the interest rate earned on the extra cost of a higher cost alternative.** If the alternative's ΔROR is greater than the MARR, the alternative would be beneficial.

Example of ΔROR

A transportation agency considers two alternatives for a project, A and B. The table shows the required investment and returned benefit of each alternative. If the agency's MARR is 8%, which alternative should be built?

Year	Alternative A	Alternative B
0	-$2,500,000	-$6,000,000
1	$746,000	$1,664,000
2	$746,000	$1,664,000
3	$746,000	$1,664,000
4	$746,000	$1,664,000
5	$746,000	$1,664,000
IRR	15.01%	11.99%

Clearly, the IRR for both alternatives is greater than the MARR. So, both projects are acceptable investments for the agency. Now the question is whether alternative B is worth the extra $3,500,000 in initial investment.

Year	Alternative A	Alternative B	Δ(A,B)
0	- $2,500,000	-$6,000,000	-$3,500,000
1	$746,000	$1,664,000	$918,000
2	$746,000	$1,664,000	$918,000
3	$746,000	$1,664,000	$918,000
4	$746,000	$1,664,000	$918,000
5	$746,000	$1,664,000	$918,000
IRR	15.01%	11.99%	9.78%

In effect, the first $2,500,000 investment in alternative B yields 15.01% IRR and the next $3,500,000 investment in alternative B yields 9.78% IRR. As the incremental rate of return is still higher than the agency's MARR, alternative B should be selected.

Payback Period

Payback period is the period of time required before the project's benefits are equal to the project's cost. In this method, the effect of interest and economic consequences after payback are ignored.

Example of Payback Period

Year	Alternative A	Alternative B
0	-$1000	-$1000
1	$200	$300
2	$500	$300
3	$800	$300
4	$1100	$300
5	$1400	$300

Payback periods for the alternatives shown in the table at a 10% discount rate are as follows:

Alternative A

$$1000 = 200 + 500 + (800 * 0.375)$$

The payback period is equal to 2.375 years.

Alternative B
Uniform annual benefits

$$\frac{1000}{300} = 3.33$$

The payback period is equal to 3.33 years.

Use of Excel's NPER function

To calculate payback period in Microsoft Excel, the NPER function can be used if the payment made each period is constant (Figure 6.3). The function uses argument NPER (rate, pmt, pv, [fv], [type]), where *rate* is discount factor, *pmt* is the payment made each period, *pv* is the present value, *fv* is the future value—or a cash balance you want to attain after the last payment is made (optional argument)—and *Type* indicates when payments are due (optional argument). If payments are due at the beginning of the period, a user should enter '1.' Otherwise, the user should not enter anything or enter '0.' The returned value is the payback period.

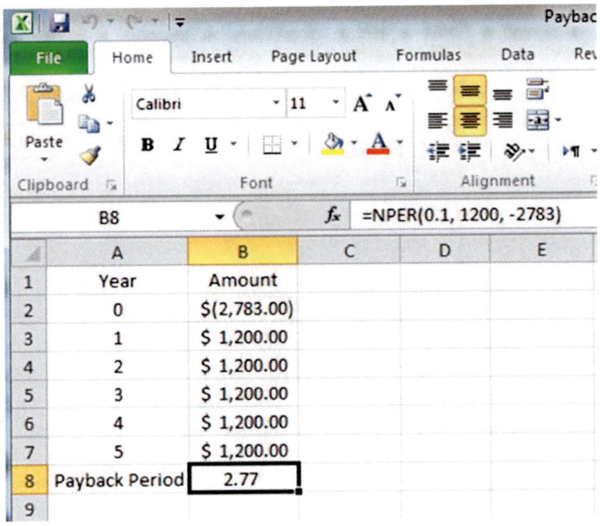

Figure 6.3: Excel's NPER Function Screen Shot

Breakeven Analysis

When a future condition is uncertain, breakeven analysis can be employed. **The computed breakeven point is the point at which expenses and revenues are equal, with no net loss or gain**. For example, breakeven analysis can be used to determine if a highway should be constructed to meet its full future demand or be constructed in multiple stages as additional demand arises. In order to conduct breakeven analysis, a breakeven point should be calculated by setting two alternatives equal to each other using NPV analysis. Then, the present worth of each alternative at each period is plotted, and the point at which the two alternatives' present worth intersect is the breakeven point (n). Finally, the plotted graphs can be used to determine the best strategy to address the uncertainty.

Example of Breakeven Analysis

Assume that a department of transportation has two alternatives for a new highway project. Alternative 1 addresses all future demand until year 20 and costs $140 million. Alternative 2 will be built in two stages: the first stage builds initial capacity and costs $100 million and the second, if required, will require an additional $120 million in year *N* to upgrade to full capacity. Determine which is the best alternative, assuming operations and maintenance costs for both alternatives are equal and interest rate is 8%. The table shows the present worth of each alternative's cost if the second stage of alternative 2 is built in year *N*.

Year (*N*)	Alternative 1	Alternative 2
0	$140,000,000	$220,000,000
4	$140,000,000	$188,000,000
8	$140,000,000	$165,000,000
12	$140,000,000	$148,000,000
16	$140,000,000	$135,000,000
20	$140,000,000	$126,000,000

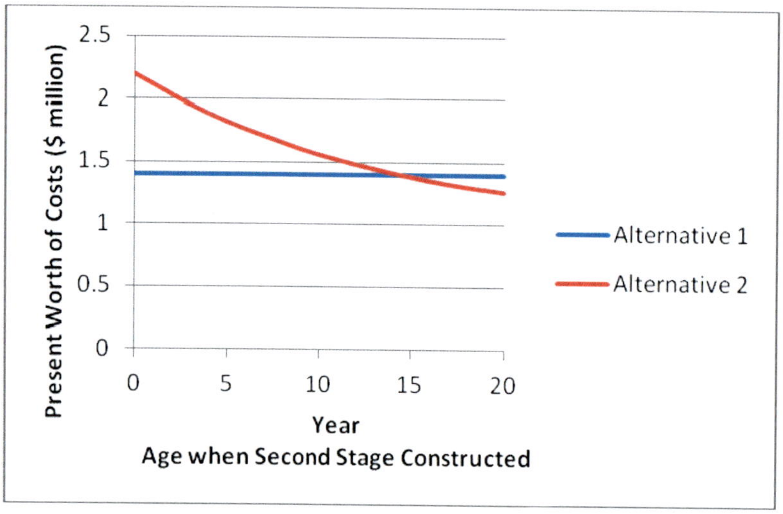

$$PW = 1{,}400{,}000 = 1{,}000{,}000 + 1{,}200{,}000(P|F, 8\%, n)$$

$$(P|F, 8\%, n) = 0.33$$

$$n = 14.3 \; years$$

As the figure indicates, if the second stage of alternative 2 is needed before year 14.3, alternative 1 is preferred, since it offers a lower present worth than alternative 2. However, if the second stage is not required until after year 14.3, alternative 2 is preferred.

The breakeven point of this example is at 14.3 years. This result means that if the second stage is deferred for 14.3 years, the present worth of the two alternatives would be equal.

Cost-Benefit Analysis (CBA)

Cost-Benefit Analysis (CBA) is a method to measure and evaluate all relative direct economic impacts of public investment projects. A useful tool for decision-making in planning and evaluation of projects, CBA can be used to determine whether and when a project should be undertaken and to rank and prioritize projects.

CBA Process

Following is a description of the CBA process.

1. **Identify project needs.** Clearly state project needs so that key relationships are identified and a wide range of alternatives can be examined. The project's objective should not be too broad, making it difficult to examine all of the trade-offs, or too narrow, excluding key relationships.
2. **Identify project constraints.** Constraints include policy and legal initiatives, and require specific assumptions about the future, such as expected regional traffic growth.
3. **Define the base case.** This is also known as the "no action" case—the continued operation of the current facility without any major investments
4. **Identify alternatives.** Identify project alternatives, which can vary from major rehabilitation of existing facilities to new construction, full reconstruction, or replacement.
5. **Define a time period.** Set the analysis period over which the life cycle costs and benefits of all of the alternatives will be measured. It should be long enough to include at least one major rehabilitation project.
6. **Define work scope.** Define the level of effort for screening alternatives. A complete analysis of all options is neither achievable nor necessary. Screening alternatives allows a wide range of initial options to be considered with only a reasonable level of effort. The level of effort is proportional to the expense, complexity, and controversy of the project.
7. **Analyze alternative traffic effects.** Analyze traffic effects that the alternative would have on the future traffic to calculate the project costs and benefits.
8. **Estimate benefits and costs.** These estimates include investment costs, hours of delay, crash rates, and other effects of each alternative relative to the base case (Table 6.3). An error in estimating costs and benefits could lead to project failure.
9. **Evaluate risk.** Look at the risks associated with uncertain costs, traffic levels, and economic values.

See Chapter 2 for a full description of costs and benefits.

10. **Conduct sensitivity analysis.** Conduct a sensitivity analysis to find factors that have an important effect on the output.

11. **Find benefit/cost ratio.** Compare net benefits with net costs and rank alternatives based on the benefit and cost ratio (B/C ratio). The benefit cost ratio is calculated by dividing total discounted benefits by total discounted costs. Options with B/C ratios greater than 1.0 are preferable. In cases where the B/C ratios of some mutually exclusive alternatives are greater than 1.0, the incremental B/C ratio should be used. In this method, the alternatives are ranked in order of investment from the smallest to the largest. Then, the incremental benefits and costs between two alternatives are calculated (ΔX-Y) where *X* is a previously justified alternative. If $(\Delta B/ \Delta C)_{X\text{-}Y}$ is greater than 1.0, alternative Y is selected. If not, alternative X will be the output of CBA.

Working with B/C ratios

12. **Make recommendations.** Recommendations are based on the B/C ratios.

Agency Costs	• Design and Engineering • Land Acquisition • Construction • Reconstruction/Rehabilitation • Preservation/Routine Maintenance • Mitigation (e.g., noise barriers)
User Costs/Benefits Associated with Work Zones	• Delays • Crashes • Vehicle Operating Costs
User Costs/Benefits Associated with Facility Operations	• Travel Time and Delay • Crashes • Vehicle Operating Costs
Externalities (non-user impacts, if applicable)	• Emissions • Noise • Other Impacts

Table 6.3: The Considered Benefits and Costs in CBA (FHWA 2003)

Example of CBA #1

In this example, a department of transportation considers five alternatives for expanding a highway. The table shows the costs and benefits of each alternative. Determine the alternative with the highest B/C ratio for this project.

	A	B	C	D	E
Costs	$6,000,000	$3,000,000	$9,000,000	$1,500,000	$15,000,000
Benefits	$11,000,000	$7,000,000	$13,100,000	$1,500,000	$12,050,000
B/C	1.8	2.3	1.4	1.0	0.8

In this example, if alternatives are not mutually exclusive (as the B/C ratio of alternative B is the largest B/C ratio), the alternative B is the best alternative.

If alternatives are mutually exclusive, the incremental B/C should be used among alternatives A, B, and C. In the first step, alternatives should be ordered from the smallest to the largest required investment (using the following table).

	B	A	C
Costs	$3,000,000	$6,000,000	$9,000,000
Benefits	$7,000,000	$11,000,000	$13,100,000
B/C	2.3	1.8	1.4

Then, the ΔCosts, ΔBenefits, and ΔB/ΔC for two cases—C-A and B-C—should be calculated using the following table.

	B-A	C-A
ΔCosts	$3,000,000	$3,000,000
ΔBenefits	$4,000,000	$2,100,000
ΔB/ ΔC	1.3	0.7

Finally, as $\Delta B/\Delta C_{B\text{-}A}$ is greater than 1.0, A is a better alternative than B. Also, as $\Delta B/\Delta C_{C\text{-}A}$ is less than 1, it shows that A is the best alternative.

Example of CBA #2

A bridge linking two towns over a river is close to failing and will be decommissioned in 5 years if repairs are not made. The state transportation agency is calculating a B/C ratio to compare the benefits of travel time savings, reduced operating expenses, crashes, and pollution to construction and maintenance costs. Following are the calculation steps taken to determine the B/C ratio.

Removing the bridge will require some users to travel further out of their way to reach destinations across the river, resulting in increased VMT overall. Assuming a lifespan of 50 years for the rebuilt bridge, the agency projects VMT in the area to be as follows:

	Total VMT	Total VHT
No-build	1,400,500	40,800
Bridge rebuild	1,275,000	39,100

Benefits

The agency estimates that travel time savings over the 50-year period would be $250 million. Vehicle operating costs, crashes, and emissions are also functions of VMT, so if the bridge helps reduce VMT, then all crashes (fatal, major, minor, property damage), operating costs, and emissions will correspondingly decrease. The transportation agency estimates that reduced operating costs will save $185 million, crash costs will decrease by $65 million, and emissions reductions will save another $45 million. Total benefit from bridge repair in present terms is $545 million.

Costs

Closing the bridge would require some funds, as would deconstruction. Bridge repair has been estimated at $100 million, with total operating and maintenance costs being $85 million over a 50-year lifespan, in present costs. Total cost in present terms for bridge repair is $185 million.

$$\frac{B}{C} = \frac{\$545{,}000{,}000}{\$185{,}000{,}000} = \mathbf{2.95}$$

This B/C ratio is greater than 1, indicating that the project returns more benefits than costs.

Life Cycle Cost Analysis (LCCA)

Engineering decision-making often entails choosing the best option among many alternatives. When alternatives require different amounts of investment and yield various levels of benefit (particularly when alternatives deviate in duration), life cycle cost analysis (LCCA) can be employed to compare the alternatives on an even playing field.

Figure 6.4: Process of Decision Making

The National Institute of Standards and Technology defines life cycle cost (LCC) as "the total discounted dollar cost of owning, operating, maintaining, and disposing of a building or a building system" over a period of time (Fuller & Petersen 1995). **LCCA considers all of the benefits and costs associated with different project alternatives over the project's lifetime, and can be applied when an agency needs to assess the total cost of a project.** It is particularly useful when deciding between various project alternatives that all meet project scope requirements, but have different initial and operating costs.

For projects with comparable benefits over the same lifetime, the alternative with the lowest LCC is usually preferred. In addition, for alternatives with different lifetimes, the alternative with the lowest Equivalent Uniform Annual Cost (EUAC) is more desirable. EUAC is determined by converting all project costs into a uniform annual recurring cost over the analysis period. In calculating the EUAC, the NPV of each alternative is shown in terms of its equivalent annual payment amount (NPV and equal payment series are discussed in detail in the next section). The required inputs for LCCA include initial expenses like equipment and right-of-way (ROW) purchase, as well as future expenses such as operation, maintenance, and replacement costs.

Note that many LCCA inputs are estimates. LCCA may fail to capture the uncertainty of future events and the impact of technological progress. This inability to account for uncertainty limits LCCA's application.

Example of LCCA

A transportation agency is deciding between flexible and rigid pavement for a new roadway. Engineers expect required surface rehabilitation after 20 years for flexible pavement and 40 years for rigid pavement. The following LCCA table was created for a 40-year analysis period, with initial and rehabilitation costs considered for each pavement type. Flexible pavement would cost $4 million initially and $2 million to rehabilitate 20 years later. Rigid pavement would cost $6 million for initial construction. Benefits for flexible pavement are reduced when travel times increase during resurfacing times.

Because rehabilitation costs are to occur 20 years in the future, they must be translated to present value before they can be compared with initial costs. Benefits are already provided in present value. Assume a 2% annual interest rate.

Present worth of flexible pavement rehabilitation costs in year 20:

$$P{=}\$2{,}000{,}000{\times}(1{+}0.02)P = \$2{,}000{,}000 \times \frac{1}{(1+0.02)^{20}} = \$1{,}345{,}943$$

	Rigid	Flexible
Initial Costs	$6,000,000	$4,000,000
Rehabilitation Costs	-	$1,345,943
Total Costs	$6,000,000	$5,345,943
Total Benefits	$11,000,000	$10,000,000
NPV	$5,000,000	$4,654,057

The rigid pavement returns a higher NPV and would therefore serve as a better choice.

Constrained Optimization

Constrained optimization is a mathematical tool used to minimize or maximize a function (f) subject to certain constraints (x). The constraints can be a combination of equations or inequalities. Every project faces some type of constraint (e.g., budget or resource), so the goal is to find the optimum, or most efficient, solution that falls within those constraints.

Constrained optimization can be used as a B/C approach in making choices to either maximize benefit within a budget or meet a given level of service with minimum cost.

For example, a transportation agency may want to decide what portion of its transportation budget should go to new road construction and how much should be spent maintenance of existing highways. Here the constraint is the transportation budget—the road construction and maintenance costs must total the budgeted amount the agency has available. For this type of problem, a function must be created that models the total utility, or benefit, the public receives from some unit measure of added roadway capacity or maintenance. The optimization process will then seek to maximize this function to give the most benefit within the constraint of the transportation budget and find the optimal amount to allocate to new roads and maintenance.

Another example is allocating monies across a variety of competing projects. Fund allocation has become an issue for all state DOTs. Constrained optimization can be used to overcome this issue. In this situation, the objective function is the following:

See Chapter 5 for discussion of budget constraints.

$$\text{maximize} \quad \boldsymbol{z(\mathrm{x}) = \sum_i b_i x_i}$$

$$\text{subject to} \quad \boldsymbol{\sum_i c_i x_i \leq B}$$

$$\boldsymbol{x_i \in \{0, 1\}}$$

Where b_i and c_i are the benefit and cost associated to each project, while B is the upper bound of the budget. x_i is a 0-1 binary decision variable (which represents the selection of project i if $x_i = 1$ and the deselection of project i if $x_i = 0$).

In order to solve constrained optimization problems, Excel's Solver add-in can be used. Following are common elements in Excel's spreadsheet model:

Use of Excel's Solver feature

1. **Inputs:** all numerical input needed to form the objective and the constraints.
2. **Changing cells:** instead of using variable names, such as x's, a set of designated cells plays the role of the decision variables. The values in these cells can be changed to optimize the objective.
3. **Target (objective) cell:** one cell called the *target value* or the *objective cell* contains the value of objective. Solver systematically varies the values in the changing cells to optimize the values in the target cell.
4. **Constraints:** Excel does not show the constraints directly on the spreadsheet. Instead they are specified in a Solver dialog box.
5. **Nonnegativity:** normally the decision variables should be non-negative.

Here are the stages of solving the constrained optimization problem in Excel:

1. **Formulate the model:** enter all the inputs, trial values for the changing cells, and formulas in a spreadsheet. In particular, the spreadsheet must include a formula that relates the objective to the changing cells, so that if the values in the changing cells vary, the

objective values vary automatically.

2. **Call solver:** designate the objective cell, the changing cells, and the constraints. Then tell Solver to find the optimal solution.

Note: based on the Solver option chosen in Excel, the solution may vary. However, the value of the objective function will be within a reasonable range of the optimal solution (performing a completely exhaustive search would be impractical).

The following example will show how to use Excel's Solver for optimizing budget allocation.

Example of Optimizing Budget Allocation

The Capital Area Metropolitan Planning Organization's (CAMPO) Transportation Improvement Program identified 14 candidate projects for fiscal years 2011–2014. These projects are a variety of roadway expansion and improvement types. The potential benefits and costs of each project are shown in the table (project costs and descriptions taken from CAMPO's MIP for FY 2011–2014). Following are the constraints for allocating the budget to these projects:

1. No more than two projects can be implemented in each of the four locations shown in the table (i.e., Austin, Georgetown, Cedar Park, and Other).
2. No more than three projects of the same type (new build, widening, and reconstruction) can be implemented simultaneously.
3. The total budget constraint is $556,780,000 (over 4 years).

Determine which projects are feasible within the listed constraints.

Proj.	Site	Road-way	Project Description (basic project type in *italics*)	Cost (Initial)	Benefit (NPV)	B/C*
1	Cedar Park	RM 1431	*Widen* a 4-lane divided arterial to a 6-lane divided arterial with wide outer lanes, raised median and sidewalk fronting public land	$26,809,766	$102,146,400	3.81
2	Other	FM 2001	*Realign* a 4-lane divided roadway	$4,899,000	$5,878,000	1.20
3	Austin	SH 71	*Build* an underpass, frontage roads and main lanes	$54,016,584	$464,265,000	8.59
4	Other	SH 195	*Widen* existing 2-lane roadway to 4-lane divided roadway	$46,191,075	$354,292,900	7.67
5	Austin	SH 130 and Cameron Rd	*Build* northbound and southbound entrance ramps and related toll integration equipment	$4,610,000	$36,100,000	7.83
6	Austin	FM 3177	*Realign* FM 3177	$4,955,552	$40,644,000	8.20
7	Other	IH 35	*Build* southbound frontage roads and convert frontage roads to one-way operation	$14,026,000	$110,390,000	7.87
8	George-town	IH 35	*Build* a 3-lane frontage road and ramps	$8,486,383	$67,295,000	7.93
9	George-town	IH 35	*Build* ramp and auxiliary lane and reconfigure ramps	$2,250,000	$16,875,000	7.50
10	Other	US 79	*Widen* roadway to a 4-lane divided arterial	$16,346,887	$137,030,000	8.38
11	Austin	US 290	*Build* 6 tolled main lanes and 6 continuous non-tolled frontage roads	$455,900,000	$3,128,500,000	6.86
12	Other	SH 71	*Build* an overpass at FM 20 and frontage roads	$16,624,199	$114,680,000	6.90
13	Other	FM 1626	*Widen* FM 1626 to a 4-lane divided roadway	$47,312,666	$402,157,000	8.50
14	Austin	Loop 1	*Build* northbound and southbound managed lanes	$253,162,143	$739,229,000	2.92

*B/C ratio includes only initial costs. A comprehensive CBA would include costs like annual maintenance.

The objective of this example is to maximize the benefit accrued to the society by selecting the proper projects, which can be represented mathematically in a standard maximization equation such as the following:

$$\text{maximize } z(\mathrm{x}) = p_1x_1 + p_2x_2 + \cdots + p_{14}x_{14}$$

where p_1 = \$102,146,400, p_2 = \$58,785,000, …,
p_{14} = \$739,229,000, and x_i represents each project i.

This function seeks to maximize the net benefit of a selection of projects by summing their benefit values (p_i) multiplied by a binary variable (x_i), which takes on a value of 1 if the project is selected, and 0 otherwise. Of course, maximization has constraints as mentioned above, including funding and resource limitations. The following equations can be written to represent each of the specific constraints introduced for this example.

$$(1)(c_1x_1 + c_2x_2 + \cdots + c_{14}x_{14} \leq B$$

$$(2)x_3 + x_5 + x_6 + x_{11} + x_{14} \leq N_{l,max}$$

$$(3)x_2 + x_4 + x_7 + x_{10} + x_{12} + x_{13} \leq N_{l,max}$$

$$(4)x_3 + x_5 + x_7 + x_8 + x_9 + x_{11} + x_{12} + x_{14} \leq N_{t,max}$$

$$(5)x_1 + x_4 + x_{10} + x_{13} \leq N_{t,max}$$

$$(6)x_1, x_2, \ldots, x_{14} \in \{0, 1\}$$

where c_1 = \$26,809,766, c_2 = \$4,899,000, …,
c_{14} = \$253,162,143;
B = \$556,780,000;
$N_{l,max} = 2$ and $N_{t,max} = 3$.

Constraint 1 ensures that the sum of the costs of the selected projects ($\sum c_i x_i$) does not exceed the budget (B). Constraint 2 ensures that no more than two projects will be implemented in Austin. Likewise, constraint 3 ensures that no more than two projects will be implemented in location "Other." *Note that two or fewer projects are being considered for Cedar Park and Georgetown; thus, constraints are not needed for these two locations.*

Constraint 4 ensures that no more than three new-build projects are selected. Likewise, constraint 5 ensures that no more than three widening projects are selected. Constraint 6 determines the binary project selection term. *Note that because only two realignment projects are being considered, no constraint is necessary.*

Stages for formulating the MS Excel spreadsheet model are as follows:

1. **Enter inputs:** Enter the various inputs such as cost, benefit, budget, maximum number of projects in each region, and maximum number of projects in each type (F4:G17, F22, F28, and F34, according to Figure 6.8, which shows the example worksheet).
2. **Change cells:** Enter any values in cells I4:I17. These values do not have to be the values shown. These are the cells where the decision variables are placed. Any values can be used initially. Solver will eventually find the optimal values.
3. **Define type of constraint for each changing cell:** As x_1 to x_{14} can be either 0 or 1, choose the bin (binary) option in the Add Constraint dialog box (Figure 6.5).

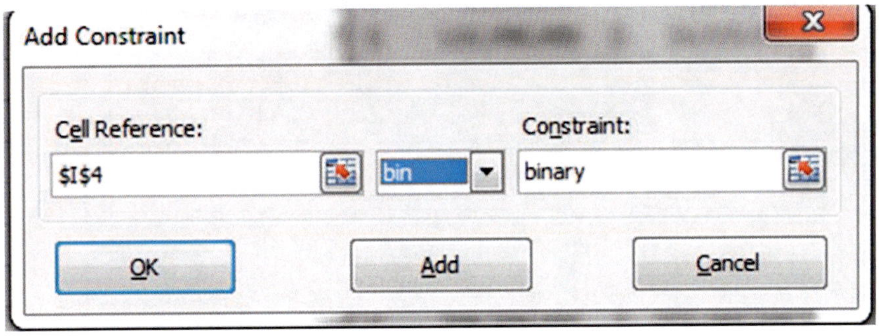

Figure 6.5: Defining Type of Constraint for x_1

Constraints include the number of projects in Austin, the number of projects in other areas, the number of widening projects, and the number of new-build projects. To define these constraints, put the left side of equation in designated cells, and define the constraints type in Solver (Figure 6.6). For example, for number of projects in Austin, put I6+I8+I9+I14+I17 in cell F24. Then, in the Add Constraint dialog box, enter F24 <= F28.

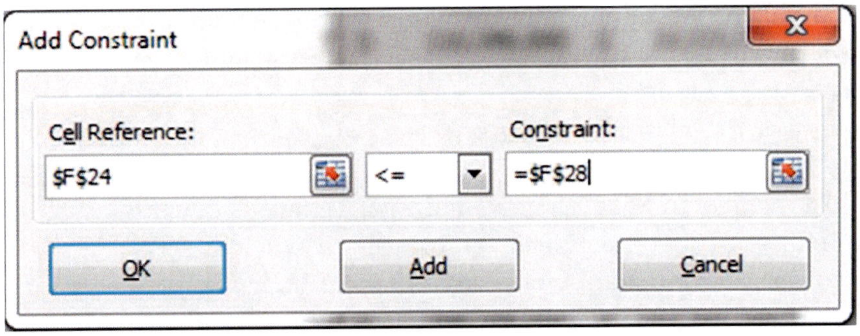

Figure 6.6: Defining Type of Constraint for the Number of Projects in Austin

4. **Enter total cost and benefit:** In cell F27, enter G4*I4+G5*I5+…+G17*I17. Cell F27 should be less than total budget, F22. For total benefit, enter F4*I4+I5*F5+…+I17*F17.

This cell is defined as an objective cell in the Solver dialog box. Figure 6.7 shows the final picture of the Solver dialog box.

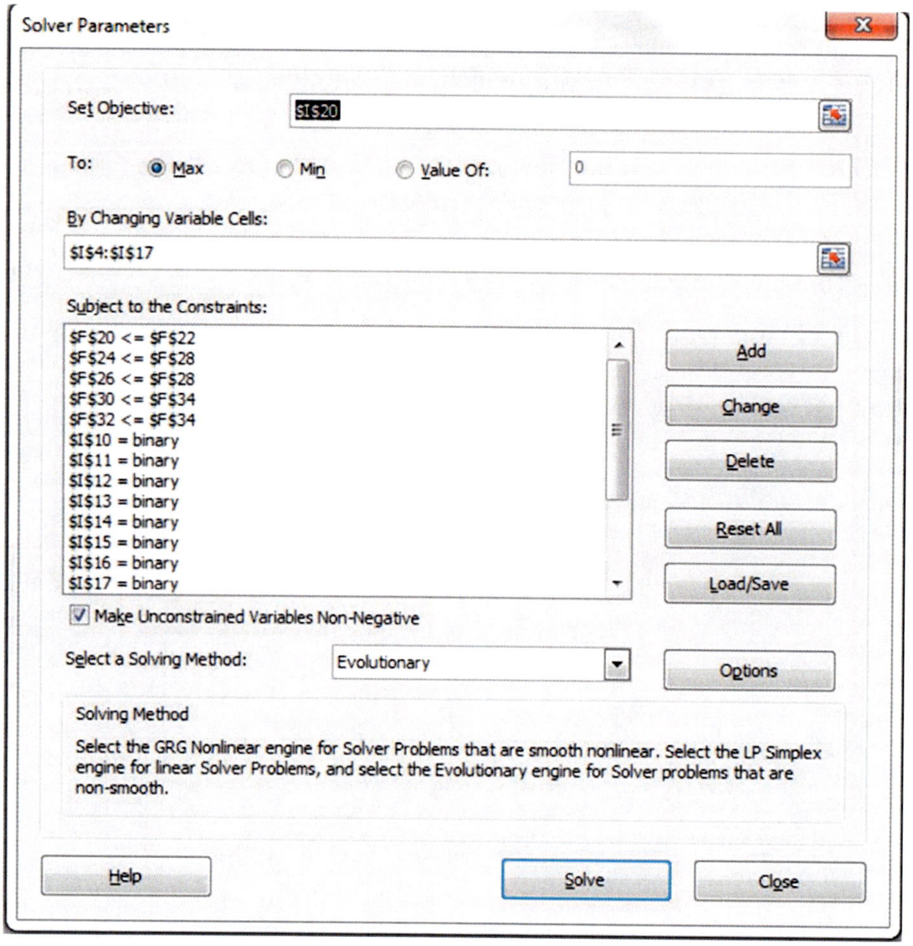

Figure 6.7: Solver Dialog Box

5. **Choose a solving method:** As the solution to this problem is not continuous (each alternative can only take on values of 0 or 1), select an Evolutionary solving method. Then click the Solve button. Tell Solver to return the values in the changing cells to their original values or retain the optimal values found. Finally, the results are put in the decision variable cells. The final results would be: $x_3 = x_4 = x_{11} = 1$, and
$x_1 = x_2 = x_5 = x_6 = x_7 = x_8 = x_9 = x_{10} = x_{12} = x_{13} = x_{14} = 0$.

As a result, the total cost is equal to $556,107,659 and total benefit is equal to $3,947,057,900. The final results are shown in Figure 6.8. Thus, the set of projects 3, 4, and 11 will maximize benefit while staying under budget.

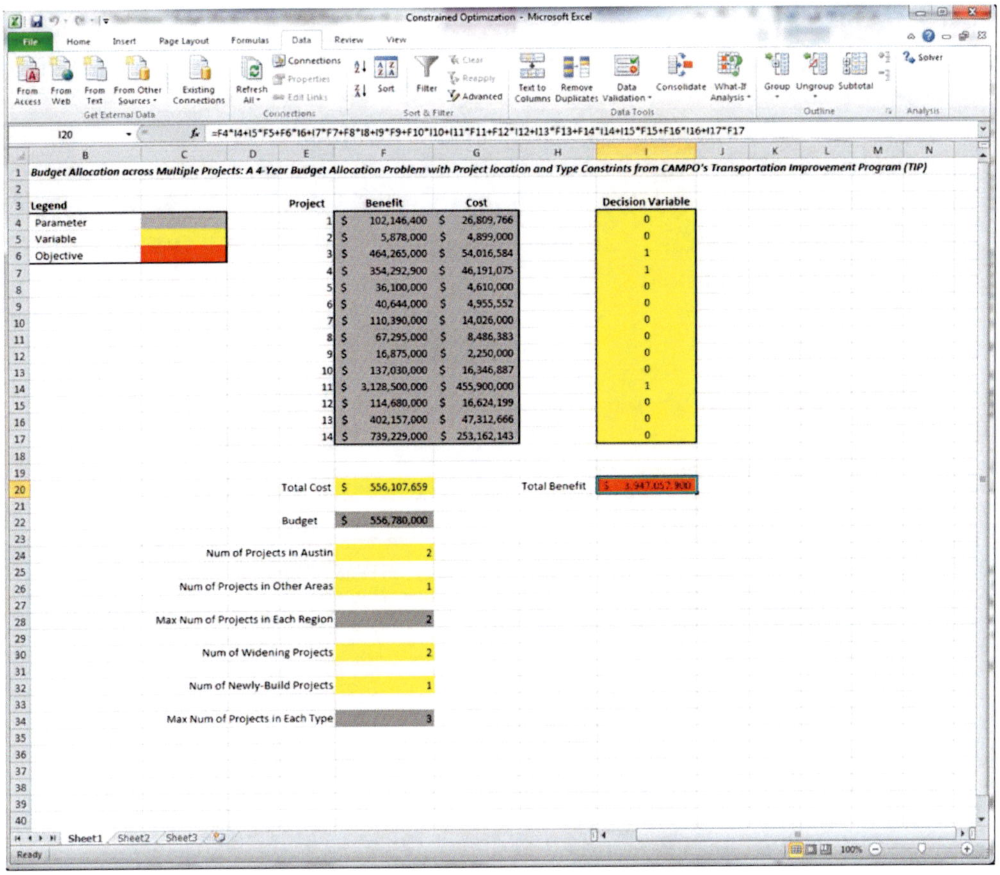

Figure 6.8: Final Results

6.3 Multicriteria Analysis

Thus far, the project evaluation techniques discussed have been purely based on economic concepts such as costs and benefits. In order to consider non-economic criteria (e.g., environmental and safety impacts) in traditional CBA, monetary value is assigned to these attributes so they can be evaluated alongside economic criteria (e.g., construction and maintenance cost). However, traditional CBA is limited in its ability to incorporate all considerations selecting between alternatives, particularly those criteria that cannot be easily measured in money terms. Social and environmental equity issues such as distributional impacts across socioeconomic groups and intergenerational sustainability are largely ignored in CBA. Furthermore, traditional CBA places emphasis on traveler benefits and often undervalues freight-related cost savings for shippers and consignees, missing non-user impacts on business productivity and competitiveness. The emphasis on traveler benefits also favors urban projects as travel time savings are greatest on high volume roadways, raising the issue of geographic equity.

Multicriteria analysis (MCA) allows alternatives analysis to be conducted across different types of criteria with various dimensions of benefits. Unlike CBA, where all variable effects are measured in monetary units, MCA allows assessment of variables on any quantitative or qualitative scale. The underlying concept involves creating a composite measure consisting of the sum of the weighted criteria scores. The best alternative is the one which scores highest on this compound measure.

Some state DOTs and MPOs are already using MCA to rate transportation projects. The Texas Department of Transportation (TxDOT) developed performance measures and an analysis tool for sustainable transportation based on a MCA method referred to as multi-attribute utility theory (MAUT). The analysis tool measures project compatibility to TxDOT goals such as congestion reduction, safety enhancement, economic opportunity expansion, transportation asset value increase, and air quality improvement.

TxDOT and KDOT applications of MCA

Example

Kansas DOT's MCA-Based Project Selection

KDOT developed a project selection approach where weighted numeric scores are calculated for each project based on the sum of three sub-scores:

- An engineering score that provides a measure of the project's impact on traffic flow (worth up to 50 points)
- A local consultant score that provides a measure of the feedback provided to KDOT district staff at meetings with public leaders, business leaders, and residents (worth up to 25 points)
- An economic impact score based on the projected change in statewide job generation and impact on present value of economic benefits (worth up to 25 points)

This composite MCA approach allows KDOT to evaluate projects based on traditional CBA considerations, feedback from the local community, and economic impact assessment. The local consultant score represents the voice of local citizens, which is valued by state DOTs but difficult to quantify into monetary terms for CBA analysis.

Compared to CBA, MCA is more flexible, but MCA is not without fault. The subjectivity of the scoring and weighting process can significantly influence outcomes. When a single decision-maker chooses the weights of the criteria, the preferences of the community may not be accurately reflected. Individual biases can be reduced by employing the Delphi technique among a group to determine weights. All members of a group of experts (or decision-makers) would be asked to assign weights to the variables along with written explanations of his/her basis for the weighting. The weights are then circulated among the other group members, and the experts (or decision-makers) are given the opportunity to revise their original weights in light of seeing others' weights and explanations.

However, beyond individual and group biases, MCA has other weaknesses. Although MCA does not necessarily require large amounts of quantitative data, the process of assigning weights to each criterion is still a considerable effort. Moreover, MCA approaches increase the likelihood of under-counting or double-counting benefits due to overlap in the variables considered. In the KDOT example, congestion impacts are likely to influence all three scores. While the scores are weighted differently, the basis on which the sub-scores are assigned has overlap.

Simple Additive Weighting (SAW)

One of the most widely used MCA methods is simple additive weighting (SAW). **SAW converts a multi-criterion problem into a single-dimension by calculating a weighted score, V_i, for each project alternative i evaluated across each criterion j** as follows:

$$V_i = \sum_{j=1}^{j=n} w_j r_{ij}$$

where w_j represents the weight for criterion j and r_{ij} represents the rating score for alternative i on criterion j. The various criteria—whether economic, environmental, social, or technical—are converted to a common scale before employing SAW. The alternative with the highest composite score V_i is selected.

Example of SAW

A transportation agency is considering three alignment options for a new route highway and has decided to compare the alignments based on the following hierarchy of criteria and their weights:

- Operations and safety considerations (0.35)
  - Congestion impacts (0.15)
  - Safety impacts (0.15)
  - Network connectivity impacts (0.05)
- Environmental considerations (0.3)
  - Noise pollution impacts (0.1)
  - Air pollution impacts (0.1)
  - Landscape (e.g., parks, wildlife and refuge) and historical site impacts (0.1)
- Cost considerations (0.25)
  - Construction cost (0.2)
  - Efficiency of construction (0.05)
- Political/community considerations (0.1)
  - Community preferences at a local level (0.05)
  - Political acceptability at a regional level (0.05)

(The weights are assumed to be previously determined but are not meant to reflect a real-world situation.)

Alternative A represents the no-build option, with no construction costs and no environmental impacts, but also no improvement in congestion, safety, or network connectivity. Alternative B represents an alignment with greater impacts to landscape and historical sites, but at a lower construction cost. Finally, Alternative C represents an option that is more expensive to construct, but has fewer impacts to the landscape than Alternative B, and is better received by the local community.

Note that the sum of the sub-criteria weights (e.g., construction cost and efficiency of construction) adds up to equal the weight of the main criteria category (e.g., cost considerations). Also, the sum of all sub-criteria weights and the sum of all main criteria weights both equal 1.

Assume the following alternative scores (converted to a common scale between 0 and 3 with the highest score the most desirable) in each criterion:

Criterion (Weight)	**Alt. A**	**Alt. B**	**Alt. C**
Congestion (0.15)	0	3	3
Safety (0.15)	1	2	2
Network connectivity (0.05)	0	3	3
Noise Pollution (0.1)	3	2	2
Air Pollution (0.1)	3	2	2
Landscape & Historical Sites (0.1)	3	2	1
Construction Cost (0.2)	3	1	2
Efficiency of Construction (0.05)	3	2	1
Community Preferences (0.05)	2	1	3
Political Acceptability (0.05)	0	3	3

The weighted scores and the composite scores (sum of weighted scores) for each alternative are shown in the following table. Each score is the product of the alternative score multiplied by the weight.

	Alt. A	**Alt. B**	**Alt. C**
Congestion (0.15)	$0 \times 0.15 = 0$	$3 \times 0.15 = 0.45$	$3 \times 0.15 = 0.45$
Safety (0.15)	0.15	0.3	0.3
Network connectivity (0.05)	0	0.15	0.15
Noise Pollution (0.1)	0.3	0.2	0.2
Air Pollution (0.1)	0.3	0.2	0.2
Landscape & Historical Sites (0.1)	0.3	0.2	0.1
Construction Cost (0.2)	0.6	0.2	0.4
Efficiency of Construction (0.05)	0.15	0.1	0.05
Community Preferences (0.05)	0.1	0.05	0.15
Political Acceptability (0.05)	0	0.15	0.15
Composite (Sum) of Weighted Scores (V_i)	*1.9*	*2*	*2.15*

SAW analysis indicates that Alternative C with the highest composite score should be selected.

It is important to note that in all MCA methods, bias is minimized when weights for criteria are chosen before the alternatives are evaluated against the criteria.

Data Envelopment Analysis (DEA)

Data envelope analysis is a decision-making model that has been used widely to compare relative performance of units within systems, such as park-and-ride facilities, bus routes, airports, and urban traffic facilities. DEA compares both inputs and outputs of a system (such as an individual park-and-ride lot or a bus route) and calculates relative efficiencies among them. The inputs/outputs called **decision-making units** (DMUs).

One study used DEA to evaluate park-and-ride facilities in Chicago, indicating how the approach helps target improvements in key areas. The research team compared 16 park-and-ride lots using the number of parking spaces and mean daily operating costs as inputs. The mean number of cars using the lot and mean daily revenue were used as outputs, or indicators of success. The results return a relative efficiency of all the lots. For example, Lot A, the most efficient, returned 100% efficiency and the least efficient lot, Lot P, returned 20% efficiency (Barnum et al. 2007).

The efficiency score of a DMU is calculated as a ratio of outputs (like daily revenue) over inputs (such as operating costs), subject to the constraints that the ratio be less than 1 and non-negative for all other DMUs. Therefore, all DMU efficiencies can be compared relative to the highest performing ratio of outputs (benefits) over inputs (costs). Two outputs can be normalized with one input to create an efficient frontier, or a graphical tool useful in determining relationships of efficiency for DMUs.

Resource
See Ozbek's 2009 study for details on DEA applications.

6.4 Sensitivity Analysis

Due to uncertainty in forecasting model inputs, the exact outcome of a project is always unknown. Rather, each input has a *probability density function* (the relative likelihood of a random value occurring at a given point) that contributes to range of possible outcomes. **Sensitivity analysis helps identify the degree to which model outputs are affected by changes in inputs, and can provide decision-makers with a range of possible outcomes rather than a single number upon which to rely.**

To conduct sensitivity analysis, a base-case scenario based on the most reliable data available is developed and different scenarios are analyzed to provide an understanding of the range of uncertainty. Aspects that contribute to future uncertainty include measurement error, background trends used for modeling future levels of gross domestic product (GDP) and fuel cost, unclear specification, future regulations and policies. After establishing a base-case and alternative scenarios, results of different models will be reported.

Sensitivity analysis can be used to determine the most important input influencing a project's NPV. In this case, sensitivity analysis begins with a base-case situation. By changing one value and holding all other values constant, an engineer can calculate how that value influences NPV, in a *single factor sensitivity analysis.* The sensitivity of the NPV to a change in a particular input can be measured as the slope of a line on a graph: the steeper the slope, the more sensitive the NPV is to change in a particular variable.

However, single-factor sensitivity graphs do not explain interactions among different variables. In some cases, the NPV may not be very sensitive to separate changes in different variables, but very sensitive to combined changes in them. This type of sensitivity analysis is called *multiple factor sensitivity analysis.* In this method, all possible combinations of different possible values for each input are developed. As the number of possible combinations of conditions can become quite large, selecting the most sensitive project factors and improving the estimates prior to conducting sensitivity analysis is important. The final analysis indicates the combinations of inputs with the most influence on the final result.

Example

Sensitivity Analysis in Action

One study on the propagation of uncertainty through travel demand models considered uncertainties in population and employment growth rates, household and employment mobility rates, location choice coefficients, and land price coefficient.

For population and employment growth rates, the high (1.5 times), low (0.5 times), and mean growth rate are considered to represent distributed sample points. As household and employment mobility rates varied from 30% to 10%, the high, average, and low rates were considered. For location choice and land price coefficients, the coefficients were varied to the 17th, 50th, and 83rd percentiles.

As a result, 81 scenarios were modeled. The long-run results showed that population and employment growth rates were the most influential factors.

Single Factor Sensitivity Analysis: A Transportation Application

Suppose a department of transportation is considering building a new highway with the following estimated parameters:

- Construction cost: $10 million
- Demand: 1,000 cars per day
- Maintenance: $500,000
- Toll: $3

Sensitivity analysis can help determine which parameter has the most influence on the NPV of the project. The analysis assumes an interest rate of 4% per year and a study period of 25 years.

Figure 6.9 shows different steps necessary for conducting sensitivity analysis for this project. In the first step, the value of toll, interest rate, construction, and maintenance cost are considered fixed, while the demand value varies. In the second step, the toll value varies, while the demand value, interest rate, maintenance, and construction costs are fixed. The same process goes for the subsequent steps, as illustrated in Figure 6.9.

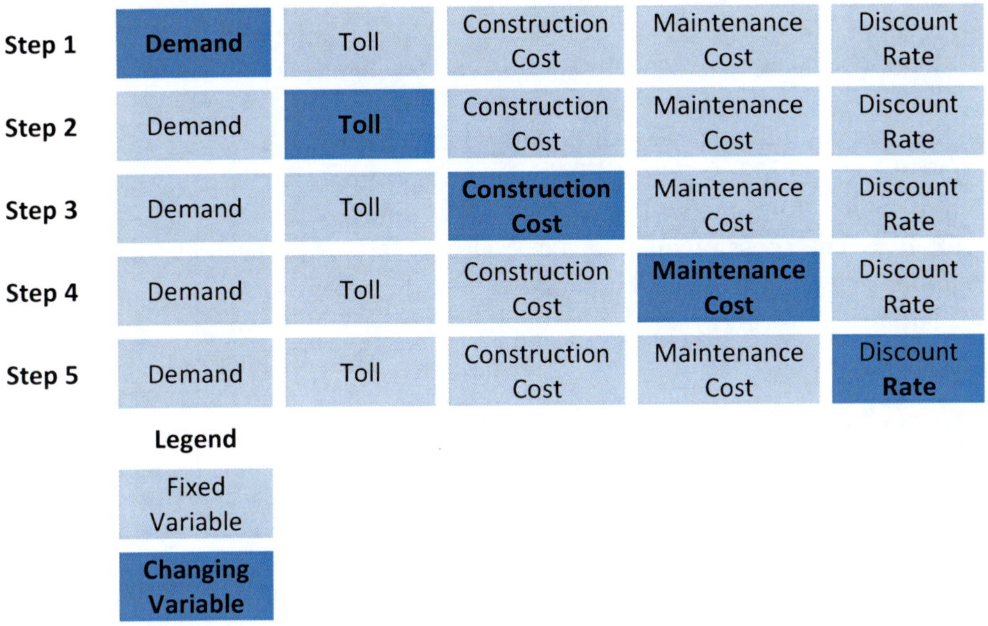

Figure 6.9: Sample Sensitivity Analysis Steps

Figure 6.10 shows the variation of NPV based on the different steps shown in Figure 6.9. As mentioned previously, the sensitivity of the NPV to a change in a particular input can be measured as the slope of a line on a graph. As the demand and toll charge have the same slope (as shown in Figure 6.10), they have the most influence on the NPV.

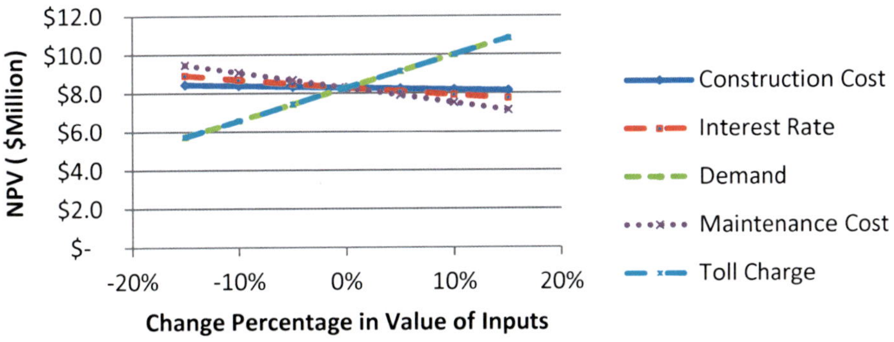

Figure 6.10: Sample Sensitivity Graph

Multiple Factor Sensitivity Analysis: A Transportation Application

Suppose a department of transportation is considering retrofitting a bridge with the following estimated parameters:

- Construction cost: $9 million
- Annual Revenue: $1 million
- Annual Maintenance: $350,000

Multiple factor sensitivity analysis can help determine which parameter combinations have the most influence on the project's NPV. The analysis assumes an interest rate of 4% per year and a study period of 25 years.

To conduct multiple factor sensitivity analysis, three different "what if" values for each input are considered: optimistic, most likely, and pessimistic. Table 6.4 shows the different values for inputs. The various combinations of the optimistic, most likely, and pessimistic factor values for revenue, construction, and maintenance cost need to be analyzed for their combined impacts on the NPV. The results for these 27 (3x3x3) combinations are shown in Table 6.5.

Table 6.4: Optimistic, Most Likely, and Pessimistic Values for Inputs ($ million)

	Optimistic	Most Likely	Pessimistic
Construction Cost	7.5	9	10.5
Revenue	1.35	1	0.75
Maintenance Cost	0.3	0.35	0.41

Table 6.5: NPV for All Combinations of Estimated Outcomes ($ million)

	Construction Cost								
	Optimistic			Most Likely			Pessimistic		
	Maintenance Cost			Maintenance Cost			Maintenance Cost		
Annual Revenue	Opti-mistic	Most Likely	Pessi-mistic	Opti-mistic	Most Likely	Pessi-mistic	Opti-mistic	Most Likely	Pessi-mistic
Optimistic	*8.9*	*8.1*	*7.2*	*7.4*	*6.6*	*5.7*	*5.9*	*5.1*	*4.2*
Most Likely	*3.4*	*2.6*	*1.7*	*1.9*	1.2	0.2	0.4	-0.3	-1.3
Pessimistic	-0.5	-1.2	-2.2	-2.0	-2.7	-3.7	-3.5	-4.3	-5.2

Note: The combinations with NPV higher than $1.2 million are italicized, while those with NPV less than 1.2 million are underlined.

As Table 6.5 displays, the combination of change in annual revenue and construction costs has the most influence on the NPV of this project. In this example, the NPV for the most likely condition for each input is equal to $1.2 million.

Monte Carlo Methods

Because forecasted inputs are often based on other underlying (and uncertain) forecasts, the number of "what if" scenarios increase quickly with the growing number of input variables. Calculating the range of possible outcomes then becomes mathematically cumbersome. **A Monte Carlo simulation is an alternative to a comprehensive sensitivity analysis with all its possible "what if" scenarios.** In a Monte Carlo simulation, a large number of discrete scenarios are run based on random input values generated by the probability density functions of the variables. The result yields a probability distribution for the outcome and provides insight into the likely range of values for the project outcome.

Resource

Commercial software such as @RISK, which can be appended to Excel, uses the Monte Carlo approach.

6.5 Summary

This chapter introduced two different approaches to project alternative evaluation and selection: traditional economic-based techniques using engineering economic concepts, and multicriteria analysis (MCA) methods. Engineering economy methods are straightforward in calculation, converting all evaluation criteria to a common monetary unit, ratio, or percentage, and the results (e.g., NPV, B/C ratio, IRR) are easily interpretable. However, for complex decisions that involve qualitative criteria in addition to quantitative criteria, MCA methods may be more appropriate, as they convert quantitative and qualitative criteria measures to a common scale to be evaluated in a composite score. Although MCA methods are flexible in nature, the final output composite scores are less easily understood, as they are not unit specific. While each project evaluation approach has its advantages and disadvantages, engineering economy analysis and MCA are not completely separate. Economic-based approaches like CBA can be a component within MCA analysis, and this chapter discussed example applications of this combined approach.

6.6 References

Ackerman, F. & Heinzerling, L. (2005). Priceless: On Knowing the Price of Everything and the Value of Nothing. New York: The New Press.

Antoniou, A. & Lu, W. (2007) Practical Optimization: Algorithms and Engineering Applications. New York, NY: Springer Science+Business Media, LLC.

Barnum, D., McNeil, S., & Hart, J. (2007) Comparing the Efficiency of Public Transportation Subunits Using Data Envelopment Analysis. *Journal of Public Transportation,* 10(2), 2007.

Capital Area Metropolitan Planning Organization (CAMPO) (2010) Transportation Improvement Program for Fiscal Years 2011–2014. Available at http://www.campotexas.org/pdfs/CAMPO_FY11_14_TIPadopted_080910.pdf.

Dalkey, N. C. & Helmer, O. (1963) An Experimental Application of the Delphi Method to the Use of Experts. *Management Science* 9, 458–467.

Department for Transport (2004) The Treatment of Uncertainty in Model Forecasting TAG Unit 3.15.5. Available at http://www.dft.gov.uk/webtag/documents/expert/pdf/unit3.15.5.pdf.

Eddington, S. R. (2006) The Eddington Transport Study, Main report: Transport's role in sustaining the UK's productivity and competitiveness. UK Department for Transport, London. Available at http://www.dft.gov.uk/about/strategy/transportstrategy/eddingtonstudy/.

Eschenbach, T. G. (2003) Engineering Economy Applying Theory to Practice. New York, NY, Oxford University Press, Inc.

Fuller, S. K. & Petersen, S. R. (1995) Life Cycle Costing Manual for the Federal Energy Management Program. Available online at http://www.wbdg.org/ccb/NIST/hdbk_135.pdf.

Kansas Department of Transportation (KDOT) (2010) Expanded Highway Project Selection Process, Briefing Paper. http://www.kansastlink.com/downloads/Project%20Selection%20Process%20White%20Paper.pdf

Kockelman, K., Fagnant, D., & Xie, C. (2010) Documentation of Project Evaluation Toolkit and Appendix D of Final Project report.

Macharis, C., et al. (2009) The Multi-actor, multi-criteria analysis methodology for the valuation of transport projects: theory and practice, *Journal of Advanced Transportation.*

Ozbek, M., de la Garza, J., & Triantis, K. (2009) Data Envelopment Analysis as a Decision-Making Tool for Transportation Professionals. *Journal of Transportation Engineering.*

Park, C. S. (2007) Contemporary Engineering Economics. Upper Saddle River, NJ: Pearson Education, Inc.

Pradhan, A., & Kockelman, K. (2002) Uncertainty Propagation in an Integrated Land Use-Transport Modeling Framework: Output Variation via UrbanSim. *Transportation Research Record No. 1805*, 128–135.

Ramani, T., Zietsman, J., Eisele, W., Rosa, D., Spillane, D. & Bochner, B. (2009) Developing Sustainable Transportation Performance Measures for TxDOT's Strategic Plan: Technical Report. *Texas Department of Transportation Report 0-5541-1.* Available online at: http://tti.tamu.edu/documents/0-5541-1.pdf

Sheth, C., Konstantinos, T., & Teodorovic, D. (2007) Performance Evaluation of Bus Routes: A Provider and Passenger Perspective. *Transportation Research Part #: Logistics and Transportation Review*, (3)4.

Sullivan W. G., Wicks, E. M., & Koelling, C. P. (2012) *Engineering Economy, Fifteenth Edition.* Upper Saddle River, NJ: Pearson Higher Education, Inc.

Transport Canada (1994) Guide to Benefit-Cost Analysis in Transport Canada. Available online at http://www.tc.gc.ca/media/documents/corporate-services/bca.pdf.

US Department of Transportation, Federal Highway Administration (FHWA) (2003) Asset Management—Economic Analysis Primer. Available online at http://www.fhwa.dot.gov/infrastructure/asstmgmt/primer.cfm.

Weisbrod, G. (2008) Models to Predict the Economic Development Impact of Transportation Projects: Historical Experience and New Applications. *Annals of Regional Science.* www.edrgroup.com/library/multi-modal/models-to-predict-the-economic-development-impact-of-transportation-projects.html.

Weisbrod, G. (2011) Incorporating Economic Impact Metrics in Transportation Ranking and Selection Processes. Presented at Transportation Research Board Annual Conference, Washington, DC.

Zerbe, R. J. (2007) The Legal Foundation of Cost-Benefit Analysis. University of Washington, Law School and Evans School of Public Affairs, Berkeley Electronic Press. Available online at http://works.bepress.com/richard_zerbe/2.

Zhang, J., Yang, C.-h., Qin, X.-h., &. Du, W. (2007) The Comprehensive Evaluation for the Synergetic Development of Urban Traffic System Based on DEA Model. *Proc., Int. Conf. on Transportation Engineering* (ICTE 2007). Reston, VA: ASCE, 2656–2661.

Zong, J. & Cao, M. (2007) Measuring the Operational Efficiency of the Six Airlines in China. *Proc., Int. Conf. on Transportation Engineering 2007* (ICTE 2007). Reston, VA: ASCE, 3082–3089.

Chapter 7. Economic Impact Analysis of Transportation Investments and Policies

7.1 Introduction

Economic impact analysis (**EIA**) unveils how transportation facilities and systems affect businesses, governments, and households. Concepts such as direct, indirect, and induced jobs are covered in this chapter, along with a discussion of terms and methods used in economic analyses. Many studies claim that transportation investments are vital for the economy and serve as engines that fuel economic growth, as seen in the following excerpts from EIA reports:

> *The transportation of freight is the life blood of the economy. Goods and materials flow in vast quantities from production sites to manufacturers and from manufacturers to customers in a highly complex, cost-minimizing system that has developed over many decades. This system has achieved high levels of efficiency and responsiveness that in turn have fueled economic growth in both domestic and international markets (Persad et al. 2011, p. 29).*

> *Using data from more than 75 state and individual airport economic impact studies, this analysis found that the 490 commercial airports in the U.S. support 10.5 million jobs, create an annual payroll of $365 billion, and produce an annual output of $1.2 trillion. These are significant contributions to the national economy. Commercial airports are often economic engines that drive the local, state, and national economies. Airports are valuable assets that contribute to the growth of jobs and economic output across the country (CDM Smith 2012, p. 1).*

An **economic impact** is any change to the amount, flow, and/or distribution of money due to changes in the production, distribution, and consumption of goods and services. Transportation is an integral part of the economy and changes are likely to impact the economy in some way. Determining whether a given transportation investment or policy makes enough of an impact to earn the description of "economic engine" requires careful data collection and analysis, as described in this chapter. An impact that cannot be directly or indirectly related to changes in the economy is considered **non-economic**. Examples include community pride and quality of life. (Weisbrod & Weisbrod, 1997)These factors are generally excluded from economic impact studies, though research exists on how to value non-economic impacts so that they can be incorporated in comprehensive studies.

7.2 Why Economic Impact Analyses Are Conducted

Prediction and Evaluation

EIAs are conducted to either

- **Predict** future changes in the economy in response to changes in future infrastructure investments or policies (all EIAs completed for transportation projects also predict economic impacts), or
- **Evaluate** changes in the economy from a project or policy implementation, past or current.

Key Terms

- ✓ **EIA:** economic impact analysis
- ✓ **EIS:** environmental impact statement
- ✓ **CBA:** cost-benefit analysis
- ✓ **IO:** input-output
- ✓ **CGE:** computable general equilibrium
- ✓ **SAM:** Social Accounting Matrix
- ✓ **BEPA:** business establishment's production accounts
- ✓ **NIPA:** National Income and Product Account

Motivations

Three major motivations for predicting or evaluating economic impacts are to

- Satisfy regulatory requirements.
- Provide public information and planning guidance.
- Develop research findings for policy development, academic study, and/or advancement of transportation economics.

Regulatory Project Assessment

One major motivation for conducting an EIA is that it is often required by law for project assessment. At the state level, several laws require or recommend an EIA as a part of the project assessment process. At the federal level, the National Environmental Policy Act (NEPA) and the associated regulations of the Council on Environmental Quality require an environmental impact statement (EIS) for transportation projects involving federal lands or funds. **Each EIS must include an economic impact analysis**. The Federal-Aid Highway Act of 1970 also requires consideration of the impacts of federal-aid transportation projects on residences, businesses, and other community entities.

EIAs are often legal requirements during project planning.

Most EISs have a standard table of contents. Typically the "social resources" or "social effects" section of the "Environmental Consequences" chapter or the "Indirect and Cumulative Effects Assessment" chapter includes the discussion of the economic impacts.

Public Information and Planning

A public or private entity may decide it needs to convey to the public the impact of an existing or planned transportation facility to "prove its worth" for public approval, or to assess how to expand or change operations.

An EIA can help determine the extent of the current impact and how to direct investments to optimize use of limited resources for future positive impacts.

For example, the California Department of Transportation Division of Aeronautics commissioned a study to evaluate the impact of aviation in California and found significant economic activity generated by those activities:

> *The aviation sector plays a substantial and vital role in California's economy. The overall contribution to the California economy (including their direct, indirect and induced impacts) amounts to nearly 9 percent of both total state employment and of total state output. In 2001, aviation contributed close to $111 billion of total output of goods and services to California's total Gross State Product (GSP) of over $1.3 trillion. Employment in California totaled 19.5 million in 2001, and 1.7 million jobs were supported by the aviation sector (Economics Research Associates and JD Franz Research, Inc. 2003, p. 2)*

Studies are also conducted for planning purposes and to compare alternative investment options. For instance, the Port of Corpus Christi in Texas requested an EIA not only to show how the public port supports employment in the community, but also to determine how to plan for the future:

A major use of an economic impact analysis is to provide a tool for port development planning. As a port grows, available land and other resources for port facilities become scarce, and decisions must be made as to how to develop the land and utilize the resources in the most efficient manner... An understanding of the commodity's relative economic value in terms of employment and income to the local community, the cost of providing the facilities, and the relative demand for the different commodities is essential in making future port development plans (Martin Associates 2004, p. 16).

A goal of many transportation facilities is to not only serve the immediate customers but also the community at large through positive economic impacts.

As a caveat, note that EIA reports typically do not put the predicted changes in perspective. For instance, the Corpus Christi economic impact study (Martin Associates 2004) mentions thousands of jobs and millions in spending, but does not cite these as a percentage of the total region's employment and spending, or compare the numbers with other similar investments that could or have been made, such as with an airport, another port, a school, roadway, or health clinic. The durations of "created" jobs are also generally neglected. Many are temporary and last only through the construction period.

Research Studies

Research studies aim to understand the significant factors influencing how and why economic indicators change in response to alterations in transportation systems, facilities and policies. Instead of being project-specific, research studies often examine a cross-section of systems or projects to find patterns in the type and magnitude of economic impacts. For instance, Texas Department of Transportation sponsored a study of the economic impacts of highway relief routes (bypasses) on 23 small and medium-sized communities in Texas, using 19 cities without highway relief routes as controls

Research results are intended to guide policy and decision-making and to advance understanding of transportation economics.

The models developed suggest that of the four sectors examined, the impact of a bypass is most negative on the per capita sales in gasoline service station. The impacts of the per capita sales in the other three sectors studies [retail, eating, and drinking establishments, and service industries] depended critically on the magnitude of the traffic diverted. When about half the approaching traffic was diverted to the bypass, all three sectors were negatively impacted. So, the better a relief route works from a traffic standpoint, the greater its adverse impact on local per capita sales (Kockelman et al. 2001, p.68).

7.3 Economic Indicators

Evaluating or predicting the economic impact of a transportation project requires measuring a change in an **economic indicator** with the amount, flow, or distribution of money in the economy. Typically, economic impact studies use one or more of the following indicators:

- spending by households and businesses
- employment, by the number of jobs
- income (wages and salaries of people living in the community)

- business sales
- exports and imports
- capital investment expenditures
- value-added (for example, a state's GDP)

Table 7.1 presents examples of economic impact studies and the indicators used. Important and common indicators are tax revenues generated, land use development, natural resources consumed, and distribution of jobs created by occupation. The indicators chosen for a study can depend on the purpose of the study, type of transportation facility, and audience for the study.

Table 7.1: Examples of Economic Indicators in Texas Economic Impact Analysis

Study	Economic Indicators
Economic Impact Analysis of the Baltimore-Washington Rail Intermodal Facility (HDR Decision Economics 2011)	• Total business sales • Value added (i.e., business output minus the cost of purchasing intermediate products) • Employment (number of jobs) • Local, state and federal tax revenue
Economic Effects of Highway Relief Routes on Small- and Medium-Size Communities: An Econometric Analysis (Kockelman et al. 2001)	• Per capita sales • Numbers of establishments • Total sales in the four highway-related • Sectors of retail, gasoline service stations, food and drink establishments, and service industries
Guide to the Value of Texas Ports(Goff et al. 1997) and *Economic Impacts of the Port of Los Angeles* (Martin Associates 2007)	• Total employment • Personal income • Business sales (revenue) • Local and state tax revenue
Estimated Economic Impact of Selected Highway Widening Projects in Texas (Buffington & Wildenthal 1998)	• Property values • Sales tax revenue • Property tax revenue • Total employment • Total output (value of goods and services)
The Economic Impact of Commercial Airports in 2010 (CDM Smith 2012) and *Economic Impact Study of the Ohio River Bridges Project* (EDR Group, Inc., 2012)	• Total employment • Annual payroll • Annual output

Example

Economic Indicators

A project for two proposed bridges to connect Louisville, Kentucky with southern Indiana and a reconstructed interchange on the Kentucky side, one of the largest transportation projects in the US called the "Ohio River Bridges Project," required an economic impact study per an Indiana state law (Indiana Code Title 8, Chapter 15.5, Chapter 4, Section 1.5 (b)) since funding came from the Indiana Finance Authority.

The impact study used the economic indicators of employment (total number of jobs), personal income (wages earned from jobs), and business output (the value of goods made and sold by businesses) to help identify the impacts specifically requested in the Indiana state law. The regional travel demand model results fed into the TREDIS model gave output indicating:

> *Overall, from 2012 to 2042, the Project is expected to generate an average of 17,796 jobs per year and a cumulative total of $27.3 billion in personal income and $78.0 billion in economic output (in constant 2012 dollars) in the regional economy.*

Measuring Economic Indicators

Local governments, counties, state agencies (such as the state comptroller's office), and federal agencies (such as the U.S. Department of Commerce, Bureau of Economic Analysis [BEA], and Census Bureau) monitor economic indicators on a continuous basis and offer a wealth of information useful for EIAs. Private sources of data are also available; while typically not free, private data can fill the gaps in public data. Measures of economic indicators are also obtained by directly surveying businesses and households.

Please note one caveat regarding data availability: data collection requires cooperation (and accuracy) on the part of businesses, consumers, and government. The government relies on households and private businesses to accurately report data. This time-consuming process often yields small samples, which can be another source of inaccuracy. For instance, input-output tables (detailed later in this chapter) are generally updated every 5 years, so unless an analyst surveys businesses and suppliers as part of the EIA, the data used could be at least 5 years old. In today's economy, with rapid technological change, the information may be outdated.

Also, analysts must avoid "double-counting" changes in economic indicators. For example, the value of increased business sales cannot be counted in personal income increases.

See the In-Depth Look section at this chapter's end for details on double-counting.

Impact Measures

An **impact measure** is anything observed to cause a change in an economic indicator. Impact measures include the following:

- Transportation costs
  - Crash savings
  - Travel time savings
- Transportation linkages
  - Addition or removal of rail service, airports, or sea ports
- Environmental quality
  - Changes in air pollution, noise pollution, and water pollution
- Community impacts
  - System changes in safety (more or fewer crashes)
  - New restrictions on access to properties
  - Removal of land from tax rolls for transportation projects
  - Business displacements

7.4 Generative and Redistributive Impacts

Transportation projects have the potential to change economic indicators. For example, removal of a freight rail bottleneck may reduce the amount of time a product is in transit, thus reducing travel costs. The savings may increase business income by improving business productivity. When travel costs fall due to system/network improvements, income for households and businesses can rise from the increased net savings, causing a **generative impact**. Generative impacts produce a net economic gain. The converse is a **degenerative impact**.

In contrast, if the savings or increases in income occur at the expense of another area's wellbeing, such as causing a bottleneck further upstream or downstream, or shifting businesses and households to a new area, then the transportation project has a **redistributive impact**. Money and income that already existed or would have existed without the project flowed to another region and so were redistributed.

What may appear as a generative impact at one geographic scale can be a redistributive impact at a larger scale. The effect of a highway bypass on a local community is one example of this. Within a town's city limits, the bypass impact may be redistributive because businesses may move outside the city limits to lie along the new highway. From the county's perspective, if the bypass results in an increase in business sales, then the impact is generative (Figure 7.1).

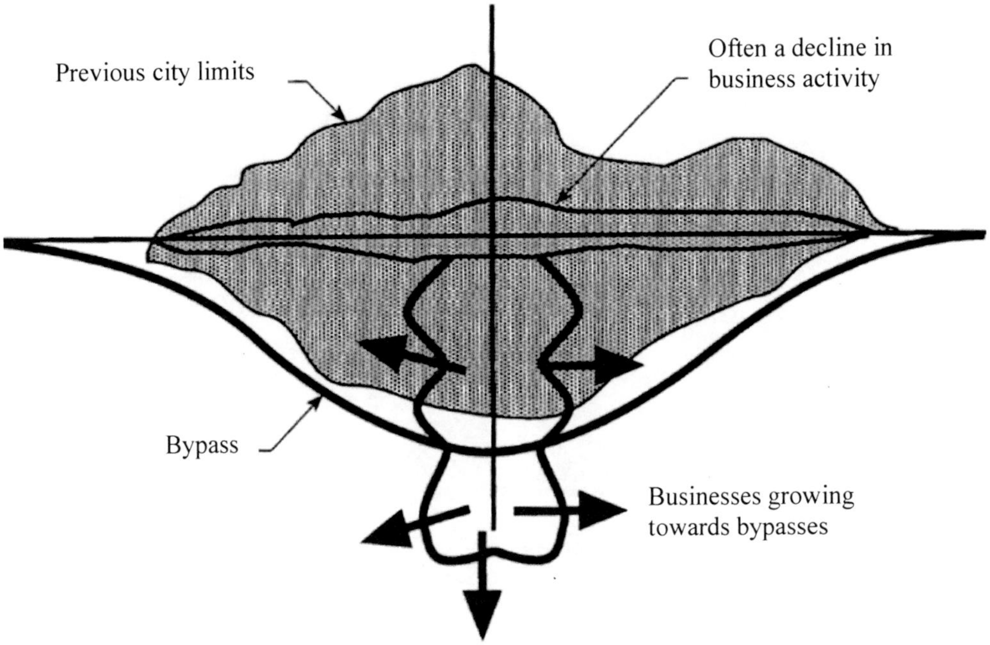

Figure 7.1: Redistribution of Businesses in Response to the Addition of a Highway Bypass (Source: Andersen et al. 1992)

Transportation infrastructure investments can yield negative local economic impacts, as the following examples illustrate.

Concept Example: Negative Local Impacts

If a port invests in making multimodal transshipments more efficient, fewer dockworkers will be needed to work the cargo. This means less labor income will be distributed within the region, perhaps even as maritime freight levels rise. The goal of the infrastructure investments was, after all, to lower transportation rates.

In a similar vein, investments to reduce congestion along major arterials can raise average speeds on those roads, making impulse shopping along those arterials a less attractive option. Thus, congestion reduction can have the unintended consequence of subduing retail sales along major arterials (Quddus et al., 2005).

Ideally, the economic goals of a project should be generative for as large a region as possible, but generative impacts at a local scale may result in redistributive impacts at a larger geographic scale. For instance, a region may wish to attract businesses from elsewhere via tax incentives and transportation system improvements. *In this case, the impact to the city or metro area would be generative, but the business relocations would not result in a net economic gain for the state or nation.*

Concept Example: Land Value

Town A adds a new major roadway and land values around the new facility rise. The roadway has a local generative impact by increasing total land values in Town A (Figure 7.2).

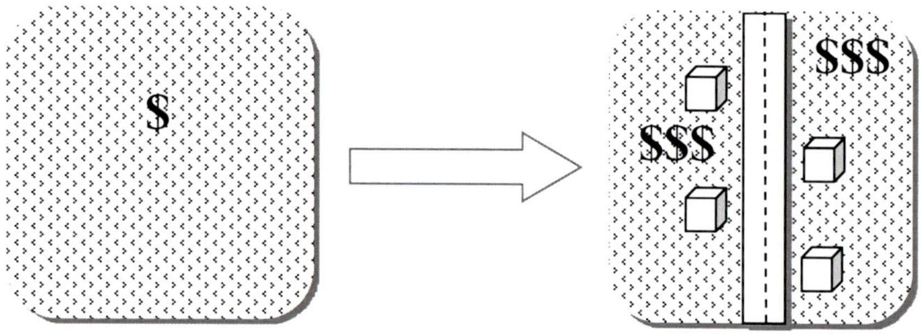

Figure 7.2: Generative Impact: Net Gain in Land Value

However, land values may fall in Town B as businesses choose Town A, with its improved network and lowered travel costs (Figure 7.3). The land value was redistributed and the overall region may not have a net gain in land values.

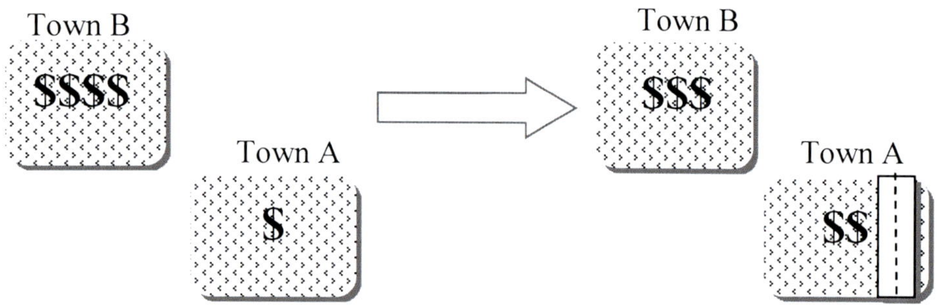

Figure 7.3: Redistributive Impact: Net Land Value Loss for Town B and Net Gain for Town A

Figure 7.2 and 7.3's dollar signs could also represent business sales or number of jobs. Regardless of the choice of economic indicator, generative and redistributive impacts must be carefully assessed to avoid double-counting changes and erroneously reporting net economic gains when only a shifting of value from one area to another occurred. EIA seeks to find the net generative impacts and net gains (or net losses). A related impact is called a **financial transfer**, or *fiscal impact*, but is not considered economic, as it refers merely to transferring funds from one entity to another, such as from the federal to the local government. Table 7.2 breaks down the three types of impacts.

Type of Impact	Description & Key Question	Examples
Generative	Changes that result in a net economic benefit in the study region because of utilization of underused resources or more efficient use of resources in the region. **Is the impact increasing the efficiency of movement of people, goods, and/or business operations?**	• Travel time savings • Infrastructure savings from more efficient land development • More efficient land use (resulting in more efficient exchange of information and goods—agglomeration benefits) • Attraction of travelers to more efficient transportation modes • Any change that results in an increase (or decrease) in the efficiency of the provision and movement of people, goods, and services.
Redistributive	Growth and/or losses that probably would have occurred in the region over time and are relocated within the study region because of the project. **Is the impact shifting economic activity, benefits, or costs from one part of the region to another?**	• Businesses move within the study region (from one part to another). • Land values decline in one area and increase in another.
Financial Transfer/ Fiscal Impact	The shifting of money from one entity to another; essentially an accounting, not an economic, impact. **Is the impact simply a transfer from the accounting books of one entity to another?**	• Direct financial impacts associated with the transfer of money from the federal government to the local government (though can be considered an economic impact if the federal funds would not arrive in region if the project did not occur) • Transfer of money from private land developer leasing public property to public entity (if raised revenues by the developer are already counted).

Table 7.2: Types of Impacts (Adapted from Cambridge Systematics et al. 1998)

7.5 Paths of Economic Analysis

The term *economic analysis* is used loosely to describe a variety of studies that could be conducted to assess a project's *economic value* or its *economic impact*. The distinction between those two types of assessments is explained in the next section.

Choosing a method of analysis depends on the final product desired, the geographic area and time period of interest, and of course, data availability and analyst expertise. Table 7.3 list typical contexts for each method.

Analysis Method	Typical Area	Typical Time Frame
Economic Value		
Cost-Benefit Analysis	Project-level to regional	Project or policy lifetime (multiple years)
Economic Impact		
Input-Output Models	Regional	Single point in time (single year)
Computable General Equilibrium Models	Regional, national, or international	Period of time (multiple years)
Economic Simulation and Forecasting	Any level	Period of time (multiple years)
Econometric Analysis	Any level	Any time frame (before/after; single or multiple periods of time)
Market and Fiscal Impact		
Market and Fiscal Impact Studies	Any Level	Period of time (multiple years)

Table 7.3: Comparison of Economic Analysis Methods

Economic Value vs. Economic Impact

A transportation project's economic impact is not the same as its economic value. **Economic value** estimation involves answering this question: Do the project's benefits outweigh its costs? **Economic impact** analysis, on the other hand, seeks to estimate a project's impacts on economic indicators.

Cost-benefit analysis (CBA) is used to calculate economic impact. CBA "attempts to capture all benefits and costs accruing to society from a project or course of actions. Used properly, CBA reveals the most economically efficient investment alternative, i.e., the one that maximizes the net benefits to the public from an allocation of resources" (USDOT 2003, p. 17). EIA does not focus on optimizing return on investment, but rather "attempts to measure the consequences that a…project….will have on considerations such as local or regional employment patterns, wage levels, business activity, tourism, housing, and even migration patterns" (USDOT 2003, p. 32).

See Chapter 6 for details on CBA and its outputs.

Many of the factors considered in each analysis are common to both, but they vary in distinct ways, as Table 7.4 shows. EIA excludes any impacts that cannot be measured as pure economic transactions (for example, personal and household time savings). Although personal time savings can be monetized, as they are for CBA, they are not considered "actual flows of money" for an EIA. Table 7.4 also illustrates how CBA is limited in terms of assessing larger-scale economic impacts (due to double-counting concerns). For example, business attractiveness and multiplier effects (such as income generated by off-highway businesses) are generally not considered.

Form of Impact	Counted in CBA	Counted in EIA
Business cost savings	✔	✔
Business-related and household out-of-pocket cost savings	✔	✔
Personal and household out-of-pocket cost savings	✔	✔
Personal and household time savings that do not result in actual out-of-pocket costs*	✔	✘
Other benefits that do not result in an actual economic transaction*	✔	✘
Attraction (relocation) of business activity into the area	✘	✔
Income generated by off-highway businesses and their suppliers	✘	✔

*These savings are monetized in CBA.

Table 7.4: Impacts Used in CBA and EIA (Source: adapted from DRT 2009)

Compared to CBA, EIA anticipates spinoffs and multiplier effects, such as investment that will save travel time, improve access, and reduce out-of-pocket travel expenditures. CBA captures these benefits using direct, monetized savings divided by project costs. Further, EIA seeks to translate travel time savings to a reduced business costs, area reinvestment, and ultimately growth in employment, wages, and GDP.

Another way to distinguish CBA from EIA is to understand how they may be used to measure social impacts. Following is an outline of how the two methods are best applied (Weisbrod 2010, p. 9).

CBA is designed to help:

- *Ensure efficient use of scarce resources*
- *Minimize cost among alternatives that achieve needs*
- *Maximize performance results for given [funds] available*

EIA is designed to help:

- *Stimulate & grow jobs and income where they are most needed (for example, distressed areas)*
- *Attract 'quality jobs' – well-paying, stable, secure, in growth industries where income can rise over time (economic vitality, sustainability, competitiveness)*
- *Ensure equity and assistance for vulnerable populations*
- *Reduce vulnerability risk from dependence on foreign suppliers (import substitution)*

The goal of EIA is to determine the changes in economic indicators, whereas the goal of CBA is to monetize costs and benefits of the project over time to find its net present value, benefit-cost ratio, or rate of return.

Economic Models

Economic models are essentially mathematical methods or equations developed under the guidance of economic theory. For example, **input-output tables** are commonly used for modeling exchange between industries and the multiplier effects of purchases and investments. They use matrix algebra, which employs as many equations as there are industry and non-industry sectors in the modeled economy. In contrast, **time-series models** generally rely on a single equation to forecast variables like employment, gas prices, or traffic counts through various points in time (as discussed in Chapter 8). Many economic models are available, each developed using different theoretical foundations and data sets. Table 7.5 lists and contrasts many of the defining characteristics of economic models.

Theoretical Tests theories and uses simulated data.	vs.	**Empirical (Applied)** Uses real-world data to solve real-world problems.
Microeconomic Study of individual businesses and industries at the disaggregated level.		**Macroeconomic** Study of interactions of economic entities at the aggregate level.
Static One time period.		**Dynamic** Multiple time periods, with future periods affected by previous periods.
Myopic Multiple time periods solved one at a time, thus not allowing economic behavior that can look ahead into the future.		**Forward-Looking** Multiple time periods are solved for, simultaneously allowing economic behavior to consider future changes.
Deterministic Predetermined equations with values fixed and no error terms.		**Stochastic** Equations with uncertain parameters and calibrated error terms.
Calibrated Equation parameters selected based on limited data and judgment.		**Estimated** Statistical methods used to estimate parameters with standard derivations for equations.
Integrated Predictive system consists of multiple model equations applied simultaneously or with feedbacks.		**Sequential** System consists of separate sub-models.

Table 7.5: Possible Characteristics of Economic Models (Source: derived partially from Partridge & Rickman 2007)

Obviously, it is beyond the scope of this reference book to delve into the details of all economic models and theories used for EIA. The more complicated economic models can seem to be "black boxes," and even seasoned economic professionals complain about the lack of transparency in proprietary and other models used for predicting economic impacts (Kockelman et al., 2009).

In addition, some popular models consist of several reasonably complicated sub-models. For instance, the website for the firm Regional Economic Models Inc. (REMI.com) describes the REMI model as

> *a structural economic forecasting and policy analysis model [that]...integrates input-output, computable general equilibrium, econometric, and economic geography methodologies. It is dynamic, with forecasts and simulations generated on an annual basis and behavioral responses to compensation, price, and other economic factors.*

The next two sections describe key concepts in and applications of two important models for relatively comprehensive EIA: the input-output (IO) model and the computable general equilibrium (CGE) model.

7.6 Input-Output (IO) Modeling

IO modeling is the most commonly applied method for economic analysis. Its developer, Professor Wassily Leontief, received the Nobel Prize for Economic Science in 1973. IO modeling tracks the flows of inputs and outputs between industries, along with outputs headed for **final demand** (for example, consumers, and international trade). The model shows that outputs of one industry become inputs to another industry, such as materials industries that produce steel, glass, rubber, and plastic products (outputs). But automotive manufacturers take these materials as inputs to build cars (outputs) to satisfy customer demand. Therefore, when you buy a car, you affect the demand for glass, plastic, steel, etc.

The IO model is frequently used to determine the change in business output, personal/household income, and employment and government revenue in response to changes in input final demand (government spending, household spending, regional or foreign trade, and business investment). This model identifies the sectors that have the greatest impacts or multiplier effects.

IO's mapping of relationships between industries allows analysts to anticipate how changes in final demand and production technologies affect other industries.

IO models can also estimate output changes in response to changes in supply (rather than changes in demand). For example, IO models are used to determine how business revenue changes resulting from transportation system improvements affect employment, household income, business profits, and government tax revenues.

Changes in economic indicators are labeled *direct*, *indirect*, and *induced*, as illustrated in Figure 7.4 (Gkritza et al., 2007).

- **Direct effects** generally are the business output (value of goods and services), jobs, and income generated on-site at a transportation facility or project. Direct effects take place only in the immediately affected industry. For example, if an automotive manufacturing facility closes, the job losses at the plant are direct effects.
- Unlike direct effects, **indirect effects** measure inter-industry transactions. In the automotive manufacturing facility example, a plant closure affects the demand for materials from the facility's former suppliers, possibly resulting in indirect job losses at the suppliers' sites. The sum of the direct and indirect effects is called **first-round effects.**
- Second-round or **induced effects** come from the changes in personal income of individuals with directly and indirectly affected jobs. Such income changes affect purchasing activity, resulting in further rounds of these demand changes. Continuing the example above, former employees of the automotive manufacturing facility now have reduced household incomes and may decrease their household expenditures, affecting business transactions at local shops and restaurants.(Martin Associates, 2004, p. 5)

Together, direct, indirect, and induced effects are called **multiplier effects**. Inter-industry transactions tables can be manipulated to derive *multipliers* of final demand (which will be addressed later in this chapter) and direct effects to very quickly gauge overall economic impacts per $1.00 or $1 million in investment. Typical multipliers vary from 2 to 4, depending on the industries directly/first affected and the size of the region.

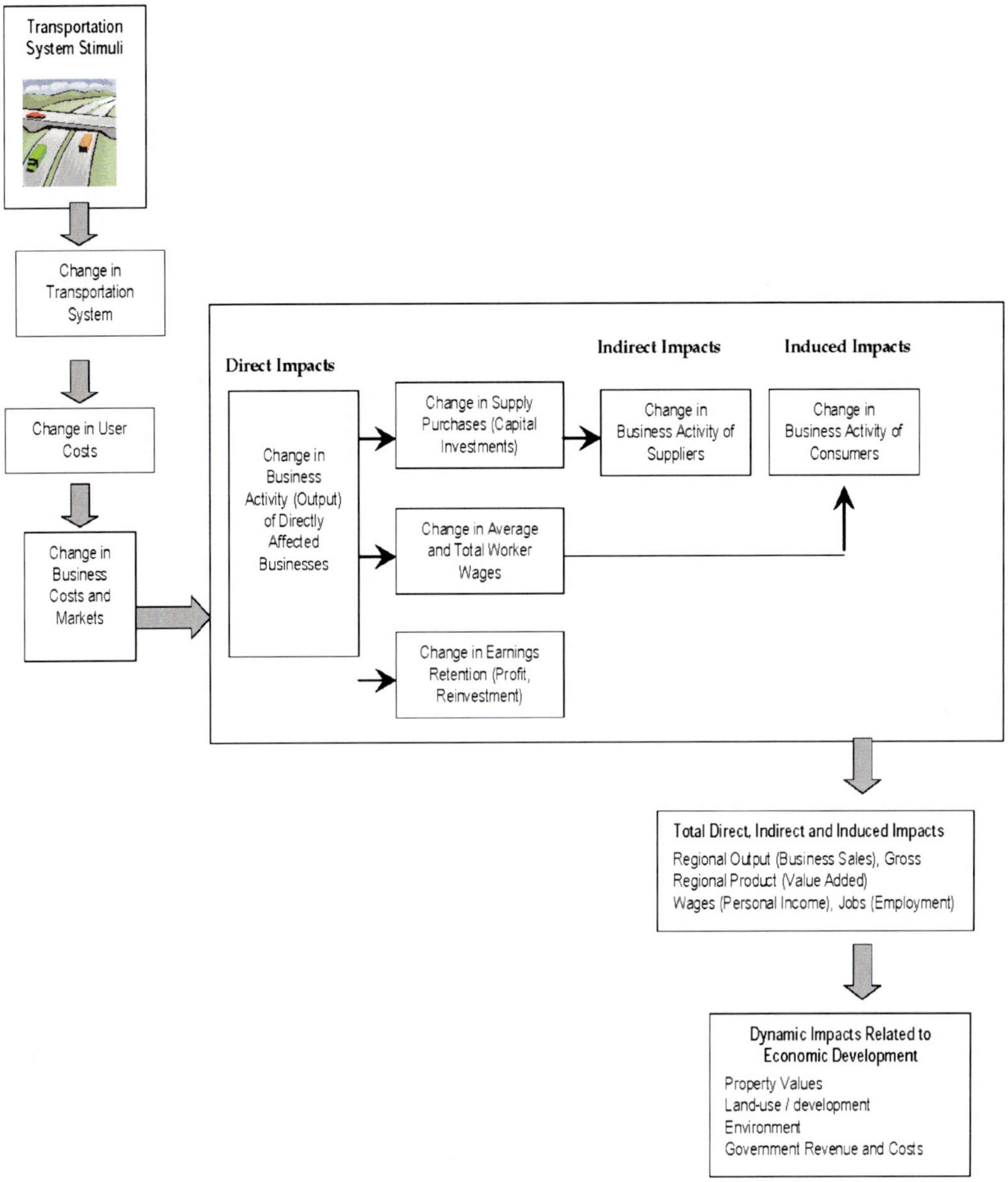

Figure 7.4: The Sources of Direct, Indirect, and Induced Impacts (Source: Gkritza et al. 2007)

The Foundation of the IO Model

The IO model's foundation is the **transactions table**, presented in Table 7.6. The outputs produced by each industry are divided into outputs purchased as inputs by other industries (**intermediate transactions**) and for final consumption or export (final demand).

		Purchases from (inputs for)		Final Demand				
		Industry Sector A	Industry Sector B	Household/ Consumer Spending (C)	Private Investment (I)	Net Exports (X-M)	Government Spending (G)	TOTAL OUTPUT
Sales from (outputs from)	Industry Sector A	150	500	50	150	100	50	1000
	Industry Sector B	200	100	400	900	250	150	2000
Value Added	Household Income	300	500	50	25	0	25	900
	Business Profit	250	750	300	50	25	25	1400
	Government Taxes/Fees	100	150	100	25	0	25	400
	TOTAL INPUT	1000	2000	900	1150	375	275	5700

Table 7.6: Example IO Transactions (Table adapted from Horowitz & Planting 2009)

Intermediate Transactions

Table 7.7's shaded gray cells show the monetary transactions between two industries or within an industry. For instance, the cell where Industry A's column intersects with Industry's A row shows that Industry A sold $150 worth of output to Industry A. The electric generation industry is one example of how an industry may purchase output from itself as an input (electricity needed to operate a power plant). In the same row but in Industry B's column, Industry A sold $500 worth of output to Industry B. In other words, Industry B purchased $500 worth of Industry A's output.

The total value of outputs produced by an industry is the sum of all its row's outputs. The total value of inputs purchased by an industry is found by summing the purchases in a column. In other words, rows represent sellers or suppliers, and columns represent buyers.

Final Demand

Changes in final demand are regularly used to estimate the economic impacts of various transportation projects. Table 7.7's four columns without shaded cells contain final demand figures for four economic sectors (C, I, X, and G):

1) **Household/consumer spending (C)** includes all purchases of goods and services for consumption. A primary engine that drives economic growth, this sector of the economy consists of two-thirds of domestic final spending (BEA 2010).

2) **Private investment (I)** is the spending by private businesses and nonprofit institutions (and in some cases, households) on new fixed assets such as buildings and equipment, or improvements to those

assets.

3) **Net export (X-M)** includes the total value of an output exported out of the region (X), less the imports coming into the region (M). If there are more imports than exports, X will be negative.

4) **Government spending (G)** on a project is a change in final demand that generally increases local demand and thus local outputs. In addition, the project may reduce costs of doing business, which may incentivize and allow additional private business investment (I) for further boosts in final demand.

Industry outputs destined for final demand are summed to find a region's or country's gross domestic product (**GDP**), which is the total market value of all final goods and services produced within a country in a defined period. GDP is frequently used as an indicator of economic health used frequently in the media and research studies. Figure 7.5 illustrates the connections between the economic entities that produce GDP.

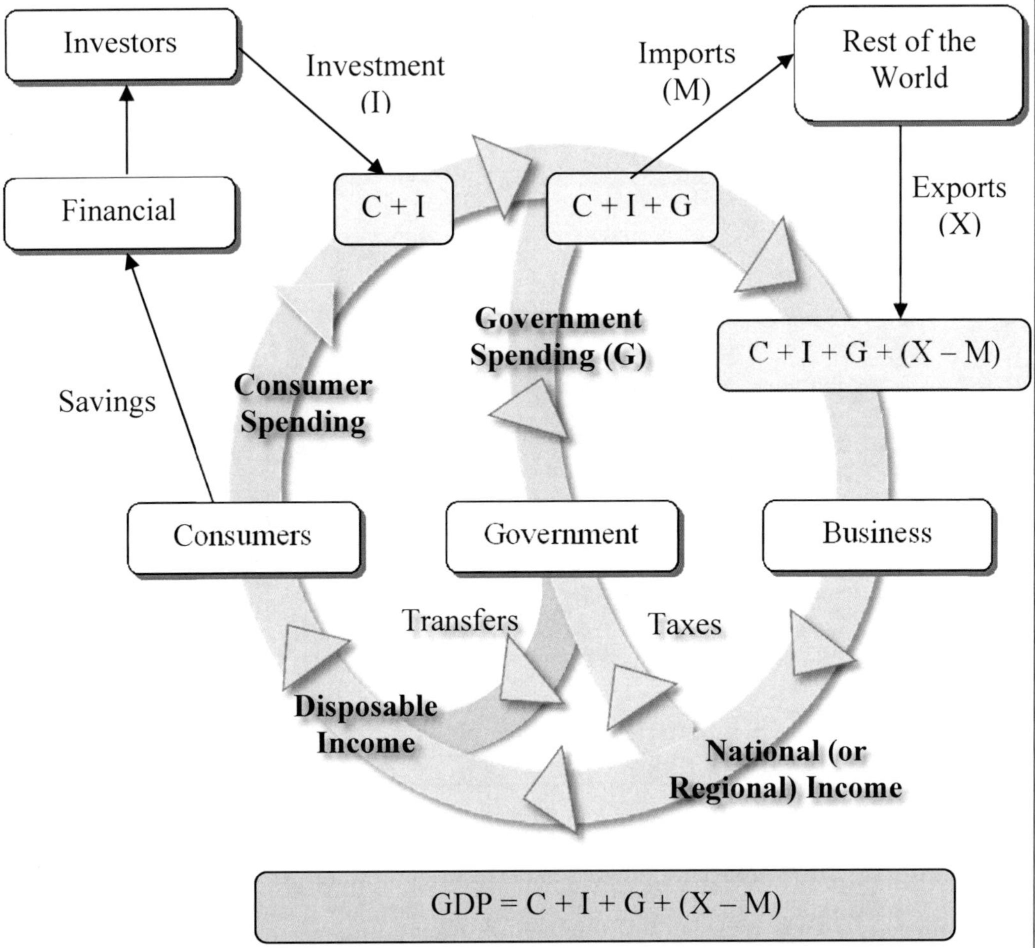

Figure 7.5: Circular Flow of Income in the Economy (Source: adapted from Baumol and Blinder 1988)

Value Added

The income approach to estimate GDP is described as **value added**. The bottom three rows of Table 7.6 contain the value added figures: industry monies devoted to wages (labor), business profit, and payments of government taxes and fees. The total compensation of local/regional employees across industry and final demand sectors is also shown as total household output. Business profit may seem like an unexpected input, but profit expectations drive the production process and such monies must be accounted for in the table and in the economic system. The government taxes and fees include taxes on production and imports less subsidies (Horowitz & Planting, 2009). Note that the expenditure approach and the income approach must yield identical GDP estimates.

Using the IO Model

As mentioned earlier, the transactions table is the basis for **IO analysis**. This analysis can be used directly to find multiplier effects of changes in final demand sectors or to find the multipliers themselves. Figure 7.6 presents the options for obtaining measures of a project's economic impact.

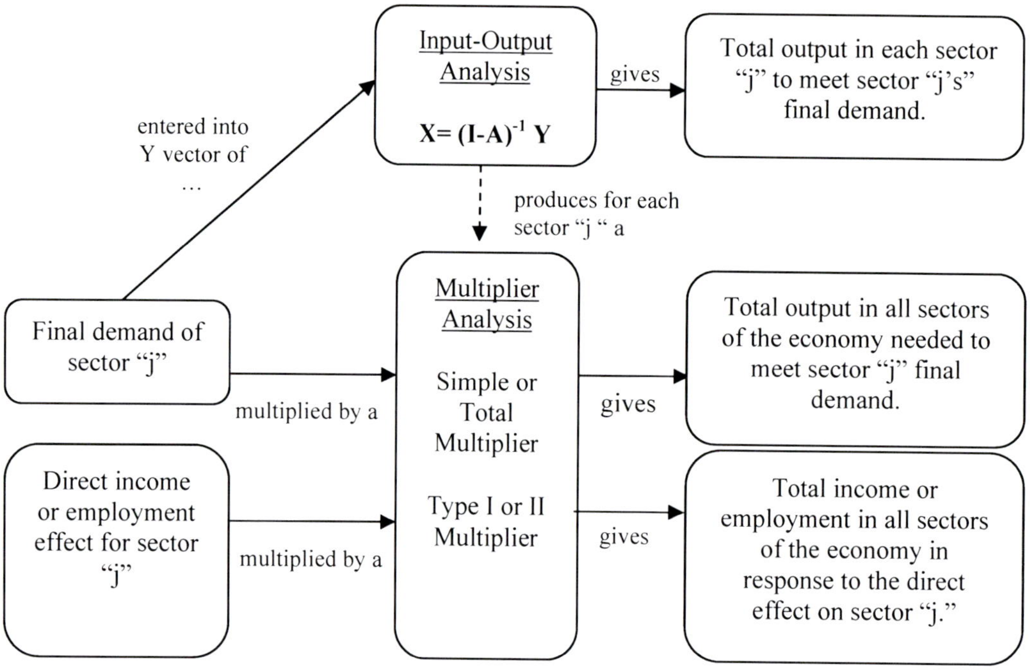

Figure 7.6: Paths to Determine Economic Impact Using IO

Multiplier Analysis

A **multiplier** is a number extracted from IO calculations that can be multiplied by output changes to estimate the total of the direct, indirect, and induced effects of transportation investments. Multipliers are specific to the industry in which the investment occurs, and are often available at the county, state, or national level from public agencies or software programs like IMPLAN (Impact analysis for Planning) and RIMS-II (Regional Input-Output Modeling System).

See this chapter's In-Depth Look section for transactions table creation and use.

The bigger the region, the bigger the multiplier, because fewer purchases of inputs are "lost" to exports. Some industries enjoy bigger multipliers than others. Some industries enjoy a high income multiplier but a relatively low jobs multiplier, suggesting that expenditures on those industries' outputs create relatively few, high-income positions. This section describes the different ways to apply and estimate multipliers.

Categories of Multipliers

Multipliers are tailor-made for each sector, with the most common economic indicators being output, employment, and income. Most multipliers are defined as either simple or total (or Type I and Type II).

The simple (Type I) and total (Type II) multipliers are also called **final demand multipliers** because their denominator is a one-unit change (for example, one dollar) in final demand (U.S. Department of Commerce, 1997) and because they are multiplied either by the new final demand or by the change in final demand. Typically, the one unit is a $1 change, but could also be one ton of the pertinent sector's product. All simple and total multipliers are derived from the *Leontief inverse matrix*, which is described later in this chapter. Table 7.7 presents real-world examples of multiplier usage.

Table 7.7: Multipliers in Action

	Simple or Type I	**Total or Type II**
Examples	(direct + indirect effects) / (change in the sector's final demand)	(direct + indirect + induced effects) / (change in the sector's final demand)
1	A local government will spend an additional $10 million on transportation projects (an increase in final demand). With a simple multiplier for industry output of 1.2, the new output is found by multiplying $10 million by 1.2 for an additional $12 million.	An employment multiplier of 7.5 per $1 million of construction spending by the local government means an additional 7.5 jobs would be created within the study area by direct, indirect, and induced effects for every $1 million spent.
2	A company plans to move to an area because of a new highway. The company will add 800 new jobs to the area. With a Type I employment multiplier of 1.6, the total number of direct and indirect jobs are determined by multiplying 800 jobs by 1.6 for a total of new direct and indirect 1,280 jobs throughout the economy. The 1,280 includes the 800 jobs directly added by the company.	With a Type II multiplier of 2.2, the company's 800 direct jobs multiplied by 2.2 results in a total of 1,760 jobs for the study area due to the induced effects of household spending. The 800 jobs are included in the 1,760 total jobs.

Because the Type II multiplier includes the induced spending, the Type II multiplier is larger than the Type I multiplier.

Example

Multiplier Analysis

Multiplier analysis is frequently used in economic impact studies. For example, in a study of the economic impact of a Baltimore-Washington Rail Intermodal Facility in the state of Maryland (HDR Decision Economics, 2011), multiplier analysis was applied using IMPLAN to anticipate the likely short-term (construction and engineering) economic impacts of the proposed project:

> *The short-term economic impacts include the effects of construction and engineering activities during the development phase (2009 – 2014) of the Baltimore-Washington Rail Intermodal Facility...Overall, the construction and engineering activities are expected to generate more than 1,300 job-years, produce in excess of $200 million of economic output, including $100 million in value added, and contribute about $25 million in Federal and State/Local taxes over the 2009 – 2014 period...the multiplier effects (i.e., indirect and induced effects combined) associated with engineering services make up a larger share of total impacts than the multiplier effects of construction activities.*

This project's construction and engineering costs are multiplied by multipliers to obtain total direct, indirect, and induced (total) effects, as shown in Table 7.8.

The study also estimated the likely long-term economic impacts from operations:

> *The long-term economic impacts include the net impacts associated with facility operations and the impacts associated with improved access and with reduced transportation and logistic costs in the study area (i.e., economic development spurred by the new facility)...In its first year of operation (2015), the new intermodal facility is expected to employ 47 people and to provide jobs for 247 workers in (outside) drayage and 30 vendors. Overall, the impacts of facility operations on output and value added constitute about one-third of the total impacts, with the drayage activities making up the balance.*

A report should list the sectors considered in the study, as shown in Table 7.9 so that the study can be replicated in a future year.

Impact Category	Direct	Indirect	Induced	Total
Construction Impacts (2009 - 2014)				
Output	$109.0	$40.8	$45.3	$195.1
Value Added	$45.9	$22.0	$27.2	$95.1
Employment	688	244	333	1,265
Taxes	-	-	-	$23.6
Federal Taxes	-	-	-	$15.3
State/Local Taxes	-	-	-	$8.3
Engineering Impacts (2009 - 2014)				
Output	$4.1	$1.7	$2.3	$8.1
Value Added	$2.3	$1.0	$1.4	$4.7
Employment	28	12	17	57
Taxes	-	-	-	$1.1
Federal Taxes	-	-	-	$0.7
State/Local Taxes	-	-	-	$0.4
COMBINED Impacts (2009 - 2014)				
Output	$113.1	$42.5	$47.6	$203.3
Employment	716	256	350	1,322
Taxes	-	-	-	$24.7
Federal Taxes	-	-	-	$16.0
State/Local Taxes	-	-	-	$8.7

Note: Dollar estimates are in millions of 2010 dollars

Table 7.8. Estimated Short-Term Economic Impacts of Baltimore-Washington Rail Intermodal Facility (Source: HDR Decision Economics 2011)

Table 4: Short-Term Impacts on Maryland, by Industry

NAICS Code	Industry	Output	Value Added	Employment
11	Agriculture, Forestry, Fishing and Hunting	$0.2	$0.1	2
21	Mining	$0.6	$0.3	2
22	Utilities	$2.1	$1.6	1
23	Construction	$109.7	$46.3	694
31-33	Manufacturing	$7.6	$2.1	18
42	Wholesale Trade	$5.5	$3.7	26
44-45	Retail Trade	$6.8	$4.6	93
48-49	Transportation and Warehousing	$2.2	$1.3	18
51	Information	$5.1	$2.1	12
52	Finance and Insurance	$7.5	$4.1	32
53	Real Estate and Rental	$13.5	$9.4	34
54	Professional - Scientific and Technical Services	$20.4	$11.6	133
55	Management of Companies	$1.0	$0.6	4
56	Administrative and Waste Services	$3.6	$2.3	50
61	Educational Services	$1.0	$0.6	14
62	Health and Social Services	$7.3	$4.5	74
71	Arts - Entertainment and Recreation	$0.8	$0.4	13
72	Accommodation and Food Services	$2.9	$1.5	45
81	Other Services	$4.2	$2.3	48
92	Government and non NAICS	$1.3	$0.6	7
	Total	$203.3	$99.8	1,322

Note: Dollar estimates are in millions of 2010 dollars

Table 7.9. Estimated Short-Term Economic Impacts of Baltimore-Washington Rail Intermodal Facility (Source: HDR Decision Economics 2011)

Example

Multiplier Analysis

A research study of the economic impacts of widening State Highway 21 in Caldwell, State Highway 199 in Parker County, and U.S. Highway 59 in Houston, all in Texas, also used multiplier analysis, with Type II multipliers derived from the 1986 Texas Input-Output Model for the New Road/Highway Construction expenditure category (Buffington & Wildenthal, 1998).

For the State Highway 21 section in Caldwell, Texas the estimated statewide Type II employment multiplier was 59.9 jobs per million dollars of construction expenditures (Table 7.9). Because these were from a statewide IO model (with relatively low spending leakages across state lines), the total number of jobs could not all be assigned to the local Caldwell area. The researchers estimated that 121 of the 364 statewide jobs were added to Caldwell (Buffington & Wildenthal, 1998). Similar estimations emerged for Parker County and Houston.

Similar calculations for total output (value of goods and services sold) yield an output multiplier of 3.69, which suggests that for every dollar spent in new road/highway construction, total output across Texas economy increases by $3.69 (Table 7.10).

Table 7.10: Employment Estimates for Texas Highway Widening Projects (Buffington & Wildenthal 1998)

Location	Construction Expenditures	Employment Multiplier	# of New Jobs in Texas	Estimated # of Jobs in Local area
State Highway 21 Caldwell	$6.095 million	59.9 jobs per million dollars of expenditures	365	121
State Highway 199 Parker County	$8.082 million	56.02 jobs per million dollars of expenditures	453	10
U.S. Highway 59 Houston	$114 million	56.02 jobs per million dollars of expenditures	421	unknown

Table 7.11: Output Estimates for Texas Highway Widening Projects (Buffington & Wildenthal 1998)

Location	Construction Expenditures	Output Multiplier	State Output	Estimated Local Output
State Highway 21 Caldwell	$6.095 million	$3.69 of output per dollar of expenditures	$22.5 million	$7.5 million
State Highway 199 Parker County	$8.082 million	$3.69 of output per dollar of expenditures	$29.82 million	unknown
U.S. Highway 59 Houston	$114 million	$3.69 of output per dollar of expenditures	$6,386 million	unknown

This review of the highway widening study reveals the ease of calculating impacts using multipliers, but difficulties remain in estimating locations of new output or employment when statewide multipliers are used.

Limitations of IO Modeling

The standard IO model neglects how rising (or falling) demand for inputs and scarce resources affects pricing, producing a negative (or positive) feedback. It also ignores the effect of financial markets and monetary policy, which can and do affect interest rates and other factors that influence multiplier effects, costs, supply, and demand.

Underlying the IO model are production functions that specify the maximum output in each sector that can be achieved under a given quantity of inputs (Samuelson & Nordhaus, 2005). These implicit functions are the foundation of the inter-industry transactions tables presented earlier.

The Leontief IO production function takes the form of the following equation:

$$X_j = \min\left(\frac{z_{1j}}{a_{1j}}, \frac{z_{2j}}{a_{2j}}, \ldots, \frac{z_{nj}}{a_{nj}}\right)$$

where X_j = industry j's total output, z_{nj} = flow from industry n to industry j, and a_{nj} = technical coefficient (the ratio of z_{nj} to X_j).

This function is a mathematical relationship between the inputs and the total maximum output. (It is also a recipe showing the amount of each input needed to attain a maximum quantity of output.)

Production functions can be graphed to show **isoquants**, the function graphed for a particular level of output. The prefix "iso-" means same, and the suffix "-quant" means quantity, so an isoquant, when graphed, shows the combination of inputs that give the same quantity of output.

In reality, Leontief's IO assumption of constant expenditure shares (a_{ij}'s) on inputs of production (for each $1 of output) implies a Cobb-Douglas production function (as discussed briefly in Section 1.5) when the z_{ij}'s are written in units of goods (for example, hours of labor, tons of steel, and kilojoules of energy). A Cobb-Douglas isoquant is smooth. While constant expenditure shares are a limitation of the model, the Cobb-Douglas is a commonly assumed functional form for production. In the case of IO equations, production is assumed to exhibit **constant returns to scale**. This means that if all inputs are increased by the same proportion (for example, all are doubled), the output increases by the same proportion. This assumption, of course, is not a guarantee. Some industries exhibit increased or decreased returns to scale.

See Chapter 1 for more on returns to scale.

Despite its shortcomings, IO modeling is widely used for estimating economic impacts. Economists have developed more sophisticated models, such as the Computable General Equilibrium model discussed next, using IO models as a foundation to capture the flows between economic sectors.

7.7 Computable General Equilibrium (CGE) Models

CGE models are large algebra problems, full of equations and variables, solved to find changes in economic indicators in response to changes in production technologies, government policies, household spending, interest rates, taxes, and other changes or system "shocks." Due to the time, resources, and complexity involved, CGE models are generally used to anticipate economic impacts across large regions or nations. For instance, CGE models were used to predict the economic impacts of the North American Free Trade Agreement (NAFTA) (Piermartini & Teh 2005) (see the CGE Modeling Example on next page).

Just as with the IO model, the CGE model measures the effects that ripple through the economy in response to change (Piermartini & Teh, 2005), but with more equations and different functions for businesses and consumers. The "computable" aspect of CGE emerged when computers were used to solve simultaneous equations efficiently. A CGE model is created by coding all of the necessary equations into computer software (Angrist & Pischke, 2010). The "general equilibrium" portion of CGE refers to the model's theoretical basis, over which there is some disagreement in the literature.

General equilibrium refers to a stable situation for prices, production, and consumption/demand across all interconnected markets being modeled in the. Such market models often seek to represent an entire economy (including households and governments), illustrated in Figure 7.7. CGE analysis examines how these markets interact simultaneously. In contrast, **partial equilibrium** refers to one market or the interconnected system of just a few markets. Prices and quantities in excluded markets are assumed constant. Spillovers into the excluded markets are ignored, if they exist.

One drawback of the general equilibrium approach (vs. partial equilibrium approaches) is a general reliance on aggregated data, resulting in a loss of detail. More detail about sub-markets can be incorporated in a partial equilibrium model (Piermartini & Teh, 2005), but such an approach is appropriate only when included markets are not strongly connected.

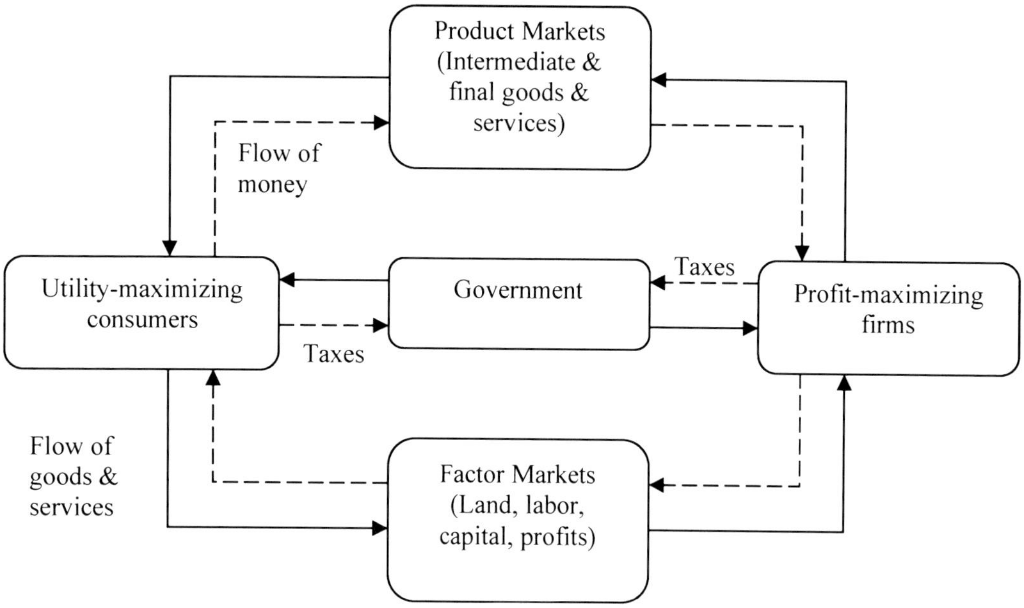

Figure 7.7: CGE Model Flows of Money, Goods, and Services (Wing 2004)

While CGE model goals are similar to IO model goals, CGE models allow for the following features:

- Elasticities of demand and substitution of inputs and products,
- Different production functions for representative businesses of each industry sector, and
- Different utility functions for (representative) utility-maximizing consumers.

In contrast, IO models assume fixed (expenditure) shares and ignore price effects.

See this chapter's In-Depth Look section for the foundation and structure of a generalized CGE model.

Example

CGE Modeling

Government agencies and academics developed CGE models to explore the possible effects of NAFTA (O'Leary et al., 2012). According to the US Congressional Budget Office's review of many of those CGE-based studies, NAFTA's net effect on the US economy would be positive and very small (US Congressional Budget Office, 1993).

O'Leary et al. (2012) summarized the CGE models and their employment estimates from several studies that predicted the impacts of NAFTA. Table 7.14 presents the conclusions made by the different CGE models, differentiated in the summary by the number of industry sectors considered in the model.

Table 7.12: NAFTA CGE Models

Author (year) and Industry Sector CGE Model	Impacts on GDP and Employment	Industry Employment Gains and Losses
Brown, Deardoff, and Stern (1991) 29-sector CGE	Real wages in the US will increase Small and diffuse increases in employment	Gains: Miscellaneous manufactures, textiles, rubber products, nonelectrical machinery, and chemicals Losses: Glass products, nonferrous metals, electrical machinery, mining and quarrying
Hinojosa-Ojeda and Robinson (1991) 7-sector CGE	Negligible change in GDP and employment	Not listed
KPMG Peat Marwick (1991) 44-sector CGE	Modest rise in real income, wages, and real rate of return on capital Job increase of 40,800 to 61,000	Gains: Miscellaneous manufactures, motor vehicles and bodies, machinery and equipment, chemicals, rubber and miscellaneous plastic products Losses: Sugar, fruits and vegetables, electronic components, computing equipment and household appliances

The CGE Model Structure

The CGE modeling problem consists of solving equations (possibly thousands) simultaneously. One relatively straightforward CGE model is called MINIMAL (as carefully documented at http://www.monash.edu.au/policy/minimal.htm), and is described in the following sections.

Developing CGE Model Linkages

Step one in model development is to assemble information on industries and economic sectors using IO tables, Social Accounting Matrices (SAMs), or similar data sets. Figure 7.8 shows how inter-industry data appears in MINIMAL. It looks like the IO tables discussed previously.

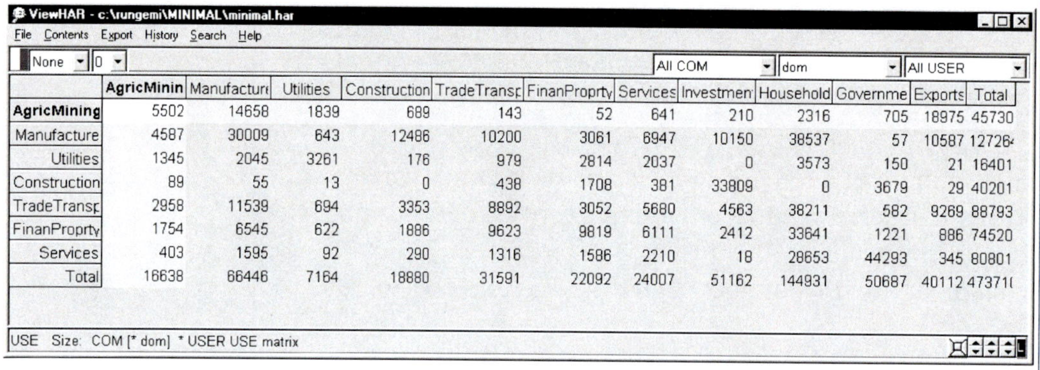
ViewHAR - c:\rungemi\MINIMAL\minimal.har

File Contents Export History Search Help

None | 0 | All COM | dom | All USER

	AgricMinin	Manufactur	Utilities	Construction	TradeTransp	FinanProprty	Services	Investmen	Household	Governme	Exports	Total
AgricMining	5502	14658	1839	689	143	52	641	210	2316	705	18975	45730
Manufacture	4587	30009	643	12486	10200	3061	6947	10150	38537	57	10587	12726
Utilities	1345	2045	3261	176	979	2814	2037	0	3573	150	21	16401
Construction	89	55	13	0	438	1708	381	33809	0	3679	29	40201
TradeTransp	2958	11539	694	3353	8892	3052	5680	4563	38211	582	9269	88793
FinanProprty	1754	6545	622	1886	9623	9819	6111	2412	33641	1221	886	74520
Services	403	1595	92	290	1316	1586	2210	18	28653	44293	345	80801
Total	16638	66446	7164	18880	31591	22092	24007	51162	144931	50687	40112	473710

USE Size: COM [* dom] * USER USE matrix

Figure 7.8: Portion of Inter-industry Data in the MINIMAL Model

In order to arrive at a general equilibrium with market-clearing prices, the consumers must maximize their utilities and industries must maximize their profits at the micro- and macro-economic levels. Figure 7.9 differentiates between the macro and micro levels. The flows of money should be tracked and balanced, such that "there are neither 'black holes' for payments to vanish in, nor mysterious fountains spitting money that agents receive" (Brocker 2004, p. 275).

To mimic profit-maximization, the model must include mathematical functions called production and cost functions relating inputs to outputs and costs for representative businesses across all industries (as introduced in Chapter 2). Likewise, utility functions relating consumer preferences for goods and their prices (as discussed in Sections 2.2 and 4.3), subject to budget constraints, must be included, for representative individuals or households.

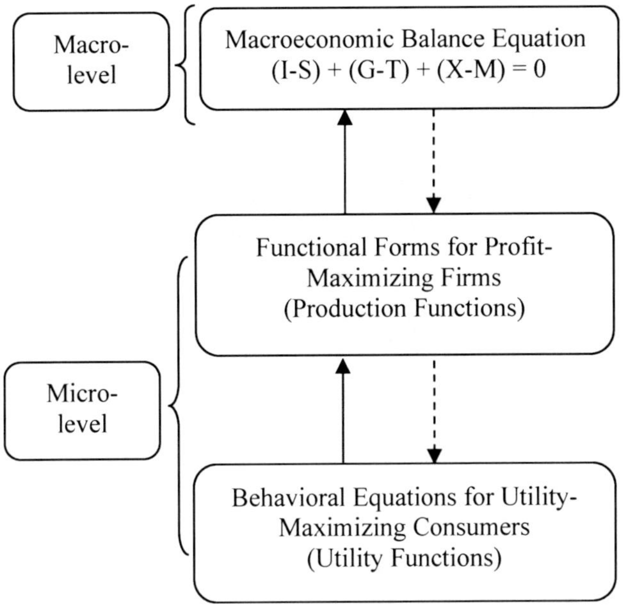

Figure 7.9 : Macro- and Micro-Level Economic Functions and Equilibrium Equations

Each function has parameters that must be specified/given. Calibration is another approach modelers can use; it refers to selecting parameter values initially from literature or expert judgment and then comparing the results of model runs with actual economic data for a particular year or time period called the **benchmark**.

See Chapter 8 for discussion of econometric methods for estimating equation parameters.

Typical CGE production and utility functions rely on Constant Elasticity of Substitution (CES), Leontief, or Cobb-Douglas specifications, because these are relatively restrained (in terms of parameter requirements), as shown in Table 7.13.

Table 7.13: Production and Utility Functions Commonly Used in CGE Models

Function Name (for output or utility)	**General Form** (where x=quantity of inputs)	**Parameters to Estimate and/or Calibrate**
Constant Elasticity of Substitution (CES)	$(a_1x_1^p + a_2x_2^p + \cdots + a_nx_n^p)^{1/p}$	The elasticity of substitution between x_i & x_j equals 1/(1-p). The "a_i" values indicate how the various inputs affect output.
Leontief (production functions only)	$\min(\frac{X_1}{a_1}, \frac{X_2}{a_2}, \ldots, \frac{X_n}{a_n})$	X are inter-industry flows and a_i's are the technical coefficients, from an IO model's direct requirements "A" matrix.
Cobb-Douglas	$\alpha x_1^{a_1} x_2^{a_2} \ldots x_n^{a_n}$	α scales the output, and a_i values reflect elasticities of production.

The elasticities of substitution that emerge from such equations indicate how businesses may substitute inputs (for example, labor for capital) and how consumers can trade off different goods and services. Most CGE models assume constant elasticity of substitution across each pair of inputs, regardless of prices and production levels, to keep things simpler. Interestingly, the Leontief and Cobb-Douglas functions essentially are special cases of the CES function. The CES becomes the Cobb-Douglas when $s = \frac{1}{1-p}$ approaches a value of 1 and it becomes the Leontief as $s = \frac{1}{1-p}$ approaches a value of 0.

A glance inside the code of CGE models reveals many, many equations. Table 7.14 presents the first of ten pages of equations used for an example CGE model.

Table 7.14: Example CGE Model Equations (Vargas et al. 1999)

Equation	Description of Equation	Parameters
$Z = \sum_i (SLACK_i + SLACK2_i)$	Objective function	
$LAB_i = \frac{\alpha_i^L PN_i X_i}{PL}$	Labor demand	α_i^L
$CAP_i = \frac{\alpha_i^K PN_i X_i}{PK_i}$	Capital demand SR	α_i^K
$CAP_i = \frac{\alpha_i^K PN_i X_i}{PK}$	Capital demand LR	α_i^K
$LAND_{ag} = \frac{\alpha_{ag}^T PN_{ag} X_{ag}}{PT_{ag}}$	Land demand	α_{ag}^T
$VA_i = a_{0i} X_i$	Composite factor demand	a_{0i}
$V_{ji} = a_{ji} X_i$	Intermediate input demand	a_{ji}

Closing the Model

In order to solve for all unknown variables in a CGE model of the economy, the number of equations must equal the number of unknown variables. Deciding which variables will be assigned values (exogenous) versus those variables solved by the model (endogenous) is called *closing the model.*

In the MINIMAL model software interface, the fourth tab in the top menu bar is for selecting exogenous variables and assigning those values. Figure 7.10 depicts a set of such variables.

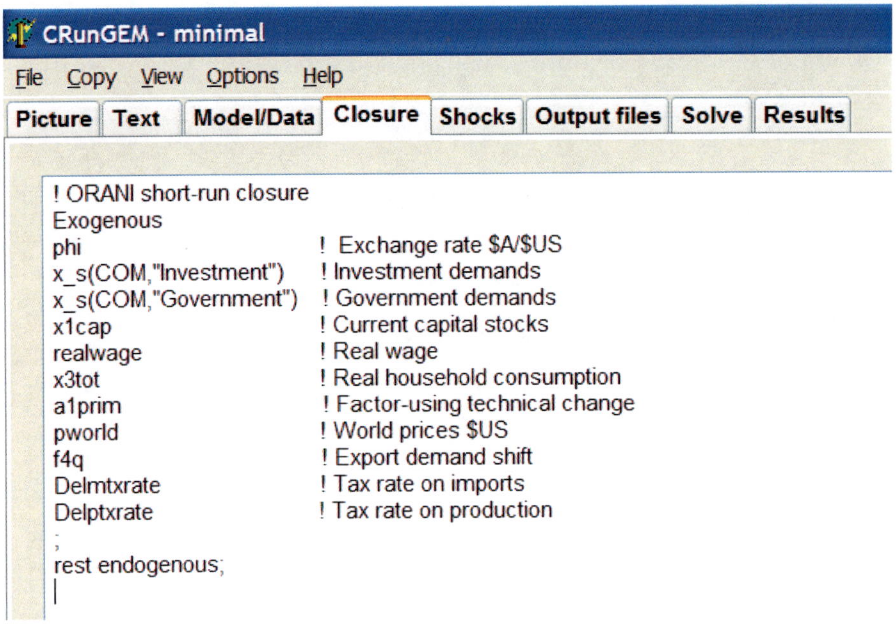

Figure 7.10: Example Closure of the MINIMAL CGE Model

Shocking the Model

A *CGE model shock* involves a change in any exogenous variable's value (from a baseline value). Figure 7.11 shows the MINIMAL interface for selecting the variable to shock.

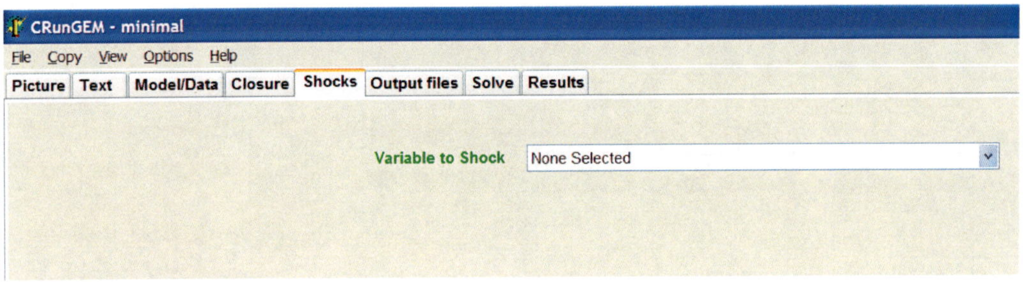

Figure 7.11: MINIMAL's Interface for Specifying an Economic Shock

Solving the Model

Solving a CGE model "involves searching for the set of prices that produces market equilibrium" (Piermartini & Teh 2005, p. 14). Running a CGE model involves solving all of the equations using one of several numerical (i.e., approximate and iterative) solution methods (as opposed to a direct, analytical solution) using a computer program. Figure 7.12 presents results from the MINIMAL model example, and Table 7.15 explains those results. Because CGE models present the relative changes from a baseline, the values, prices, and quantity results are given as percentage changes (from the baseline scenario).

In solving a set of equations, more than one equilibrium or set of values could potentially solve the model. Multiple equilibrium solutions can be a problem, and few CGE models test for this potential error(Wing, 2004). Wing (2004, p.14) notes that work is ongoing to address this issue but "without the ability to test for—or remedy—the problem of multiple equilibria, most applied modelers proceed on the assumption that the solutions generated by their simulations are unique and stable."

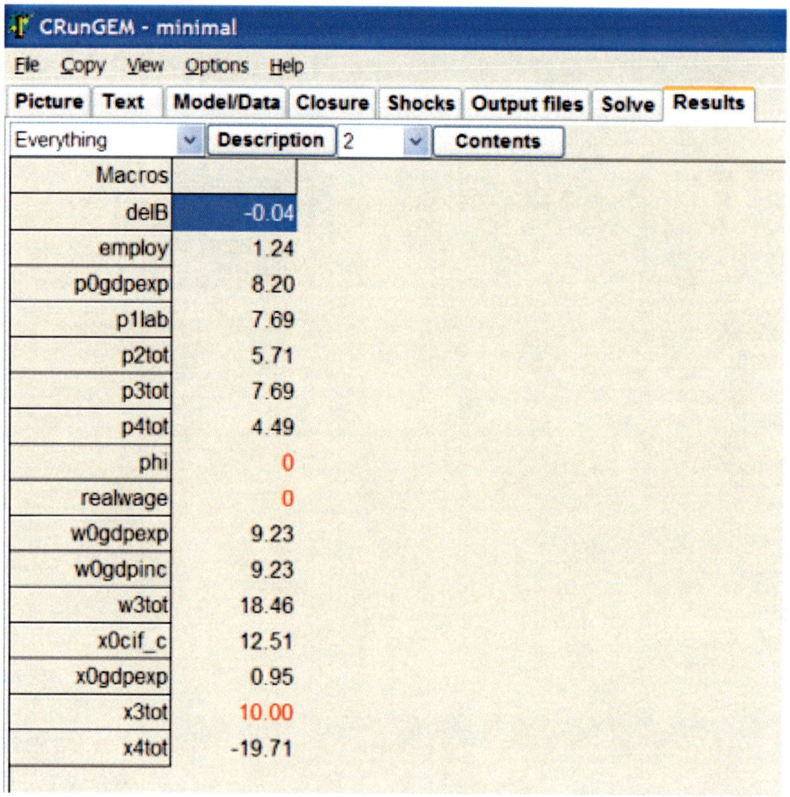

Figure 7.12: MINIMAL's CGE Model Results (Wing 2004)

Table 7.15: Explanation of MINIMAL CGE Model Results

Variable	Description	% change
delB	Balance of trade/GDP	-0.04
employ	Aggregate employment	1.24
p0gdpexp	GDP price index, expenditure side	8.20
p1lab	Economy-wide wage rate	7.69
p2tot	Investment price index	5.71
p3tot	Consumer price index	7.69
p4tot	Export price index	4.49
phi	Exchange rate, (local $)/(foreign $)	0.00
realwage	Wage rate deflated by CPI	0.00
w0gdpexp	Nominal GDP from income side	9.23
w0gdpinc	Nominal GDP from expenditure side	9.23
w3tot	Nominal total household consumption	18.46
x0cif_c	Import volume index, CIF prices	12.51
x0gdpexp	Real GDP from expenditure side	0.95
x3tot	Real household consumption	10.00
x4tot	Export volume index	-19.71

Strengths and Limitations of CGE Models

The strength of CGE modeling is its ability to approximate a wide variety of key interactions across an entire economy to produce more realistic estimates of changes in economic indicators.

However, CGE models rely on a variety of imperfect assumptions, including the following (Samuelson & Nordhaus, 2005; Piermartini & Teh, 2005; Conrad, undated):

- **Homogenous goods and services**, meaning goods and services produced within each industry sector are the same; there is no differentiation of goods or services.
- All markets are perfectly competitive, meaning that
  - Consumers and producers are perfectly informed about prices.
  - Businesses maximize profits (revenues less expenses).
  - Consumers maximize their utility (they rationally choose the best set of goods).
  - Prices for outputs move to balance/match supply and demand at all times.
  - All goods are produced under constant or decreasing returns to scale.
  - No externalities or other market failures exist.
  - A single household or a single business cannot affect prices, so all are **price-takers** (they take the price from the market; they do not set the market price like a monopoly would).
- Production and production functions exhibit constant elasticity of substitution across inputs.
- CGE models usually apply to a single region (just like the vast majority of IO models), so they neglect local production and consumption and price differences, along with travel costs and traffic patterns.

To overcome some such limitations, a few CGE models have incorporated monopolistic (imperfect) competition (with increasing returns to scale and product differentiation, technological progress, and other enhancements of the theory assumptions listed above) to better reflect real-world market conditions. More and more flexible and spatially complex models are emerging. Models such as PECAS and RUBMRIO include travel costs, and land use-transportation models like MUSSA and TRANUS incorporate land prices (though they ignore industries and production).

7.8 Summary

EIAs often rely on complex models and abstractions of reality, but their outputs/estimates can and generally do provide insight into a region's economy and opportunities for better decision-making. The basic IO model and its descendants provide an organized framework for anticipating and quantifying real impacts of transportation investment and policy choices. Consultant reports typically give only a brief overview of study methodology, since methods and data are often considered proprietary. This shielding of how the numbers were generated can elicit skepticism. In addition, the validity or accuracy of future-year forecasts cannot be evaluated until after the time period.

Government expenditures for transportation projects, depending on the source of funds, could also have potentially been used for other government goods or services with larger economic impacts. Most economic impact studies do not compare across a variety of different government spending options. EIA methods like IO and CGE do not determine whether the creation of new jobs in an area is generative or redistributive. Additional studies have to be conducted to put those changes into a wider perspective of how the impacts are potentially a result of a shifting from one geographic location to another.

Recommended Reading

The following studies may prove most helpful for readers seeking more information about EIAs:

- For EIA details and applications, see *Regional Economic Impact Analysis and Project Evaluation* (Davis 1990).
- The Transit Cooperative Research Program's Report 35, *Economic Impact Analysis of Transit Investments: Guidebook for Practitioners* (Cambridge Systematics 1998) is a useful resource for selection of methods to conduct EIA for transit investments. It is available at http://onlinepubs.trb.org/onlinepubs/tcrp/tcrp_rpt_35.pdf. A recent update of this work can be found in Weisbrod and Reno (2009).
- The FHWA also lists offices and websites for more information about economic analysis of highway infrastructure and operations available: http://www.fhwa.dot.gov/infrastructure/asstmgmt/econlinks.cfm.

7.9 An In-Depth Look

Double-Counting

In economics, inadvertently double-counting economic impacts will in most cases overstate the net economic benefit of a project. This section explains how to determine if double-counting is occurring.

The following useful guidelines (Weisbrod & Weisbrod 1997) indicate when combining impacts are acceptable, either for EIA or other methods, like CBA. These three types of impacts are combinable:

1. User impacts (travel time savings, safety, reliability, etc.)
2. Government fiscal impacts (public revenues and expenditures)
3. Societal benefits (air quality, noise impacts and other environmental conditions, social conditions)

In general, changes in most economic indicators should NOT be added together because most reflect changes in others. For instance, an increase in business revenue may result in an increase in personal income. Adding the increase in personal income with the business revenue would be double-counting the benefits. Changes in the following economic indicators, or any similar indicator, should not be added together(Weisbrod & Weisbrod, 1997):

- Employment (number of jobs)
- Personal income (wages)
- Property values
- Business sales

Table 7.16 presents examples of double-counting.

Impact Measure or Economic Indicator 1	Impact Measure or Economic Indicator 2	Why It's Double-Counting
Crash cost savings	Reductions in government expenditures	Less money spent on managing crash.
Travel cost savings	Increase in business profits	Less money spent on transportation, resulting in more profit.
Travel cost savings	Increase in property values	Property values also reflect travel costs.
Increase in business sales (from improved transportation system)	Increase in GDP	GDP already reflects business sales.
Increase in business sales	Increase in business income	Income already reflects business sales.
Increase in business sales	Increase in wages	Wages tend to rise with business profits and income.
Increase in property values	Increase in business income	Property value increases come from rising business incomes.
Changes in government revenue	Changes in property value	Government revenue increases (decreases) because of increased (decreased) tax revenue from increased (decreased) property values.
Changes in property values	Changes in air quality	Air quality benefits (cost) realized in values of property.
Changes in wages	Changes in property value of land	Property values rise because of rising incomes and profits.
Value of development losses (buildings, infrastructure) from disasters	Financial losses (decline in equity values)	Decline in equity value reflects the direct loss.
Cost of loss to property/capital	Lost value-added from the services from the capital	The cost of the loss is determined from the lost value-added.

Table 7.16: Examples of Double-Counting Across Impact Measures and Economic Indicators

Creating the IO Model

The Leontief transactions table values originate from businesses' responses to an economic census, a local survey of businesses, or something similar. Keep in mind that the data is only as accurate as the information provided by the businesses.

IO Model Preparation

To conduct an IO analysis, the analyst must follow these three steps:

1. Define the economic area for the (1) direct impact and (2) for the indirect and induced impact.

At least two economic areas must be defined. The area of the direct impact must be the location of the actual transportation project. Only the jobs, business revenue, and personal income earned for work on the transportation project site count as direct impacts. If the project is a service or a program rather than infrastructure construction, then only the economic indicator factors within the operations of the service or program are counted as direct impacts.

Because IO involves the relationship between industries, the model must set the boundaries of the economic areas. Only those industries within the area of interest are used for determining the indirect and induced impacts. As discussed earlier in the chapter, the size of the evaluation area sometimes affects whether an impact is considered generative or redistributive. The type and purpose of the transportation project will guide the selection, as well as consideration of how the selection will affect the determination of generative and redistributive economic impacts. Most sources of data for IO analysis aggregate the data at the county level, so economic areas will either be a county, multiple counties, a state, or multiple states.

When using the Regional Input-Output Modeling System (RIMS II), the choice of area is the county, several counties, or component economic areas (micropolitan or metropolitan statistical areas). As the Figure 7.13 map shows, Texas has 13 Bureau of Economic Analysis (BEA) areas.

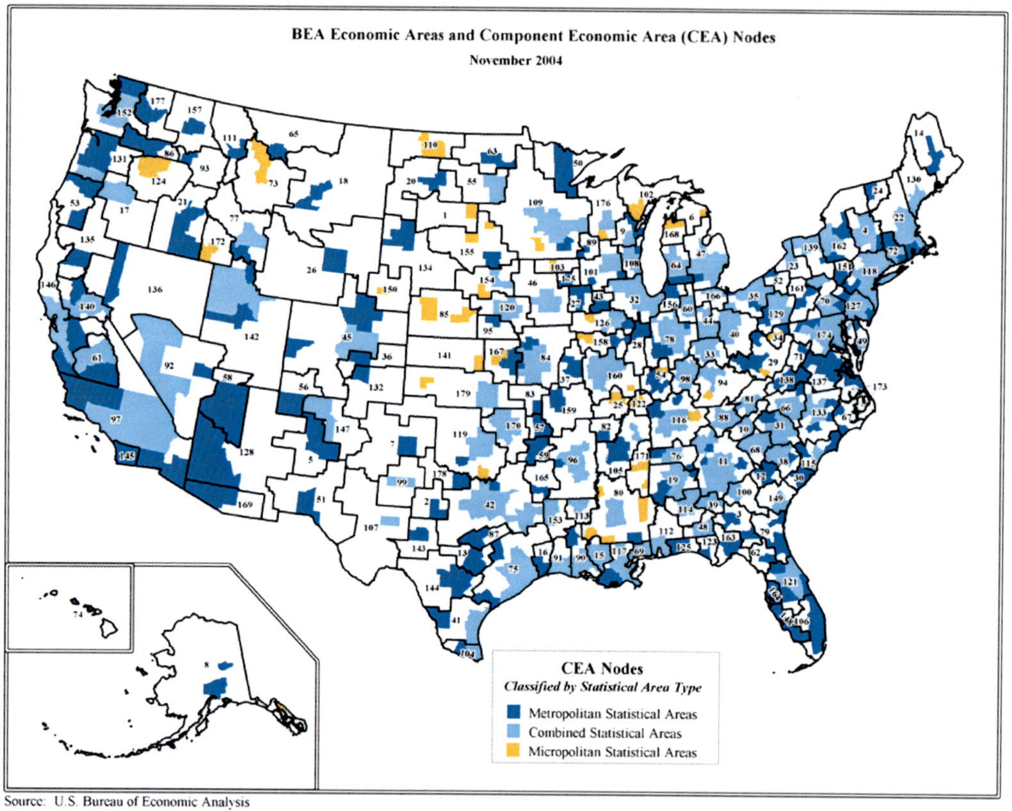

Source: U.S. Bureau of Economic Analysis

Figure 7.13: BEA Areas (Source: BEA 2010)

2. Select the industry sectors in the economic area of interest.

Only the industry sectors within the defined economic area should be included in the IO model. However, the study does not have to include all the industry sectors within the area. Relevant industries are determined by the study's objectives. The selection also depends on the preferred level of detail. For instance, a study could use the wide-ranging manufacturing sector or recognize the different types of manufacturing sectors, such as electronics, shoes, and furniture.

3. Gather data on the inter-industry flows within a defined time period.

Once the area and industries are selected, a table is created showing the flows in either monetary or physical terms (number of goods) between industries. Typically the flows are presented in dollars (which is a potential issue, because price changes do not reflect the change in quantity of goods exchanged between industries). The resulting IO tables (or inter-industry transaction tables) can be quite large, depending on the number of sectors surveyed, and are rarely reported in economic impact studies. Rather, they reside within the computer as an intermediate step in determining the multipliers.

Example

Corpus Christi Port Study

For the Corpus Christi, Texas Port study (Martin Associates 2004), the following impact areas were defined:

- **Direct impact area**: The private and public terminals along the Inner Harbor and within the Port of Corpus Christi Port District.
- **Indirect impact area**: the Corpus Christi region and the state of Texas.
- **Induced impact area**: the Corpus Christi region.

The following industry sectors were selected:

- **Direct sectors**: Surface Transportation, Maritime Service, Shippers/Consignees, Port of Corpus Christi Authority.
- **Indirect sectors**: All sectors providing services for businesses with direct impacts at the Port of Corpus Christi.
- **Induced sectors**: Only examples were given, such as Construction, Home Furnishings, Restaurant, Social Services, Business Services, and Educational Services in the Corpus Christi region.

As with the areas, a report should list the sectors so that the study can be replicated in a future year.

Creation of U.S. Transactions Tables

The BEA surveys businesses across the U.S. annually to prepare national and regional accounts (using its RIMS II software program) that track inputs and outputs. An in-depth description of how they create U.S. IO tables is provided at http://www.bea.gov/methodologies/index.htm. An abbreviated explanation follows.

The BEA collects information on each business establishment's production accounts (BEPAs) and organizes the information needed for the Leontief transactions table into a "T" account table for each business surveyed (Table 7.17).

Debit	Credit
Purchases from	Sales to
Industry sector A	*Industry sector A*
Industry sector B	*Industry sector B*
Wages and Salaries	Sales to Households
Profits	Government Purchases
Other Value Added	Other Final Demand
Total Expenses and Profit	Total Revenues

Table 7.17: Business Establishment Production Accounts (BEPAs)

Businesses with similar products are grouped by the North American Industry Classification System (NAICS) codes. The NAICS groups businesses across various levels, as illustrated in Table 7.18.

NAICS level	NAICS Code	Example
Very general	2-digit	Construction
General	3-digit	**Construction of Buildings**
Somewhat general	4-digit	**Nonresidential Building Construction**
Specific	5-digit	Commercial and Institutional Building Construction
Very specific	6-digit	Commercial and Institutional Building Construction

Table 7.18: NAICS Codes

Once surveyed companies are grouped by NAICS codes, the BEPA "T" table values (Table 7.19) are added together across all companies in the same industry. The resulting table contains the same "T" table information, but at an industry level (rather than the company level.) Next, a National Income and Product Account (NIPA) table is created (or, in the case of a region, a Regional Income and Product Account, or RIPA), as shown in Table 7.19.

Debit	Amount	Credit	Amount
Wages and Salaries		**Sales to Households**	
Industry sector A		*Industry sector A*	
Industry sector B		*Industry sector B*	
Profits		**Government Purchases**	
Industry sector A		*Industry sector A*	
Industry sector B		*Industry sector B*	
Other Value Added		**Other Final Demand**	
Industry sector A		*Industry sector A*	
Industry sector B		*Industry sector B*	
Total Charges Against GNP	**SUM**	**Total Contributions to GNP**	**SUM**

Table 7.19: National Income and Product Account

The NIPA (or RIPA) table is then organized to create the IO Transactions Table (as shown in Table 7.7 in this chapter's main section). Table 7.20 presents the expanded table in a generic way, with variables for cell values, rather than specific numbers.

- C values stand for household consumption.
- I values signal (private) investment.
- X-M values represent net exports (the total value of an output exported out of the region [X], less the imports coming into the region [M]).
- G values refer to government activities.
- L, N, and T signal labor expenses, industry profits, and taxes paid.
- The z_{ij} values indicate the flow of money between row industry i and column industry j. **Notation note: the first subscript "i" in flow z_{ij} refers to the row and the second subscript refers to the column.**

Table 7.20: Notation for Transactions Table

		Purchases from (inputs for)		Final Demand (sum = Y_i)				
		Industry Sector A	Industry Sector B	Household/ Consumer Spending (C)	Private Investment (I)	Net Exports (X-M)	Government Spending (G)	TOTAL OUTPUT
Sales from (outputs from)	Industry Sector A	z_{11}	z_{12}	C_1	I_1	X_1-M_1	G_1	X_1
	Industry Sector B	z_{21}	z_{22}	C_2	I_2	X_1-M_2	G_2	X_2
Value Added	Household Income (Labor, L)	L_1	L_2	L_C	L_I	L_X	L_G	L
	Business Profit (Income, N)	N_1	N_2	N_C	N_I	N_X	N_G	N
	Government Taxes/Fees (T)	T_1	T_2	T_C	T_I	T_X	T_G	T
	TOTAL INPUT	X_1	X_2	C	I	X-M	G	X

Table 7.21 presents this transactions table again (from Table 7.7), but with final demands and value-added components merged to simplify the presentation.

Table 7.21: Demand-Collapsed Transactions Table—The Make Table (Source: part of this table is from Miller & Blair 1985)

		Purchases from (inputs for)			
		Industry Sector A	Industry Sector B	Final Demand	TOTAL OUTPUT
Sales (outputs from)	Industry Sector A	150	500	350	1000
	Industry Sector B	200	100	1700	2000
	Value Added	650	1400	650	2700
	TOTAL INPUT	1000	2000	2700	5700

Table 7.22 shows how to calculate total output using the IO table. Macroeconomic equilibrium requires that the **total inputs** (industry inputs plus value added) equal the **total output** (industry outputs plus final demand).

Table 7.22: Algebraic Formulation of Total Output

Sector	Total Output Equations	Table 12 Examples
Industry A	$X_1 = z_{11} + z_{12} + Y_1$	$X_1 = 150 + 500 + 350 = 1000$
Industry B	$X_2 = z_{21} + z_{22} + Y_2$	$X_2 = 200 + 100 + 1700 = 2000$

Social Accounting Matrices

An alternative to the transactions table is called the Social Accounting Matrix (SAM). According to IMPLAN's website, "SAM accounts are a better measure of economic flow than traditional input-output accounts because they include 'non-market' transactions" such as taxes and unemployment benefits. They can be used for analysis similar to input-output and multipliers can be derived from them.

A SAM emphasizes the spending and generation of personal income so the focus "shifts from how regional output is produced to also address how regional income is generated and distributed" (Vargas et al. 1999). Whereas traditional transactions tables show only income to the factors of production (the value-added components of capital and labor), the SAM shows the owners of the factors of production (households, government, or business firms) to illustrate "the circular process of demand leading to production leading to income, which in turn leads back to demand" (Rutherford & Paltsev 1999, p. 5).

As with an IO table, the SAM total will equal the sum of its columns. In other words, no money disappears or magically appears.

Creation of Direct Requirements Tables

An IO model finds the direct, indirect, and induced effects by mathematically transforming the interactions table to create a direct requirements table. This table is a recipe for interindustry sector outputs and input flows, and the set of linear equations relating a sector's output to its inputs from other sectors is the basis of IO analysis.

The calculation of a new total output in response to a change in final demand is solved by using the matrix equation $\mathbf{X} = (\mathbf{I}-\mathbf{A})^{-1}\,\mathbf{Y}$ where **X** is the vector of gross (total) outputs, **Y** is the vector of final demands, **A** is the matrix of IO **technical coefficients** (and is referred to as the direct requirements table), and **I** is the identity matrix. To go from the transactions table to the set of $\mathbf{X} = (\mathbf{I}-\mathbf{A})^{-1}\,\mathbf{Y}$ equations requires first computing the technical coefficients (a_{ij}) in matrix A (the direct requirements table).

To calculate the technical coefficients, the value of input in each cell (z_{ij}) is divided by the total output for the sector in the column (X_j). This ratio of input to industry output is **the technical coefficient** (a_{ij}), also known as an **input-output coefficient** or **direct input coefficient**.

$$a_{ij} = \frac{z_{ij}}{X_j}$$

This share of output, a_{ij}, is a "unitless" number because both z_{ij} and X_j have units of dollars, thus cancelling each other out. Table 7.23 shows how the calculation is completed, with the value added and final demand sectors included, and with household income and spending separated from other value-added and final-demand sectors (so that induced effects from household spending can later be computed).

Table 7.23: Calculation of Technical Coefficients (Source: part of this table is from Miller & Blair 2009)

		Purchases from (inputs for)		**Final Demand**		
		Industry Sector A	Industry Sector B	Household Spending	Other Final Demand	**TOTAL OUTPUT (X_j)**
Sales from (outputs from)	Industry Sector A	$\frac{150}{1000}$	$\frac{500}{2000}$	$\frac{50}{900}$	$\frac{300}{1800}$	1000
	Industry Sector B	$\frac{200}{1000}$	$\frac{100}{2000}$	$\frac{400}{900}$	$\frac{1300}{1800}$	2000
Value Added	Household Income	$\frac{300}{1000}$	$\frac{500}{2000}$	$\frac{50}{900}$	$\frac{50}{1800}$	900
	Other Value Added	$\frac{350}{1000}$	$\frac{900}{2000}$	$\frac{400}{900}$	$\frac{150}{1800}$	1800
	TOTAL INPUT	$\frac{1000}{1000}$	$\frac{2000}{2000}$	$\frac{900}{900}$	$\frac{1800}{1800}$	5700

The resulting table of technical coefficients (Table 7.24) is referred to as the **direct requirements table** (the "A" matrix). The technical coefficients give the proportions of each input needed to produce $1 of the producing industry's output. For instance, industry sector A needs $0.15 of inputs from industry sector A, $0.20 from industry sector B, and $0.30 from labor (household income) in order to produce $1 of output.

Table 7.24: The Direct Requirements Table—"A" Matrix (Source: part of this table is from Miller & Blair 2009)

		Purchases from (inputs for)				
		Industry Sector A	Industry Sector B	Household Spending (endogenous)	Other Final Demand (Y_i)	**TOTAL OUTPUT (X_i)**
Sales from (outputs from)	Industry Sector A *(Row 1)*	0.15	0.25	0.0556	0.1667 (Y_1)	X_1
	Industry Sector B *(Row 2)*	0.20	0.05	0.4444	0.7222 (Y_2)	X_2
Value Added Sectors	Household Income *(Row 3)*	0.30	0.25	0.0556	0.0278 (Y_3)	X_3
	Other Value Added *(Row 4)*	0.35	0.45	0.4444	0.0833	---
	TOTAL INPUT	1.0	1.0	1.0	1.0	---

Derivation of $X= (I-A)^{-1} Y$

A typical goal of IO analysis is to determine what the new total output, *X*, will be in each sector after a change in final demand *Y*. This is accomplished by using the matrix equation **X= (I-A)$^{-1}$ Y.** Here, Table 7.24's technical coefficients are used. Other Final Demand is used as the exogenous demand vector *Y*, so Table 7.25's equations apply. If household spending is to be included to find final demand *Y*, Table 7.26's equations will govern.

Table 7.25: Inter-Industry Flow Equations with Endogenous Household Income

Sector	General Equations	Equations With Table 7.24 Coefficients
Industry A	$X_1 = a_{11}X_1 + a_{12}X_2 + a_{13}X_3 + Y_1$	$X_1 = 0.15X_1 + 0.25X_2 + 0.0556X_3 + Y_1$
Industry B	$X_2 = a_{21}X_1 + a_{22}X_2 + a_{23}X_3 + Y_2$	$X_2 = 0.20X_1 + 0.05X_2 + 0.4444X_3 + Y_2$
Household	$X_3 = a_{31}X_1 + a_{32}X_2 + a_{33}X_3 + Y_3$	$X_3 = 0.30X_1 + 0.25X_2 + 0.0556X_3 + Y_3$

(Y = Other Final Demand)

Table 7.26: Inter-Industry Flow Equations with Exogenous Household Income

Sector	General Equations	Equations With Table 7.24 Coefficients
Industry A	$X_1 = a_{11}X_1 + a_{12}X_2 + Y_1$	$X_1 = 0.15X_1 + 0.25X_2 + Y_1$
Industry B	$X_2 = a_{21}X_1 + a_{22}X_2 + Y_2$	$X_2 = 0.20X_1 + 0.05X_2 + Y_2$

(Y = Household Spending + Other Final Demand)

Solving for each sector's output, *X*, after a change in final demand values (*Y*) requires that we rearrange equations to solve for *Y* (Table 7.27).

Table 7.27: Inter-Industry Flow Equations Rearranged to Solve for Final Demand (Y)

Sector	General Equations	Equations With Coefficients
Industry A	$X_1 - a_{11}X_1 - a_{12}X_2 - a_{13}X_3 = Y_1$	$X_1 - 0.15X_1 - 0.25X_2 - 0.0556X_3 = Y_1$
Industry B	$X_2 - a_{21}X_1 - a_{22}X_2 - a_{23}X_3 = Y_2$	$X_2 - 0.20X_1 - 0.05X_2 - 0.4444X_3 = Y_2$
Households	$X_3 - a_{31}X_1 - a_{32}X_2 - a_{33}X_3 = Y_3$	$X_3 - 0.30X_1 - 0.25X_2 - 0.0556X_3 = Y_3$

With Table 7.27's three equations and three unknowns (X_1, X_2, and X_3), we can solve for total output (X_1, X_2, and X_3) using basic algebra if the final demands (*Y*) are given. However, in most IO analyses, the number of industries is much greater than three and matrix algebra is used. Such matrix algebra requires that the technical coefficients be placed in the "A" matrix and *X* and *Y* in their own column vectors, as shown in Table 7.28.

Table 7.28: Matrices for Technical Coefficients (a_{ij}), Total Output (X), and Final Demand (Y)

$$\begin{array}{l} \text{Industry A} \\ \text{Industry B} \\ \text{Household} \end{array} \quad \mathbf{A}= \begin{bmatrix} a_{11} & a_{12} & a_{13} \\ a_{21} & a_{22} & a_{23} \\ a_{31} & a_{32} & a_{33} \end{bmatrix} \quad \mathbf{X}= \begin{bmatrix} X_1 \\ X_2 \\ X_3 \end{bmatrix} \quad \mathbf{Y}= \begin{bmatrix} Y_1 \\ Y_2 \\ Y_3 \end{bmatrix}$$

The equation relating these three matrices is **(I-A)X=Y**, where *I* is the **identity**

matrix (the square matrix of zeros with ones down the diagonal). The matrix is of the same size as A, or 3x3 in this example. The resulting equations are shown in Table 7.29 and the *I*, *A*, and *I-A* terms are shown in Table 7.30. Typically, the goal is to solve for each X_i, the total output in each sector *i*.

Table 7.29: Inter-Industry Flow Equations Solved for Final Demand (Y)

Sector	General Equations	Equations With Coefficients
Industry A	$(1 - a_{11})X_1 - a_{12}X_2 - a_{13}X_3 = Y_1$	$(1 - 0.15)X_1 - 0.25X_2 - 0.0556X_3 = Y_1$
Industry B	$- a_{21}X_1 + (1 - a_{22})X_2 - a_{23}X_3 = Y_2$	$- 0.20X_1 + (1 - 0.05)X_2 - 0.4444X_3 = Y_2$
Household	$- a_{31}X_1 - a_{32}X_2 + (1 - a_{33})X_3 = Y_3$	$- 0.30X_1 - 0.25X_2 + (1 - 0.0556)X_3 = Y_3$

Table 7.30: Formation of the I-A Matrix

$$\mathbf{I} = \begin{bmatrix} 1 & 0 & 0 \\ 0 & 1 & 0 \\ 0 & 0 & 1 \end{bmatrix} \quad \mathbf{A} = \begin{bmatrix} a_{11} & a_{12} & a_{13} \\ a_{21} & a_{22} & a_{23} \\ a_{31} & a_{32} & a_{33} \end{bmatrix} \quad \mathbf{I - A} = \begin{bmatrix} 1 - a_{11} & - a_{12} & - a_{13} \\ - a_{21} & 1 - a_{22} & - a_{23} \\ - a_{31} & - a_{32} & 1 - a_{33} \end{bmatrix}$$

$$\mathbf{I} = \begin{bmatrix} 1 & 0 & 0 \\ 0 & 1 & 0 \\ 0 & 0 & 1 \end{bmatrix} \quad \mathbf{A} = \begin{bmatrix} 0.15 & 0.25 & 0.0556 \\ 0.20 & 0.05 & 0.4444 \\ 0.30 & 0.25 & 0.0556 \end{bmatrix} \quad \mathbf{I - A} = \begin{bmatrix} 0.85 & -0.25 & -0.0556 \\ -0.20 & 0.95 & -0.4444 \\ -0.30 & -0.25 & 0.9444 \end{bmatrix}$$

The goal is to solve for each *X*, the total output in each sector. In matrix algebra, as in algebra, solving for ***X*** involves multiplying both sides by the inverse of **I-A** matrix:

$$(I-A)^{-1}(I-A)X = (I-A)^{-1}Y \quad \text{results in} \quad \boxed{X = (I-A)^{-1}\,Y}$$

Finding $(\mathbf{I-A})^{-1}$ involves matrix operations explained in any linear algebra textbook. Only square matrices have inverses, which is why the number of rows must equal the number of columns in matrix **A**.

Solving for X, the Economic Impact

The $\mathbf{X} = (\mathbf{I-A})^{-1}\,\mathbf{Y}$ equation provides the X_i values or total output for each sector when *Y*, the final demand for each sector, is set exogenously (outside the model). This important equation is used to answer this question: *What happens to a sector's output when final demand (household spending, government spending, business investment and/or exports) changes?*

Continuing with the matrices **I-A**, **X,** and **Y** defined above, final demand in industries A and B (excluding household spending, which is endogenous to the model) could be set as \$600 and \$1,500, respectively. Such values may come from spending on a small transportation project that increases payments to the construction industry by \$300 (\$600-\$300) and the steel industry by \$200 (\$1500-\$1300) for work done on site. Table 7.31 shows such investment in local industries is estimated to result in higher output and/or expenditures that exceed prior conditions (old **Y** and old **X**).

Table 7.31: Comparison of Total Output Under New and Old Final Demand, Households Endogenous

						Final Demand		Total Output
Industry A		1.3815	0.4394	0.2881		300		1000
Industry B	$(I-A)^{-1}=$	0.5663	1.3815	0.6834	Old **Y**=	1300	Old **X**=	2000
Household		0.5888	0.5053	1.3313		0		900

Industry A		1.3815	0.4394	0.2881		600		1487.98
Industry B	$(I-A)^{-1}=$	0.5663	1.3815	0.6834	New **Y**=	1500	New **X**=	2412.00
Household		0.5888	0.5053	1.3313		0		1111.18

The results given in the new *X* matrix show that when the final demand for Industry A changes from $300 to $600 and Industry B's final demand rises from $1,300 to $1,500, Industry A's total output is expected to grow from $1,000 to $1,488. Overall effects are listed in Table 7.32.

Table 7.32: Identifying Direct, Indirect, and Induced Effects in IO Analysis Results

Direct, Indirect, and Induced Effect	$1,111
Direct Effect	$500
Indirect and Induced Effect	$611

If only direct and indirect effects are of interest, then the matrices should not include the household income row and household spending column. Instead, the final demand, **Y**, will need to include the household spending, so final demand for Industry A with household spending for Industry A becomes $350 and for Industry B becomes $1,700. When final demands rise to $650 and $1,900 in these two industries, their total output levels jump to $1,442 and $2,304.

Table 7.33: Comparison of Total Output under New Final Demand and Old Final Demand, Households Exogenous

Calculation for Total Output Due to Direct and Indirect Effects

					Final Demand		Output
Industry A		1.2541	0.3300		350		1000
Industry B	$(I-A)^{-1}=$	0.2640	1.1221	Old **Y**=	1700	Old **X**=	2000

					Final Demand		Output
Industry A		1.2541	0.3300		650		1442
Industry B	$(I-A)^{-1}=$	0.2640	1.1221	New **Y**=	1900	New **X**=	2304

A comparison of Table 7.32 and Table 7.33 results shows what intuitively should be the case: removing the households' induced effects reduces total output for each sector, and thus the total impact on the economy (Table 7.34).

	Output (Direct and Indirect Effects)		Total Output (Direct, Indirect, and Induced Effects)	
Sector	**Old**	**New**	**Old**	**New**
Industry A	1000.00	1500.00	1000.00	1487.98
Industry B	2000.00	2000.00	2000.00	2412.00
Household	---	---	900.00	1111.18
Total Output Across All Sectors	3000.00	3500.00	3900.00	5011.16
Difference Between Old and New Output		$500.00		$1111.16

Table 7.34: Comparison of Results

In summary, the process for finding the new sector's output in response to a change in final demand involves

- Collecting information about inter-industry flows to develop an IO transactions table.
- Using the transaction table values to estimate the technical coefficients to create the direct requirements table (the "**A**" matrix).
- Setting the **Y** vector to equal the new final demand or the change in final demands.
- Solving for the new total output (or change in total output) in each industry sector (**X** matrix) by using $X = (I-A)^{-1} Y$.

If the new total outputs include induced effects along with the direct and indirect effects, then household income and spending must be included in the **A** matrix. Table 3.5 summarizes how to find the multiplier effects.

Type of Multiplier Effect	How to Find Effects
Direct	The change in final demand in the sector.
Direct & Indirect	The new total output calculated from $\mathbf{X} = (\mathbf{I}-\mathbf{A})^{-1}\,\mathbf{Y}$ when household income and spending are *excluded* from the **A** matrix.
Direct, Indirect & Induced	The new total output calculated from $\mathbf{X} = (\mathbf{I}-\mathbf{A})^{-1}\,\mathbf{Y}$ when household income and spending are *included* in the **A** matrix.

Note: Instead of totals, the matrices could include the changes ΔX and ΔY

Table 7.35: How to Find Multiplier Effects

The IO analysis presented in this section is the most common. Variations exist and allow, for instance, the induced effects of government spending (by including government taxes and spending instead of household income and spending), or job and income direct effects.

The $(I-A)^{-1}$ matrix is also used to estimate economic multipliers and to determine how a change in each sector of final demand changes total output, employment, or income throughout the economy.

How to Derive Multipliers

As mentioned earlier, the $(I-A)^{-1}$ **Leontief inverse matrix** provides the information needed to find simple and total multipliers. Simple and total output multipliers for each sector are found by summing the values down a column of the $(I-A)^{-1}$ Leontief inverse matrix. The simple output multiplier for Industry sector A is 1.5181 as shown in Table 7.12 and the total output multiplier is 2.5366, shown in Table 7.13.

Table 7.36: Simple Output Multiplier

		Industry A	Industry B
Industry A	$(I-A)^{-1}=$	1.2541	0.3300
Industry B		0.2640	1.1221
Simple Output Multiplier	Sum	1.5181	1.4521

Table 7.12 shows that a $1 increase in final demand for Industry A products requires $1.254 more production from Industry A and $0.264 more output from Industry B. The simple output multiplier of 1.5181 is the total increase in Industry A output in response to a $1 increase in final demand for Industry A.

Table 7.37: Total Output Multiplier

		Industry A	Industry B	Household Income
Industry A		1.3815	0.4394	0.2881
Industry B	$(I-A)^{-1}=$	0.5663	1.3815	0.6834
Household		0.5888	0.5053	1.3313
Total Output Multiplier	Sum	2.5366	2.3262	2.3028

When local households and their labor are also considered in the model, Table 7.13 shows how a $1 increase in final demand for Industry B product(s) uses $0.4394 more Industry A output, $1.3815 more from Industry B, and $0.5053 more household production (in the form of labor). The total output multiplier of 2.3262 is the total increase in output in Industry B in response to a $1 increase in final demand for Industry A.

Total output multipliers are larger than simple output multipliers because they include the induced effects of household spending and income (labor).

CGE Model Foundation

The empirical foundation of the model is an IO table, a SAM, or some similar way of organizing empirical data about the inter-industry relationships in an economy.

The theoretical foundation of the model (general equilibrium) is a bit of a controversy. Depending on which article about CGE is read, CGE is based on one of three theoretical foundations (excerpts from the literature proclaiming the foundation of CGE are given to show the variety of opinions):

1. **Walras general equilibrium.** CGE models are founded on Walrasian equilibrium and are "simulations that combine the abstract general equilibrium structure formalized by Arrow and Debreu" (Wing 2004, p. 2). The accounting rules for CGE "are the cornerstones of Walrasian general equilibrium" (Wing 2004, p. 4).
2. **Arrow-Debreu model of general equilibrium.** "In principle, a CGE model is a member of the class of macroeconomic models which has as its theoretical underpinning the application of an Arrow-Debreu general equilibrium framework" (Conrad, undated, p. 4).
3. **Macroeconomic balancing or equilibrium.** "CGE models are *not* and have *never been* general equilibrium models in the Arrow-Debreu or Walras tradition…CGE models are macro balancing models, and not modern microeconomic models" (Mitra-Kahn 2008, p. 72).

This Reference will present only the basics of all three theoretical foundations, leaving the pursuit of further investigation to the reader. The debate essentially reduces to a question of whether CGE models are primarily founded on microeconomic equilibrium (the Arrow-Debreu and Walras general equilibrium) or on macroeconomic equilibrium. A structural review of the model suggests that the CGE model may be founded upon all three (the micro- and macroeconomic equilibriums).

Walras General Equilibrium

Leon Walras was a French economist who authored *Elements of a Pure Economics*, published in the late 1800s. This text was updated several times into the early 1900s and eventually translated to English in 1954. Kenneth Arrow, an American economist, and Gerard Debreu, a French economist (of the Arrow and Debreu theory), began their seminal 1954 article with a concise description of Walras's theory of a general equilibrium economic system(Arrow & Debreu, 1954). The original article with the description of the Walras general equilibrium is provided in Figure 7.14.

ECONOMETRICA

VOLUME 22 July, 1954 NUMBER 3

EXISTENCE OF AN EQUILIBRIUM FOR A COMPETITIVE ECONOMY

BY KENNETH J. ARROW AND GERARD DEBREU[1]

A. Wald has presented a model of production and a model of exchange and proofs of the existence of an equilibrium for each of them. Here proofs of the existence of an equilibrium are given for an *integrated* model of production, exchange and consumption. In addition the assumptions made on the technologies of producers and the tastes of consumers are significantly weaker than Wald's. Finally a simplification of the structure of the proofs has been made possible through use of the concept of an abstract economy, a generalization of that of a game.

INTRODUCTION

L. WALRAS **[24]** first formulated the state of the economic system at any point of time as the solution of a system of simultaneous equations representing the demand for goods by consumers, the supply of goods by producers, and the equilibrium condition that supply equal demand on every market. It was assumed that each consumer acts so as to maximize his utility, each producer acts so as to maximize his profit, and perfect competition prevails, in the sense that each producer and consumer regards the prices paid and received as independent of his own choices. Walras did not, however, give any conclusive arguments to show that the equations, as given, have a solution.

The investigation of the existence of solutions is of interest both for descriptive and for normative economics. Descriptively, the view that the competitive model is a reasonably accurate description of reality, at least for certain purposes, presupposes that the equations describing the model are consistent with each other. Hence, one check on the empirical usefulness of the model is the prescription of the conditions under which the equations of competitive equilibrium have a solution.

Figure 7.14: Arrow-Debreu's Description of Walras's General Equilibrium Theory

Walrasian equilibrium models the competitive economy with the following assumptions:

- Each company maximizes profits.
- Each consumer maximizes their utility (preferences), subject to their budget constraint.
- Prices are independent of choices and are taken as given.
- All markets clear (i.e., reach equilibrium; price becomes static until another buyer and seller willing to exchange emerges in the market)

According to the Walrasian theory, prices reach equilibrium when "there is no good for which there is positive excess demand" (Varian 1978). This is a distinctly different case, then, of equilibrium occurring as supply equals demand. Consider cases for which there is no demand for an undesirable good or service; the demand would be zero, but the supply would be greater than zero. Under Walrasian equilibrium, the price remains stable until the consumer's utility or the company's profit maximization preferences for the goods or services changes; then the price changes again.

Walras realized the market was not static, but rather "like a lake agitated by the wind, where the water is incessantly seeking its level without ever reaching it." (Mitra-Kahn 2008, p. 51) The English translation of Walras's use of the term "tátonnement" to describe the approach towards equilibrium is "groping" (Mitra-Kahn 2008, p. 48). Perhaps, under the assumption that the CGE model is based on Walrasian equilibrium, it should be called *computable groping equilibrium*.

With a market described as constantly reaching for a general equilibrium, the challenge of using the theory is to make the claim that all markets clear (all reach an equilibrium price). With that claim serving as the basis for the CGE model of the economy, there must be a good reason to believe this is not too far-fetched an assumption. Economists develop what are called *proofs of existence* to show that indeed it is possible to reach a particular theoretical equilibrium. Arrow's and Debreu's 1954 *Econometrica* article presented the proof "that settled the issue" (Weintraub 2002). The mathematical proof for competitive equilibrium's existence is described the Arrow and Debreu article as well as many economics textbooks.

Arrow-Debreu General Equilibrium

Arrow and Debreu set out to prove in their 1954 *Econometrica* article that a competitive equilibrium, like the one described by Walras, can exist under specific conditions. The goal of the proof was to show that there is a *price vector*, a set consisting of a price for each market, which clears all markets. Under general equilibrium, the price is stable for all markets at once. The Arrow-Debreu model of general equilibrium is based on the Walras general equilibrium with specifications for the following conditions under which a market-clearing price vector can occur(Arrow & Debreu, 1954):

1. Every individual has initially some positive quantity of every commodity available for sale.
2. There are some types of labor in which (1) each individual can supply some positive amount of at least one such type of labor, and (2) each of this type of labor has a positive usefulness in the production of desired commodities.

The Arrow-Debreu model "presents an integrated system of production and consumption which takes account of the circular flow of income" (Arrow & Debreu 1954, p. 60).

Macroeconomic Balancing or Equilibrium

The equilibrium under this theoretical foundation is that the following macroeconomic balancing equation will equal zero.

$$(I - S) + (G - T) + (X - M) = 0$$

Where:

I= Investment

S= Savings
G= Government spending
T= Taxes
X= Exports
M= Imports

The argument for macroeconomic equilibrium as the theoretical base for CGE is that the model is ultimately driven by the macroeconomic balance equation. Microeconomic analysis according to general equilibrium theory is a part of the model, but in such a way that the microeconomic results must conform to the macroeconomic equilibrium (Mitra-Kahn, 2008).

7.10 References

Andersen, S. J., Harrison, R., Euritt, M. A., Mahmassani, H. S., Walton, C. M., & Helaakoski, R. (1992). *Economic Impacts of Highway Bypasses.* The University of Texas at Austin, Center for Transportation Research, Austin, Texas.

Angrist, J. D., & Pischke, J.-S. (2010, Spring). The Credibility Revolution in Empirical Economics: How Better Research Design is Taking the Con out of Econometrics. *Journal of Economic Perspectives, 24*(2), 3-30.

Arrow, K. J., & Debreu, G. (1954, July). Existence of an Equilibrium for a Competitive Economy. *Econometrica, 22*(3), 265-290.

Baumol, W. J., & Blinder, A. S. (1988). *Economics: Principles and Policy* (4th edition ed.). Harcourt Brace Jovanovich, Publishers.

Brocker, J. (2004). Computable General Equilibrium Analysis in Transportation Economics. In D. A. Hensher, K. J. Button, K. E. Haynes, P. R. Stopher, & D. H. al. (Ed.), *Handbook of Transport Geography and Spatial Systems* (pp. 269-289).

Buffington, J. L., & Wildenthal, M. T. (1998). *Estimated Economic Impact of Selected Highway Widening Projects in Texas.* Texas A&M University, Texas Transportation Institute, College Station, Texas.

Bureau of Economic Analysis. (2010, November). *National Economic Accounts.* Retrieved from NIPA Handbook: http://www.bea.gov/national/index.htm

Cambridge Systematics, I. R. (1998). *Report 35: Economic Impact Analysis of Transit Investments: Guidebook for Practitioners.* Transit Cooperative Research Program. Washington, D.C.: Traansportation Research Board.

Center for Economic Development and Research. (2011). *The Economic Impact of General Aviation in Texas.* Texas Department of Transportation.

Conrad, K. (undated). *Computable general equilibrium models in environmental and resource economics.* Retrieved November 2010

Dorfman, R. (1973, December). Wassily Leontief's Contribution to Economics. *The Scandanavian Journal of Economics, 75*(4), 430-449.

EDR Group, Inc. (2012). *Economic Impact Study of the Ohio River Bridges Project.* Prepared for Indiana Finance Authority and Indiana Department of Transportation. Boston, MA: EDR Group, Inc.

Eisele, W. L., & Frawley, W. E. (2000). *A Methodology for Determining Economic Impacts of Raised Medians: Final Project Results.* Texas Transportation Institute, College Station, Texas.

Gkritza, K., Labi, S., & Sinha, K. (2007). *Economic Development Effects of INDOT Transportation Projects.* Joint Transportation Research Program.

Goff, Z. A., McFarland, W. F., Edge, B., Basilotto, J., & Graalum, S. (1997). *The Value of Texas Ports.* Texas A&M University, Texas Transportation Institute.

HDR Decision Economics. (2011). *Economic Impact Analysis of the Baltimore-Washington Rail Intermodal Facility Final Report.* Prepared for CSX Intermodal Terminals, Inc. Silver Spring, Maryland: HDR Decision Economics.

Horowitz, K. J., & Planting, M. A. (2009). *Concepts and Methods of the U.S. Input-Output Accounts.* Bureau of Economic Analysis.

Horridge, M., & Powell, A. (2001). *MINIMAL A Simplified General Equilibrium Model.* Monash University, Centre of Policy Studies and the Impact Project.

Juri, N. R., & Kockelman, K. M. (2005). *Evaluation of the Trans-Texas Corridor Proposal: Application and Enhancements of the RUBMRIO Model.* Retrieved from http://www.ce.utexas.edu/prof/kockelman/.

Kockelman, K. M. (2007). *RUBMRIO.* Retrieved November 2010, from Random-Utility-Based Multiregional Input-Output Model: http://www.ce.utexas.edu/prof/kockelman/RUBMRIO_Website/homepage.htm

Kockelman, K., Srinivasan, S., & Handy, S. L. (2001). *Economic Effects of Highway Relief Routes on Small- and Medium-Size Communities.* The University of Texas at Austin, Center for Transportation Research.

Kockelman, K., Zhou, B., & Tirumalachetty, S. (2009). Meeting Minutes. *NCTCOG's Land Use Modeling Workshop.* Arlington: NCTCOG.

Martin Associates. (2004). *The Local and Regional Economic Impacts of the Port of Corpus Christi.* Prepared for the Port of Corpus Christi.

Miller, R. E., & Blair, P. D. (1985). *Input-Output Analysis.* Englewood Cliffs, New Jersey: Prentice-Hall, Inc.

Mitra-Kahn, B. H. (2008). *Debunking the Myths of Computable General Equilibrium Models.* The New School for Social Research. New York: Schwarz Center for Economic Policy Analysis.

Mohring, H. (1961, June). Land Values and the Measurement of Highway Benefits. *Journal of Political Economy, 69*(3), 236-249.

O'Leary, C. J., Eberts, R. W., & Pittelko, B. M. (2012). *Effects of NAFTA on US Employment and Policy Responses.* OECD Publishing.

Partridge, M. D., & Rickman, D. S. (2007). *CGE Modeling for REgional Economic Development Analysis.*

Piermartini, R., & Teh, R. (2005). *WTO Discussion Papers No. 10: Demystifying Modelling Methods for Trade Policy.* Geneva, Switzerland: World Trade Organization.

Quddus, M., Carmel, A., & Bell, M. G. (2005). *Working Paper 12: The congestion charge's impact on retail- The London experience.* London: Greater London Authority.

Rutherford, T., & Paltsev, S. (1999, August). *From an Input-Output Table to a General Equilibrium Model: Assessing the Excess Burden of Indirect Taxes in Russia.* (M. I. Technology, Producer) Retrieved November 2010, from http://web.mit.edu/paltsev/www/docs/russia.html

Samuelson, P. A., & Nordhaus, W. D. (2005). *Economics* (Eighteenth ed.). New York, New York: The McGraw-Hill Campanies, Inc.

Texas Department of Transportation. (2007). *US 290 Environmental Impact Statement and Schematic Design.*

The Perryman Group. (2006). *Moving Into Prosperity The Potential Impact of the Trans-Texas Corridoron Business Activity in Texas: An Analysis of the Effects in Key Trans-Texas Corridors and the State of Texas.* Waco, Texas.

U.S. Department of Commerce. (1997). *Regional Multipliers: A User Handbook for the Regional Input-Output Modeling System (RIMS II).* 3rd edition, Washington, D.C.

US Congressional Budget Office. (1993). *A Budgetary and Economic Analysis of the North American Free Trade Agreement.* Washington, D.C.: US Congressional Budget Office.

Vargas, E., Schreiner, D., & Tembo, G. a. (1999). *The Web Book of Regional Science.* (R. R. Institute, Producer, & West Virginia University) Retrieved November 2010, from Computable General Equilibrium Modeling for Regional Analysis: http://www.rri.wvu.edu/WebBook/Schreiner/contents.htm

Varian, H. (1978). *Microeconomic Analysis.* New York: W.W. Norton & Company, Inc.

Varian, H. R. (1996). *Intermediate Economics.* New York: W.W. Norton & Company.

Weintraub, E. R. (2002). *How Economics Became a Mathematical Science.* Durham, NC: Duke University Press.

Weisbrod, G. E., & Alstadt, B. B. (2007). Progress and Challenges in the Application of Economic Analysis for Transport Policy and Decision Making. *Research Roundtable on Infrastructure Planning and Assessment Tools.* OECD International Transport Forum.

Weisbrod, G., & Weisbrod, B. (1997). *Assessing the Economic Impact of Transportation Projects: How to Choose the Appropriate Technique for your Project.* Washington, D.C.: Transportation Research Circular.

Wing, I. S. (2004). *Computable General Equilibrium Models and Their Use in Economy-Wide Policy Analysis.* Massachusetts Institute of Technology. Cambridge, Massachusetts: MIT Joint Program on the Science and Policy of Global Change.

Chapter 8. Econometrics for Data Analysis

8.1 Definitions and Steps for Econometric Analysis

Econometrics is the application of statistics to the analysis of economic and other behavioral data (like household incomes, VMT, and local retail sales). Econometrics allow analysts to discern relationships among variables of interest (such as gas prices, household size, and vehicle ownership) in order to forecast future trends. The field of econometrics is well developed, with hundreds of textbooks offering various levels of detail and mathematical sophistication for data applications in a variety of fields. Examples include

- Professor Bruce Hansen's *Econometrics* text at http://www.ssc.wisc.edu/~bhansen/econometrics/Econometrics.pdf, and
- Ken Train's *Discrete Choice Methods with Simulation* athttp://elsa.berkeley.edu/books/choice2.html.

The goal of this chapter is not to repeat what can be readily obtained elsewhere, but to provide an overview of key econometric methods for transportation applications. It begins with basic data examination and data set definitions, and leads to least-squares and maximum likelihood estimation techniques for linear and nonlinear regressions of various types of response variables (like annual VMT per vehicle) on sets of explanatory data that may help explain changes to the response variables (like household size and income, neighborhood density, and vehicle age).

The standard econometric investigative approach is to

1) specify questions of interest (e.g., what is the response of businesses and households to new highway capacity?),

2) acquire data sets of interest (e.g., sales data from a tax office, household travel survey data from an MPO, and/or property values from an appraisal district),

3) specify the econometric model that makes the most sense for the data (e.g., ordinary least squares regression for home prices and a multinomial logit for mode choice), and

4) select statistical software to estimate the parameters (β's) that quantitatively weight the predictive variables (X's) to estimate response variables (Y's). We can then forecast the response of others (e.g., other households and firms) in new, related settings (e.g., another highway corridor in a different city).

Clearly, econometrics offers a useful set of tools.

Some Caveats

While mathematical models are meaningful tools for understanding behavior and anticipating future outcomes, they are all abstractions of reality. Human behavior is highly variable, and many outside-the-model factors intervene (e.g., economic recessions, technological advances, and restrictive work schedules). Related to this

Key Terms

- ✓ **OLS:** ordinary least squares
- ✓ **LS:** least squares
- ✓ **MLE**: maximum likelihood estimation
- ✓ **MSE**: mean squared error
- ✓ **WLS**: weighted-least squares
- ✓ **FGLS**: feasible generalized least squares
- ✓ **IV:** instrumental variables
- ✓ **SEM:** system of equations model
- ✓ **FE:** fixed effects modeling
- ✓ **FD:** first difference modeling
- ✓ **RE:** random effects modeling
- ✓ **ARIMA:** autoregressive integrated moving average
- ✓ **SES:** simultaneous equations systems
- ✓ **SUR:** seemingly unrelated regression

notion, validating models of complex urban phenomena is difficult, unless all other factors at play (those not included in the set of explanatory variables) are held constant—a nearly impossible task in free markets. As statistician George Box is credited for saying, "essentially, all models are wrong, but some are useful." As an example, traffic count models may always predict the direction of volume *changes* when capacity is added, and often give a good indication of the magnitude of traffic change, but their absolute levels (before and after the roadway's expansion) may be off by 25% or more (Wang & Kockelman 2007). Thus, pivoting off of known numbers, whenever possible (e.g., existing land use patterns and traffic counts), is generally very helpful in generating predictions.

Resource

For a deeper discussion of the discrepancy between economic theory and the complex and unpredictable processes of real-world behavior, see David Orrell's *Economyths: Ten Ways Economics Gets It Wrong.*

Another point of caution relates to behavioral assumptions underlying many choice models. While economic laws attempt to characterize rational, benefit-maximizing behavior, and these laws form the basis for many econometric models, decisions at the individual level are often somewhat irrational. This is largely due to limitations on information and decision-makers' cognitive abilities under time- and resource-constrained settings.

Despite such concerns, econometric modeling is still preferred to subjective argument and natural biases and cognitive limitations that exist in all humans. As the saying goes, knowledge is power. Data sets and econometric methods are improving, and common sense remains paramount in any application. Common sense is important to construct key questions, identify the most appropriate data sets, specify the most robust models, and evaluate results.

Some Terminology

Before launching a discussion of regression techniques, we'll define key terms, as shown in Table 8.1. Explanatory variables combine with model parameters (and error terms) to predict other, response variables. And analyses often rely on extensive data sets, holding information on 1,000 or more observational units (e.g., travelers) and various variables of interest (e.g., a traveler's number of trips per day, age, gender, and income).

Key Term (and Synonyms)	**Basic Definition**
Observational Unit	The entity (or time period) used as the unit for data collection and analysis. Examples include households or persons, towns or states, vehicles or road segments, businesses or parcels. Examples of time periods (for tracking traffic counts, gas prices, and regional VMT, for example) include days, months, and years.
Variable	Any characteristic whose value or category changes across observational units in the data set. Examples include the employment and retail sales of different cities and the size and income of different households.
Explanatory Variable (Predictor, Regressor, Independent)	Any variable whose value is used to predict or explain the model's response variable. In lab experiments or stated-preference surveys, these variables can generally be set/controlled by the analyst, and so are exogenous or independent. In cause-and-effect settings, these variables cause the response. Typically indicated as X_1 through X_p (for *p* distinct variables).

Response Variable (Dependent, Explained)	This is the outcome variable, or the effect. Typically indicated as a Y. (Examples include travelers' chosen modes, link traffic counts, and county-level retail sales.)
Discrete Variable	These variables are normally integer by nature (e.g., number of vehicles owned or crash counts) or coded as integers (e.g., Y = 0 & 1, for no & yes responses). They may be ordered (such as no-injury crashes to fatal crashes) or unordered (as in mode choice outcomes).
Continuous Variable	These variables allow for all values in a range on the number line, such as household incomes or parcel sizes.
Indicator Variable	Indicator variables are simple binary variables, with 0/1 values. The value 1 applies if the observational unit has the characteristic; if not, the value is 0 (e.g., $X_{Male} = 1$ for male respondents and 0 for females.)
Error Term	A model's random component, typically indicated as ε. Such unknowns serve as a catch-all for missing/unobserved attributes of each observational unit (and its context). Error terms allow for unobserved variability or heterogeneity across observational units that have the same observed X values (so their outcomes, Y's, can differ).
Residual	The residual for each observational unit is the difference between its observed and *estimated* response values. It is an estimate of the error term (as defined above).
Parameter (Coefficient)	Model parameters interact (either multiplicatively, as a coefficient, or otherwise) with explanatory variables to render response variable values. They are commonly denoted as β (e.g., $Y = \beta_0 + \beta_1 x_1 + \varepsilon$), but other notation can also apply (e.g., σ, θ, and γ). A key objective of data analysis is to estimate these unknown parameter values, in order to understand which variables are most important in a model and predicting others' behavior.
Sample Size (n)	Sample size is the number of observations included in a statistical sample.
Mean (μ)	Mean is the expected value of a random variable. For a data set, it describes the center location of a set of data. The sample mean takes the form $\widehat{\mu_x} = \bar{x} = \frac{1}{n}\sum_{i=1}^{n} x_i$, where n is sample size.
Median	Median is another useful statistic for measuring the center of a set of data. The sample median is the numerical value that separates the higher half of the sample from its lower half.

Key Term (and Synonyms)	**Basic Definition**
Variance (σ^2)	Variance measures the variability, or spread, of data around its center of gravity (the distributional mean). Sample variance is the square of standard deviation and takes this form: $\widehat{\sigma_x^2} = s^2 = \frac{\sum_{i=1}^{n}(x_i-\mu_x)^2}{n-1}$.
Correlation (ρ)	Characterizes the direction of any potentially linear relationship between two variables. Positive correlation means X and Y tend to rise together; negative means one tends to fall as the other rises. Zero correlation means no obvious and simple linear relationship appears to exist (but a sinusoidal, symmetrically convex, or other relationship may exist). Sample correlation takes this form: $\rho_{xy} = \frac{\sum_{i=1}^{N}(x_i-\mu_x)(y_i-\mu_y)}{\sqrt{\sum_{i=1}^{N}(x_i-\mu_x)^2}\sqrt{\sum_{i=1}^{N}(y_i-\mu_y)^2}}$

Table 8.1: Useful Econometric Terms

Regression	The association between a set of explanatory variables and one or more types of response variables. Estimating a model's regression parameters by maximizing some measure of model fit, in order to predict one or more response variables.
Null hypotheses (H_0)	Null hypotheses are the hypotheses that are formulated with the hope of rejecting.
p-Value	The probability that, if the null hypothesis is correct, a new sample of data could have yielded a parameter estimate further from the hypothesized value(s).
t-statistic	Used in hypothesis testing, it is a ratio between the departure of an estimated parameter from its notional value and its standard error. If the null hypothesis is correct, t-statistic follows a standardized student distribution with (n-k) degree of freedom where n is the number of observations and k the number of parameters.
R^2	A common model fit statistic. The R^2 measures how well a linear regression model fits the observed data (i.e., how close the estimated response values, $\hat{Y}_i$'s, come to the actual y_i's)
Likelihood	A data set's likelihood is a function of the data and all parameters to be estimated in a model. It is the probability of observing the outcomes (for a set of data) given those parameter values.
Maximum Likelihood Estimation (MLE)	MLE is a method of estimating the parameters of a model, such that they maximize the likelihood function.
Likelihood Ratio Index (LRI)	LRI is a goodness-of-fit statistic for regression models estimated using maximum likelihood methods. LRI is used to compare the fit of two models: a more complete model (with more parameters) and a reduced version of this model (e.g., a simple "no-information" model with no X values). LRI = $1 - \frac{LogLikelihood\ (more\ complete\ model)}{LogLikelihood\ (simpler\ model)}$
Probability Distribution	Presented either as a graph, table, or function, a probability distribution gives the probability of the value of a random variable in a population. There are different types of equations that provide probability distributions. Examples include the normal (the familiar "bell-curve"), log-normal, logistic and Gumbel distributions.

Table 8.1: Useful Econometric Terms (continued)

A meaningful concept to mention here is the difference between **correlation** and **causation**. Just because two variables exhibit correlation does not mean that a change in one will cause a change in the other. Similarly, just because an analyst chooses to put one variable on the left side of a regression equation and call it the "response variable," the variables on the right are not automatically the appropriate causal variables. They may be helpful in prediction, but not in generating the response that policymakers seek. It may be simple correlation, rather than causation (e.g., household VMT per year predicting household income or city population predicting GDP per capita). Another possibility is that the Y actually affects the X more than the X affects the Y, and/or they vary simultaneously, in response to similar, underlying factors or in some form of behavioral balance (like supply and demand). Such nuances are discussed in more detail in this chapter's section on simultaneous equations estimation.

Understanding the Data: Use of Summary Statistics

Transportation engineers and planners must examine their data set *before* applying any real econometric techniques. Such examinations help identify outliers (e.g., ages of 120 years) and incorrect values (e.g., annual incomes of $200 million). Furthermore, summary statistics provide a quick snapshot of the data sample (e.g., the share of drive-alone trips during peak commute hour or percentage of heavy duty trucks on highways).

Summary statistics tables typically list the mean (average) value of each variable in the data set, along with the standard deviation, and minimum and maximum values. Other basic descriptors are also meaningful, including medians (the 50th percentile value) and the first and third quartile values (which divide the lower and upper halves of ordered values into equal parts). Table 8.2 presents example of summary statistics.

Example

Summary Statistics Tables

Table 8.2 shows such a table for a study of the costs and condemnation likelihood of acquired parcels in recent Texas Department of Transportation (TxDOT) ROW data. The table indicates that 15% of these acquisitions went through condemnation (an expensive and time-consuming process), 90% involved a partial (rather than whole) taking, 66% came from Texas's northern region, and the average share of land taken was 20% of a parcel's total area. Perhaps most importantly, the average cost of acquisition was $236,000 per parcel, with substantial variance, as characterized by a standard deviation of $791,000. The histogram of acquisition cost and its log-transformation are presented in Figures 8.1 and 8.2. **Such numbers impact TXDOT's bottom line, project-completion schedules, and the net benefits of projects pursued.**

Variables	Description	Average	Std. Deviation	Min	Max
Condemnation	Parcel possession type is condemnation (y=1) else negotiated (y=0)	0.15	0.36	0	1
TotalCost	Total acquisition cost (in year 2011 dollars)	236,000	791,000	72.89	1.61E+07
LnTotalCost	Natural log of total cost	10.58	1.98	1.86	7.21
TakenSF	Land area of part acquired (square feet [sf])	49,400	159,000	7.00	2.55E+06
RemainderSF	Land area of remainder parcel (sf)	14,400,000	474,000,000	0	1.96E+10
TimeTrend	Trend variable for year of acquisition (1=2008, 2=2009,…, 4=2011)	2.31	0.69	1	4
North	X=1 when parcel is located in northern region of Texas, else X=0	0.66	0.47	0	1
South	X=1 when parcel is located in southern region of Texas, else X=0	0.15	0.36	0	1
East	X=1 when parcel is located in eastern region of Texas, else X=0	0.14	0.35	0	1
West	X=1 when parcel is located in western region of Texas, else X=0	0.04	0.20	0	1
PopDensity	Population density (at county level) per square mile (U.S. Census, 2010)	795.7	852.8	1.24	2692
Agriculture	X=1 when parcel land is in agricultural use, else X=0	0.09	0.29	0	1
Residential	X=1 when parcel land is in residential use, else X=0	0.26	0.44	0	1
Commercial	X=1 when parcel land in commercial, retail and service uses, else X=0	0.35	0.48	0	1
OtherUse	X=1 when parcel land in other use (e.g., ecclesiastical, industry, education, multi-use, and special use), else X=0	0.06	0.24	0	1
Vacant	X=1 when parcel land is vacant, else X=0	0.24	0.42	0	1
PartialTaking	X=1 when parcel was acquired with a remainder	0.90	0.30	0	1
Ratio	Fraction of taken area to total area	0.20	0.31	0	1
Individual	X=1 for individual ownership type, else X=0	0.51	0.50	0	1
Corporation	X=1 for corporation and partnership ownership type, else X=0	0.38	0.49	0	1
OtherOwnership	X=1 for other ownership types (e.g., federal or state agency & municipality), else X=0	0.10	0.30	0	1

Table 8.2: Example Summary Statistics (for 1,710 TxDOT ROWIS properties, as acquired between 2008 and 2011, with data from Xiong and Kockelman [2012])

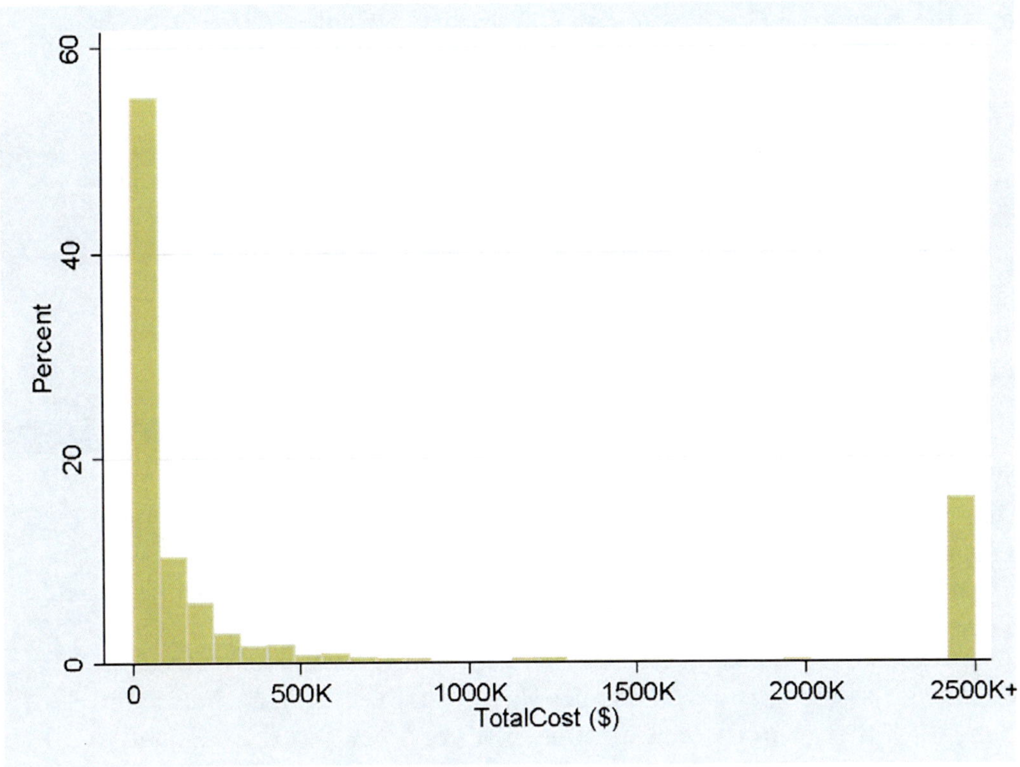

Figure 8.1: Histogram of Total Acquisition Cost

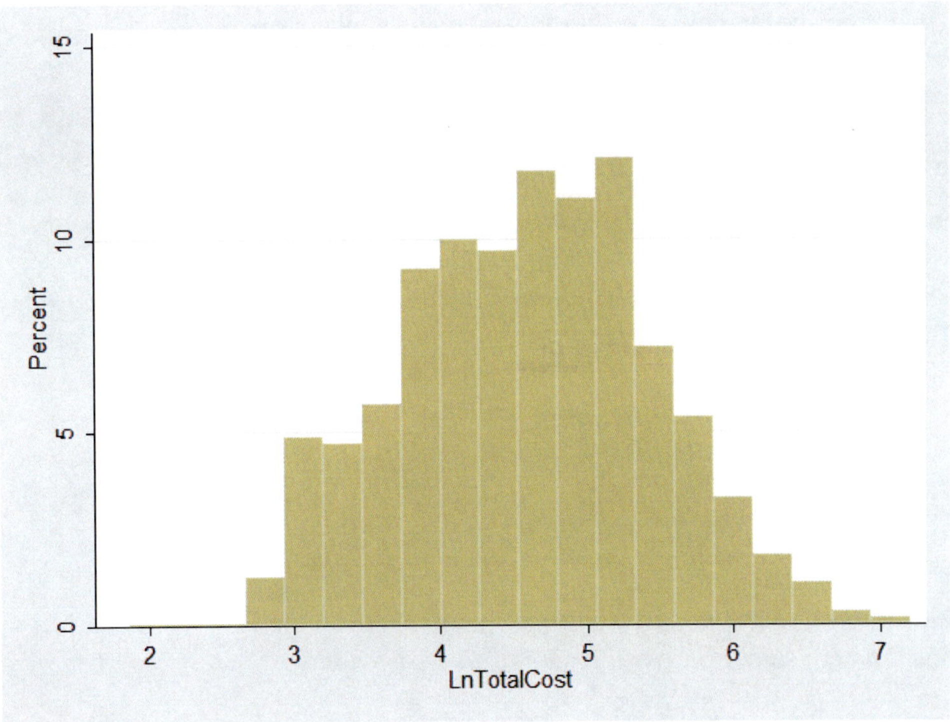

Figure 8.2: Histogram of LnTotalCost

Equations for the sample mean and standard deviation (SD) of each variable (X_j) are as follows:

$$\text{Mean}(X_j) = \overline{x_j} = \frac{\sum_{i=1}^{n} x_{ji}}{n}$$

$$\text{SD}(X_j) = \sqrt{\frac{\sum_{i=1}^{n}(x_{ji} - \overline{x_j})^2}{n-1}}$$

where $i=1,2,\ldots,n$ & n is the sample size. The $\sum$ symbol means sum, so for the mean and SD means summing the values across all observational units.

Another basic attribute of interest is the **correlation coefficient**,ρ, which measures the strength of linear association between two paired variables (e.g., parcel size and parcel cost). Correlation can vary from -1 to +1. A positive correlation means that the two variables tend to rise or fall together (e.g., travel time and cost) and a negative correlation indicates that when one rises the other falls (e.g., distance to the downtown and land values as discussed in Chapter 4).

Resource

All of these summary statistics can be generated easily in MS Excel, using the Data Analysis toolpak add-in. Plus, Excel can generate histograms for each variable.

Correlation is also the square root of the R-squared (**R^2**) goodness-of-fit statistic that emerges when we regress a single variable on another single variable, using a straight-line assumption (e.g., $Y = \beta_o + \beta_1 X + \varepsilon$, where ε is the error term). An R^2 of 1.0 means a perfect fit or one-to-one relationship (with no errors) between two variables, implying that we can perfectly predict the Y variable without error by simply knowing the X variable (or set of X variables). For example, if the correlation between ROW acquisition costs (Y) and acquiring property areas (X) is 0.43, then the R^2 will be 0.18, implying that 18% of a parcel's ROW acquisition costs can be predicted by knowing the parcel area. Adding relevant explanatory variables (X's) to the regression (e.g., land use, ownership type) will increase the R^2, yielding better ROW acquisition cost estimates.

8.2 Data Sets for Regression Models

Regression is a priceless tool in a number of data contexts because it can identify the predictive variables that have statistically-significant associations with the response variable. Whereas correlation measures the association between two random variables, without causal order (in other words, without assigning one variable the independent variable and the other the response variable), regression recognizes correlation of response variables with predictive variables, but incorporates more variables, and provides the parameters characterizing the relationship.

Regression in transportation economics is regularly used to

- Value homes, businesses, and vehicles (using least-squares hedonic models);
- Estimate production, profit, preference, and consumption functions (to understand how business and consumer behavior responds to changes in predictive variables);
- Predict travelers' selection of alternatives (using logit regression techniques

across network routes, transportation modes, trip destinations, or residential locations); and

- Forecast the future (of gas prices, populations, and truck volumes).

In any setting, analysts require data to form predictions, and the choice of model specification (e.g., ordinary least squares or logit) regularly depends on the type of data set used (e.g., cross-sectional or time-series, aggregate or disaggregate), as well as the unit of observation (e.g., road segments along a corridor).

Cross-Sectional, Time-Series, and Panel Data Sets

Cross-sectional data sets are the most common: they offer a snapshot of values across a set of observational units at a specific moment in time, such as the Austin household travel surveys in 2006, or the state Workforce Commission's number of jobs and population in each county in January 2011.

Table 8.3 shows an example of cross-sectional data, where the units are acquired properties (parcels) for roadway construction or expansion, with various variables tracked: total amount paid (to property owners), property type (i.e., land use type), taking type, taking amount (taken area, in square feet), remainder (area, in acre), and ownership type. A useful approach might be to model the first of these six as the response variable, and the latter five as predictor/explanatory variables.

Table 8.3: Example of First Three Records in a Cross-Sectional Data

ROW CSJ	Parcel #	Total Amount Paid	Property Type	Taking Type	Taking Amount	Remainder	Ownership Type
0009-02-056	16	$5,000	Commercial	Partial Taking	168 sq ft	2.68 Acres	Corporation
1400-04-027	102	$95,000	Vacant Lot	Whole Taking	7,505 sq ft	0	Individual
2243-01-011	2	$1,470	Agriculture	Partial Taking	39,465 sq ft	31.22 Acres	Corporation
…	…	…	…	…	…	…	…

Serial or **time-series data sets** track the behavior of a single observational unit over an ordered series of points in time. Examples include the number of retail jobs in Bexar County every year for the past 50 years, or the average price of gas in Fort Worth from month to month over the past 5 years.

Finally, **panel data sets** are a marriage of cross-sectional and time-series data. They require information on a set of observational units over a set of times slices, such as the number of retail jobs in each county or city in a state over each of the past 50 years.

Concept Example: Panel Data Set

Researchers Eisele and Frawley (2000) used a panel data set and studied the impact of raised medians on adjacent businesses by following property values along the corridor (relative to city and county trends). They found that property values dipped during the construction phase in many of the cities, but rose overall, across corridors with added medians.

Other examples of panel data sets include the miles traveled by each household in a data set each day for a month, or parcel attributes for a set of properties over a 20-year period. Panel data sets can be balanced or unbalanced, depending on where the set of observational units is perfectly maintained over time, or units are lost and added. The latter is more challenging to estimate, but both types of data are common and all styles of analysis are feasible.

In nearly all regression contexts, more than one variable is tracked across units and/or over time. This approach allows analysts to predict one or more response variables as a function of other attributes (e.g., gas tax revenues as a function of population and employment, or vehicle ownership as a function of household size, income, and presence of children).

Table 8.4 shows an example of panel data, with three observations in time for each of two distinct properties. (The complete data set would be much longer than this, because regression model parameters cannot be confidently estimated with three data points. The standard rule of thumb is to have *at least* 5 data points per parameter [e.g., per β], though more than 100 records per parameter is typical.) Notice how some attributes remain constant over time (e.g., acreage). As long as the attribute varies across observational units (properties in this case), the regression should be able to pick up this variable's effect.

Table 8.4: Example of Panel Data (only first two observational units shown)

one observational unit across several rows {

Property ID	Year	Tax Appraisal	Proximity to Transit (feet)	Acreage
1039	2009	$200,000	500	1.2
1039	2010	$225,780	500	1.2
1039	2011	$228,050	500	1.2
1040	2009	$52,000	250	0.4
1040	2010	$56,400	250	0.4
1040	2011	$48,000	750	0.4
…	…	…	…	…

Example

Panel Data Regression

Kweon and Kockelman (2005) examined the impacts of speed limit increases on yearly crash counts across 6,000-plus similar road segments in Washington State over a 4-year period. This panel data set had discrete integer counts as the response variable (rather than a continuous variable, like traffic flows), so the researchers specified fixed-effects and random-effects Poisson and negative binomial regression models. Regressing crash counts, by severity level, on vehicle exposure (VMT/day), design attributes (e.g., length and degree of horizontal curvature, grade, median width, and number of lanes), and speed limits. The researchers estimate that a 10-mph speed limit increase is associated with significant increases in deadly and injurious crashes, while having a slightly negative effect on total crash counts (everything else constant). Tighter horizontal curves, fewer lanes, and lower traffic volumes were also associated with higher fatal and injurious crash rates (all else held constant).

8.3 Specifying the Model

Econometric models are mathematical equations that relate the parameters, explanatory variables and error term to a response variable. The type of data, the quantity of data, the characteristics of the underlying the data, and other factors play a role in choice of econometric models for data analysis. The choice of model dictates the parameter estimation methods that can be used. Following are some critical initial questions: What is my response variable? (Are the values discrete or continuous? Ordered or categorical?) What kind of data do I have— cross-sectional, time-series, or panel? And are the records in my data set independent or dependent over time and space, or within households and other group settings?

Table 8.5 lists common models for various types of response variable.

Continuous Y
Least-squares Regression (weighted & unweighted, linear, & nonlinear)
Discrete Y: Binary
Binary Logistic Model
Binomial Logit or Probit Models
Discrete Y: Categorical
Multinomial Logit and Probit
Nested Logit
Discrete Y: Ordered
Ordered Probit & Ordered Logit
Discrete Y: Integer
Poisson & Negative Binomial Models

Table 8.5: Response Variable Types for Different Regression Models

The classic linear regression specification is presented as Equation 8.1, with labels for each component using Table 8.1's statistical terms.

Dependent Variable ↓ (y)   Explanatory Variable ↓ (x_1)

$$y = \beta_0 + \beta_1 x_1 + \cdots + \beta_k x_k + \varepsilon \qquad (8.1)$$

Parameters ↗ (β_0, β_1)   Error Term ↑ (ε)

The explanatory variables and error term should correlate with the dependent variable. As alluded to earlier, the model's error term (or error terms, in certain cases) reflects unobserved information that affects the response variable. For example, weather patterns will affect crash counts and traffic volumes, mode choices, and so forth, but such details are generally not available to analysts, in part because they vary from day to day, hour to hour, and location to location, and are not recorded for most sites. The error term, ε, is the difference in observed and modeled/expected values of *Y*; it is often estimated as a residual, once the parameters of the model's primary equation are known ($y = f(x,\beta)$). Correlation between the error term and explanatory variables leads to biased parameter estimates (unless corrected and/or controlled for). Such correlation means that the associated X_1 variable's coefficient (β_1) will try to pick up the effects of the missing variable(s), resulting in bias (either too high or too low, upward or downward) in the estimated coefficients.

Parameter Estimation

Two main methods are used for parameter estimation in the classic linear model. The first is called **least squares** (LS) and the second is **maximum likelihood estimation** (MLE). LS minimizes the sum of squared error predictions, while MLE maximizes the likelihood of the data set's observed Y values. In other words, a parameter estimated by MLE "is simply the value of β that maximizes the probability of drawing the sample actually obtained" (Kennedy 1998, p. 30).

The mathematics behind each method differs. But, in most cases, application of either approach (LS or MLE) leads to very similar, if not identical, estimates of the slope parameters,β. A key difference is that the MLE approach requires specification of a probability distribution for the error terms, whereas LS enjoys being distribution free (i.e., the analyst does not have to assume a probability distribution for the error terms).

Why is there so much emphasis on the error term in parameter estimation? Well, the key objectives of model estimation are 1) identification of variables that significantly influence the response variable and 2) accurate prediction of Y. Error terms (ε's) indicate how far off such models will be in prediction. The hope is that these errors are small.

Econometricians seek models and methods that provide **unbiased, consistent, and efficient parameter estimates**. Bias is simply the (expected or average) difference between the estimated parameter ($\hat{\beta}$) and its true (unknown) value: $E[\hat{\beta} - \beta]$, where $E[\ \ ]$ denotes the expected or mean value of the embedded function (here the simple difference between the estimator and its true value). Each data set sample is in essence a random experiment or outcome, so each data sample's set of β estimates will differ from sample data set to sample data set. Thus, $\hat{\beta}$ is a random variable (across data sample sets) and so follows some probability distribution. The form of this probability distribution (e.g., normal or uniform) depends on the data sets drawn to estimate a single $\hat{\beta}$.

Ideally, our estimates should cluster around their target, the true β, without bias.

Related to the notion of unbiasedness is **consistency**. The set of estimates $\hat{\beta} = (\hat{\beta}_0, \hat{\beta}_1, \cdots, \hat{\beta}_k)$ (referred to as vector $\hat{\beta}$) is said to be consistent (or asymptotically consistent) if all values in the vector converge to their true values as the sample size gets larger (towards infinity, in theory). In such contexts, bias and uncertainty become negligible, if we can obtain a large enough sample of data points. An **efficient** estimator is one that estimates the unknown parameters with maximum precision or minimum uncertainty. Overall, analysts seek an estimator offering a minimum **mean squared error** (MSE), where MSE=$E[(\hat{\beta} - \beta)^2]$. In other words, MSE is the expected or mean squared difference between the (unknown/random) estimator and the true β. Some estimation methods are able to achieve efficiency as sample size grows; these are called *asymptotically efficient* estimators.

Ordinary Least Squares (OLS) Regression

The most common form of regression is **OLS**. This popular specification requires a response variable that is linear and continuous in nature, and the regression equation is as follows:

$$y = \beta_0 + \beta_1 x_1 + \cdots + \beta_k x_k + \varepsilon$$

The set of parameter estimates (vector $\hat{\beta}$) that minimizes the sum of squared residuals (the squared error-term estimates), SSR=$\sum_{i=1}^{n}(y_i - x_i'\hat{\beta})^2$, is the OLS estimator. For y regressed on a set of predictive variables x, the estimated parameter is found with the following formula:

Resource

This type of regression can be estimated in MS Excel.

$$\hat{\beta}_{ols} = \frac{(\sum_{i=1}^{n} x_i' y_i /n)}{(\sum_{i=1}^{n} x_i' x_i /n)}$$

where $x_i = (1, x_{i2}, \ldots, x_{ik})$ is the value of the single predictive variables for the ith observation and n is the number of observations in the sample. The x_i' refers to …

Consistency requires that the ε's be zero and be uncorrelated with the predictive variables. Often econometricians prefer the more concise block-matrix notation for $\hat{\beta}_{ols}$:

$$\hat{\beta}_{ols} = (X'X)^{-1}X'Y$$

where X is an n by k matrix with the ith row containing the predictive variables for the ith observation, x_i, Y is an n by 1 vector containing the stacked response variable for all n observations, and the exponent -1 indicates matrix inverse operation.

As sample sizes rise, the asymptotic variance of the $\hat{\beta}_{ols}$ estimator (with covariance term) is a matrix: $\hat{\sigma}^2(X'X)^{-1}$, where $\hat{\sigma}^2$ is an estimate for the homoscedastic error terms' (constant) variance σ_ε^2 and each X matrix is simply the table of covariate values (with only 1's in the first column, to interact with the β_0 term, and rows signifying data records). The variance terms (or squared standard error term for each estimator, $\hat{\beta}_j$) can be found along the main diagonal of this matrix; the off-diagonal terms are the covariances (correlation times standard deviations) of the various pairs of parameter estimates: $cov(\hat{\beta}_l, \hat{\beta}_j)$.

MLE Estimation of the Classic Linear Model

In contrast to OLS (which focuses on minimizing squared residuals), MLE estimators are the set of parameter estimates ($\hat{\beta}_i$'s) that maximize the response variables' likelihood function. This likelihood is (or is proportional to) the probability of the observed response values, given the observed covariate values x's and their associated coefficients (β's):

$$Likelihood = \prod_{i=1}^{n} f(y_i | x_i;\ \beta)$$

where f is the probability mass function or density function (for example, a normal density function) of the dependent variable y and $\prod$ is the symbol for multiplying all these f values (across all n observations).

The MLE estimators are consistent if the model is correctly specified, if the unknown parameters can be identified uniquely, and if the log-likelihood function is continuous in those unknown parameters. The asymptotic distribution of the resulting MLE estimators is then approximately normal, centered on the true β's, with a covariance matrix that relies on the (vector) derivatives of the log-likelihood (cross multiplied and inverted).

When the error terms are independent of one another and follow the same normal distribution (centered on zero), the MLE parameter estimates will equal those of the OLS approach. Both sets of parameter estimates under this setting will enjoy exactly normal distributions (rather than only asymptotically).

MLE Estimation of the Binary Probit Model

Another basic application of MLE techniques is a binary probit model, for indicator responses ($y = 0,1$). This type of model is useful in generating a model for a discrete two-outcome response variable (e.g., yes or no, car or bus, peak or off-peak travel), instead of a continuous response variable. A general linear model can be formed by defining a latent (i.e., unobservable) response variable (y^*) to be a linear function of x:

$$y^* = x\beta + \varepsilon$$

where the observed y (yes or no, car or transit) equals 1 if $y_i^*>0$, and equals zero otherwise. Here, the error terms, ε, are assumed independent across observations and normally distributed, with mean zero and variance σ^2. Because y^* is unobservable, σ is typically set to 1, so that unique parameter estimates can be obtained/identified.

MLE estimators are used to generate many transportation estimates that are helpful to planners.

Application of such models can be found in land use (e.g., y=1 for developed parcels and 0 for undeveloped parcels), vehicle ownership (y=1 if the household purchased a vehicle that year), crash modeling (y=1 if the occupant suffered a severe injury), mode choice (y=1 if the traveler chose a motorized mode), and many other transportation applications.

To form the likelihood function, we must first define the probability of $y = 1$, as follows:

$$P(y_i = 1) = P(y_i^* > 0) = P(x_i\beta + \varepsilon_i > 0) = P(\varepsilon_i > -x_i\beta) = \Phi(x_i\beta),$$

where $\Phi(\cdot)$ denotes the standard normal cumulative distribution function (or normal CDF).

In this binary indicator-response model setting, the probability $P(y_i = 0)$ is simply 1- $P(y_i = 1)$. Hence, this model's likelihood function can be written as $\mathcal{L} = \prod_{i=1}^{n} \Phi(x_i\beta)^{y_i} [1 - \Phi(x_i\beta)]^{1-y_i}$. Note that use of a different error term distribution (like the Gumbel, rather than a normal) will lead to a slightly different model name and likelihood function (e.g., a binary logit, rather than a binary probit).

Maximizing the natural logarithm (ln) of the likelihood function is equivalent to maximizing the likelihood. The **log-likelihood** can be written as a sum, instead of a multiple: $LogLik = \sum_{i=1}^{n} y_i \ln[\Phi(x_i\beta)] + (1 - y_i)\ln[1 - \Phi(x_i\beta)]$.

Search routines (in Excel, R, SPSS, and other mathematical and statistical programs, described in Section 8.8) then iteratively seek the set of unknown parameters (the $\hat{\beta}_{mle}$ values) that maximize this function. Statistical programs also provide the matrix of (estimated) variance and covariance terms for these parameter estimates. Taking square roots of the variance terms delivers the standard error terms for each parameter, providing a set of t-statistics ($t = \frac{\hat{\beta}}{SE_{\hat{\beta}}}$).

Statistical Significance vs. Practical Significance

The t-statistics allow analysts to ascertain whether each estimate differs in a statistically significant way from a value of zero. That is to say, a t-statistic close to zero indicates that the explanatory variable X is highly unlikely to have any impact on the response variable Y (because its associated slope parameter is statistically close to zero). On the other hand, if the t-statistic is greater than 2, the parameter and explanatory variable associated with the t-statistic is considered highly **statistically significant**. However, with very large data sets (e.g., n>10,000), almost all parameter estimates will prove statistically significant. That does not mean they offer *practically* significant magnitudes of effect on response. A statistically significant variable's effect on the response variable can be very small, which is why researchers will often include measures of practical significance (e.g., elasticities or standardized coefficients) in addition to t-statistics.

Section 8.9 discusses how to determine statistical and practical significance.

Bayesian Estimation

Both OLS and MLE are classical estimation methods, using the frequentist approach that most traditional statistical testing is based on. They assume that each true parameter takes a single value. In contrast, **Bayesian methods** assume that each parameter follows a distribution, as suggested by Bayes' rule: $p(\beta|y) = \frac{p(y|\beta)\cdot p(\beta)}{p(\beta)}$, where $p(\beta|y)$ is the joint posterior distribution of unknown parameters, $p(y|\beta)$ is the likelihood function, or density of the response vector y conditional on the regressor and the parameters; $p(\beta)$ is the prior or starting distribution for parameters of interest, and $p(y)$ is the probability distribution of y and can be treated as a constant here. In other words, Bayesian statisticians do not view a set of data as a unique set of frequencies, but rather as a distribution (posterior) updated with each previous set of assigned probabilities (priors).

Bayesian techniques require more computing effort than most frequentist approaches (such as OLS, 2SLS, and MLE), but many Bayesian specifications have been coded in user-friendly software packages like BUGS (Bayesian Inference Using Gibbs Sampling), BACC (Bayesian Analysis, Computation and Communication), and James LeSage's MATLAB toolbox (a reservoir for spatial econometric algorithms as well as non-spatial models, available at http://www.spatial-econometrics.com/). To write their own codes, analysts often use MATLAB©, Gauss©, R, and Ox©.

Choice of Estimation Methods: Error Terms' Distributions

Selection of a parameter estimation method also depends on the distribution of model error terms. For example, OLS estimation presumes that all error terms have equal variance and are independent, regardless of the size of the X's and Y's. The assumption of constant variance is called **homoskedasticity**. If, instead, error terms rise and fall as a function of X's or other factors, it is useful to specify a model that allows for **heteroskedasticity** (non-constant variance in error terms).

Concept Example: Variance in Error Terms

> For example, in a crash, the travel speed of a vehicle impacts the uncertainty or variance in occupant injury severities. A vehicle speeding on an interstate highway has the potential for a greater range of injury severities than a vehicle driving slowly across a parking lot. *Heteroskesdastic* model specifications account for this increased variance in outcome, while a *homoskedastic* model specification assumes constant variance in injury severity, no matter the vehicle speed.

If a model neglects to reflect such heteroskedasticity, the model's estimates of the standard deviations of the errors (referred to as standard errors) are compromised but slope-parameter estimates remain unbiased and consistent. For example, the standard errors for the parameter estimates may be too low (which will bias the associated t-statistics high, making parameters and their explanatory variables appear more statistically significant than they actually are).

Analysts should carefully examine the statistical relationships between variables to check for existence of heteroskedasticity. For models involving a continuous response variable, one easy way to check for the presence of heteroskedasticity and/or serial correlation in error terms is to run an OLS model and then plot the residuals versus each predictor variable value and the response value, to see whether these estimates of errors show any patterns. The graphs in Figure 8.3 illustrate an example where the residuals are plotted against the predicted response values.

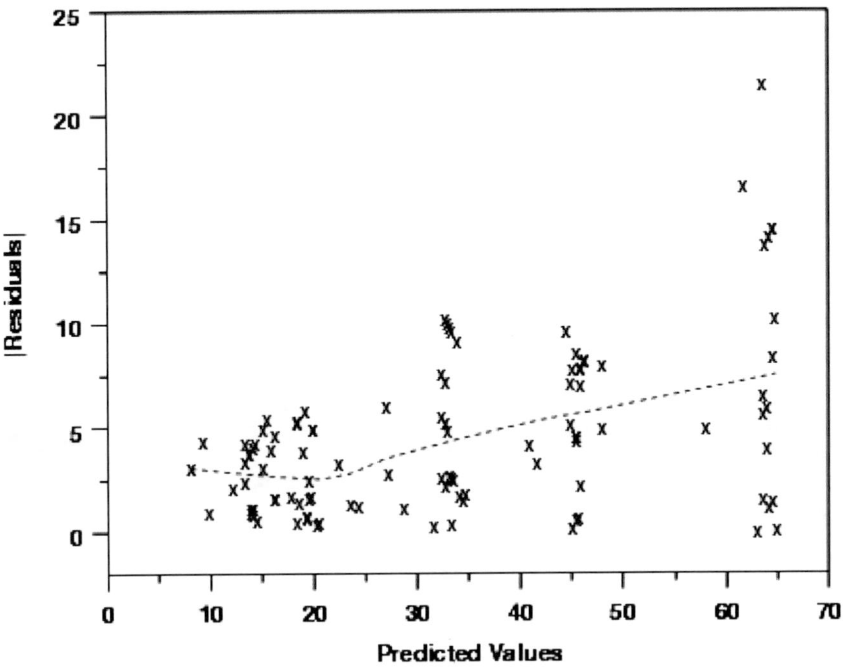

Figure 8.3: Heteroskedastic Residuals (Top) versus Homoskedastic Residuals (Bottom)

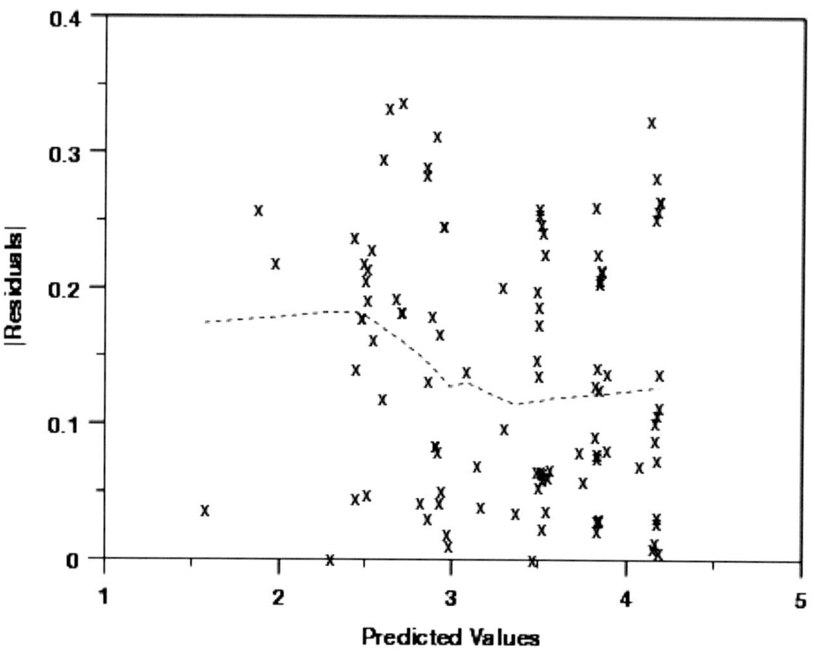

The first graph illustrates a situation where the residuals (or estimates of the error terms) increase as predicted values increase, signifying the presence of heteroskedasticity. The second graph shows a case where residuals are evenly distributed regardless of the predicted values, suggesting homoskedasticity.

To address heteroskedasticity, many analysts will create a model for the variance term, as a function of factors that influence such outcomes. They may weight each data record and apply **weighted-least squares** (WLS) regression, so that each data point's weight value is inversely proportional to its variance (e.g., use household sizes as the inverse weight terms when estimating household VMT values).

Choice of Estimation Methods: Error Term Correlation

Another situation of interest is when error terms rise or fall together, across correlated or complementary data records. For example, members of the same family may be present in the mode-choice data set. The presence of common family values (such as a social consciousness towards eco-friendly life styles or distaste for non-motorized/active modes) may create trends or correlation in intra-family choices, even after controlling for individuals' ages, genders, education levels, and other explanatory variables. **Autocorrelation** of their error terms emerges because the records have something in common, not controlled for by the explanatory variables. As noted above, respondents may be may be close in space (neighbors or residents of the same city, for example), or may reside in the same household. Data points may be close in time and/or observational units repeated multiple times in a single data set (e.g., trip after trip by the same person in a one-day travel diary). Finally, several response variables of interest may depend on similar unobserved factors (e.g., y_1= HH VMT, y_2= HH spending, and y_3= number of vehicles owned) and those associations will create correlation across

Resource

Many econometric textbooks tackle these issues, including these listed in this chapter's References: Greene (2008), Wooldridge (2002), and Washington et al. (2003).

many error terms.

Common sense can also help detect possible autocorrelation in a data set: Could some of the records potentially share unknown factors in common with other data records? (For example, drivers in the same household may trade off use of the household's one car, thus affecting each other's mode choice; and people in the same neighborhood presumably have more similar preferences and travel choices than those far away, thanks to the presence of similar unknowns, like access to hiking trails and popular restaurants.)

Exploiting such correlation via thoughtful model specification and analysis is important in achieving unbiased, consistent, and efficient estimates and predictions. Time-series analysis, discussed later in this chapter, takes care of many forms of serial autocorrelation. Spatial econometric techniques seek to tackle the two-way autocorrelations that exist over space. Fixed- and random-effects models address the repeat-observation issue common to many trip-based or shipment-based data sets. And systems of regression equations exploit shared information across different outcomes, as in Srinivasan and Kockelman's (2002) study of the sales impacts of highway bypasses on different industry sectors across Texas towns over time (with each industry's sales having its own equation). Figure 8.4 illustrates many of the methods that exist for analyzing data sets exhibiting correlation across error terms.

Common Modeling Mistakes

The issue of correlation also arises between unobserved factors (contained in the ε error term) and predictor variables (X). As briefly mentioned earlier in this chapter, most modelers assume this away (often improperly), because such correlations produce bias in estimates (as the control variable's coefficient attempts to absorb the effects of the missing variable(s)). Econometric texts explain how the **Hausman test** can be used to detect the presence of such correlation. In addition, theory and intuition help identify the correlated variables that may be concealed in the error terms.

Concept Example: Unobserved Factors

For example, a model of vehicles miles traveled (VMT) per household as a function of household size and income neglects the effects of neighborhood attributes. If larger and/or higher-income households are more likely to live in suburban or exurban neighborhoods, where driving distances are longer and alternative-mode options fewer, the model will attribute most of these location effects to the household size and income variables (both of which also have positive effects), resulting in β values that are probably biased high.

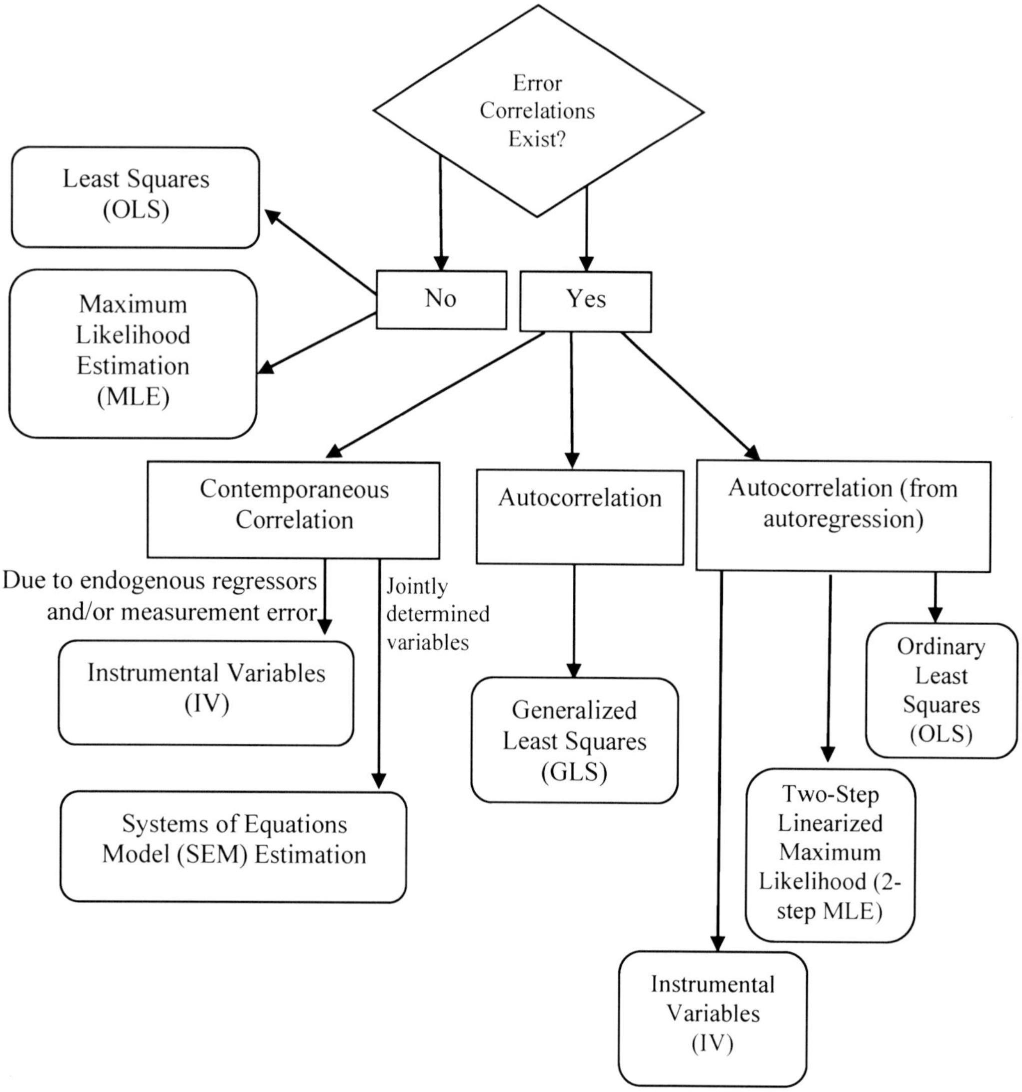

Figure 8.4: Regression Techniques and Tools for Cross-Sectional Data Sets

Related problems can occur when explanatory **variables are measured with error**. For example, if many less-educated people consistently overstate their educational attainment in interview surveys, and education has a positive effect on household VMT, the coefficient on education may be biased high. (The regression line steepens in this example, with low X values falsely being shown as higher.) Such errors also add noise to the data, producing lower t-statistics or less confidence in the parameter estimates of the misstated variable(s).

Use of **endogenous predictor variables** is another concern that many analysts ignore. For example, vehicle ownership is not truly exogenous when predicting a household's VMT, though modelers regularly imply this in their model specifications. In reality, those who expect to travel more tend to acquire more vehicles. Thus, the relationship is two-way, rather than strictly cause-and-effect from right side ($x\beta + \varepsilon$) to left side (Y). Multi-equation models, also called **simultaneous equations systems**, that allow for both VMT and vehicle ownership to be predicted simultaneously (with correlated error terms) make the most sense in such contexts (see, e.g., Mannering and Winston's 1985 model of vehicle ownership and use).

In contrast, **exogenous** variables are those whose values are essentially pre-set, upstream of the behavior being modeled. Gender and age generally qualify here. Vehicle ownership would not be an exogenous variable in predicting household income because the reverse causation typically holds (i.e., income is much more likely to be a determinant of vehicle ownership than ownership a determinant of income).

8.4 Nonlinear Parameter Estimation Methods

While common specifications assume model equations that are linear in unknown parameters (β's), many behaviors are characterized by nonlinear equations. One example is population growth, which is better modeled as exponential growth rather than linear. Another example is the Cobb-Douglas production function, popular for characterizing business production levels.

Concept Example: Cobb-Douglas Production Function

This function can describe production levels, such as a transit agency's seat-miles per year for Y, as a function of employees and buses. The following equation uses K for capital (in this case, buses) and L for labor (employee-hours worked):

$$Y = \beta_0 K^{\beta_1} L^{\beta_2} \varepsilon$$

This equation can be linearized by taking the natural log of both sides. In that case, OLS or WLS may apply (though the error terms are unlikely to be truly homoskedastic, before or after such a transformation).

Alternatively, we can pursue nonlinear least-squares techniques (available in commercial statistical software), MLE, or other estimation techniques.

Generalized Least Squares (GLS) Estimation

As noted earlier, OLS assumes error terms are homoskedastic and independent. **GLS** is a modification of OLS that relaxes those two assumptions by permitting various forms of heteroskedasticity and correlation (Washington et al. 2003). GLS is a general estimation method that includes WLS, feasible GLS, and more general LS techniques.

As noted earlier, WLS regression estimates β parameters from a weighted sum of the squared residuals (giving more weight to observations with less uncertainty/less variance in the in their error terms). Weights are simply inversely proportional to the variances (e.g., the number of shipments used to compute an average load-per-container term).

Feasible generalized least squares (FGLS) is an OLS-followed-by-WLS regression iterative technique where squared residuals from the initial OLS and later GLS estimation are used to improve parameter estimates in an iterative way:

1. Using a diagonal weight matrix constructed from the first-round OLS's squared residuals, conduct GLS.
2. Perform another GLS estimation using the squared residuals obtained from step 1's GLS estimator.
3. Repeat step 2 until the weight matrix converges.

By reflecting underlying correlations and/or heteroskedasticity, FGLS estimation can produce unbiased estimates of standard errors on parameter estimates and more efficient (lower variance) parameter estimates.

More general forms of GLS allow for serial and other forms of autocorrelation, with parameters in the covariance matrix to be estimated using OLS residuals or other techniques), as discussed below.

Instrumental Variables (IV) Estimation

IV estimation methods are most useful in cases where the setup shows correlation between explanatory variables and error terms, where errors occur in the measurement of independent variables, and/or where autocorrelation can be addressed using autoregression techniques.

In these various contexts, another variable or two (or more) are sought to serve as IVs, in order to estimate the β parameters without bias. IVs are needed that can predict changes in the problematic X variable(s) without being correlated with the error term. Figure 8.5 illustrates how the error term and explanatory variables interact within the equation, with and without the IV.

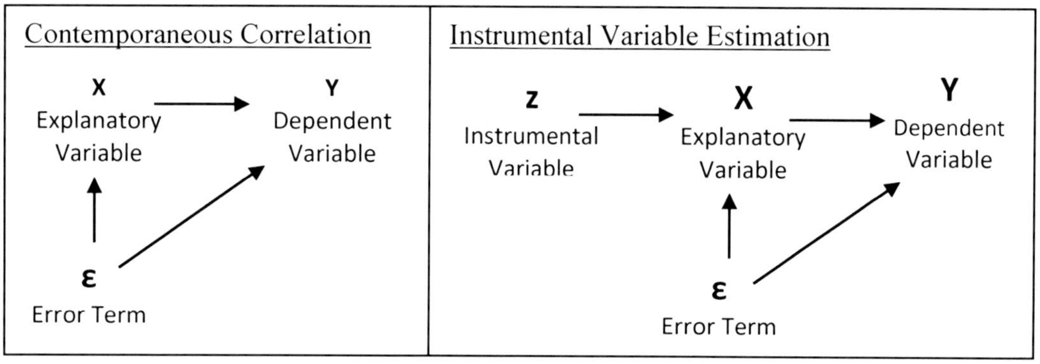

Figure 8.5: Visual of Explanatory and IV Relationship (Source: Cameron & Trivedi 2010, p. 178)

See this chapter's In-Depth Look section for a detailed discussion of SEMs.

Systems of Equations Models (SEMs)

SEM refers to a mathematical model consisting of a set of interrelated equations that must be solved simultaneously. Several different estimation methods are used to find the βs.

8.5 Panel Data Models and Parameter Estimation Methods

The standard linear regression model for panel data takes the following form, with "i" for the observational units (e.g., counties or cities) and "t" for the time period of the observation (e.g., years 1990 through 2010):

$$Y_{it} = \beta \ X_{it} + \alpha_i + \varepsilon_{it}$$

where α_i is an error-term component that is constant across time periods, and ε_{it} is the time-variant component of the error term. X_{it}does not contain a constant term. Parameter estimation methods for panel data mainly differ in how they address the two different error components.

As with models for cross-sectional data, regressions using panel data can have nonlinear forms. For example, Mannering and Winston (1985) specified a dynamic model of household vehicle ownership and use (VMT per year) based on data collected before, during, and after the 1979 energy crises. Choice of vehicle type served as a discrete response variable, while vehicle use was a continuous response with correlated error terms, resulting in a nonlinear panel data model. They found that households exhibited strong brand loyalties, including a distinct preference for U.S. manufacturers.

See the In-Depth Look section for a recent example of equations modeling used to illuminate ownership behaviors.

Dargay (2001) examined the effects of income on car purchases using pseudo-panel data constructed from 1970 to 1995 UK Family Expenditure Survey data. She concluded that that car ownership responds more strongly to rising family income than falling income and that income elasticity declines with rising car ownership.

Common methods for estimating parameters from panel data are presented in Figure 8.6 and are quite different from the methods used for cross-sectional data. Because panel data contains data for the same observed unit over different time periods, autocorrelation is expected over time points for each observational unit's error term (the ε_{it}). Model adjustments can remove such autocorrelation. For instance, the first differences method basically subtracts out each observation unit's common effects, with the drawback that time-invariant variables (like gender) are not included in the model.

Figure 8.6 divides linear regression methods for panel data by whether a contemporaneous correlation exists, or whether the individual-specific (and time-invariant) latent effect, α_i, is correlated with any of the observed covariates,X_{it}. Note that the error term ε_{it} is assumed to be uncorrelated with X_{it} and α_i.

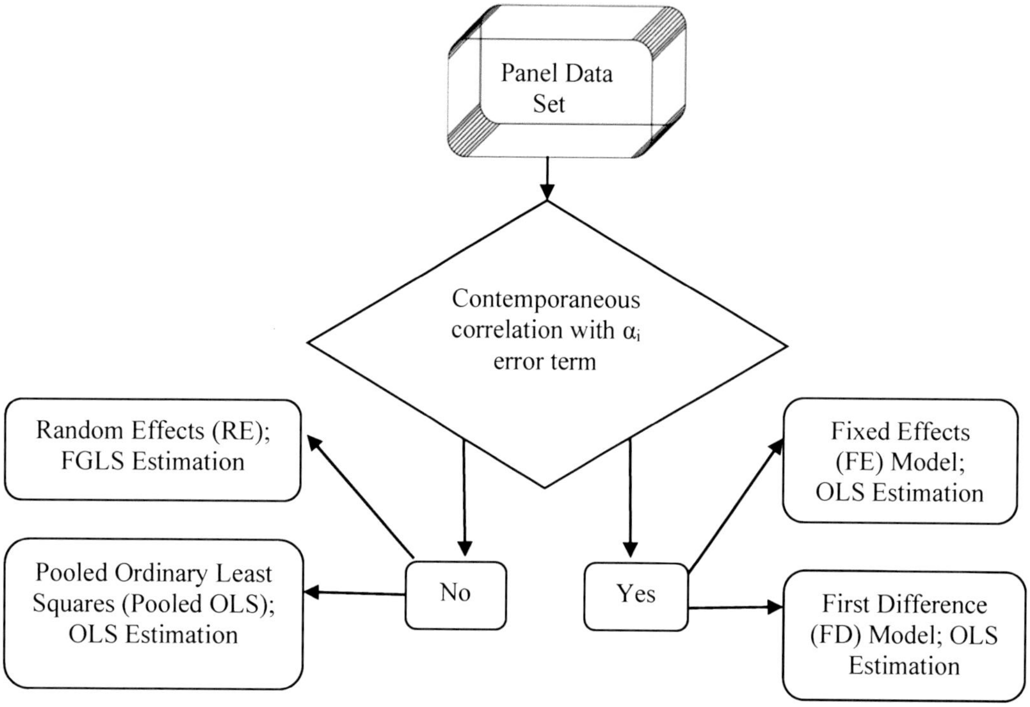

Figure 8.6: Model and Parameter Estimation Options for Panel Data Sets

As evident in Figure 8.6, OLS estimation for cross-sectional data can also be applied to panel data using pooled OLS, fixed effects, and first difference techniques. These and other methods are summarized below.

Pooled OLS

If we assume no liaisons between α_i and X_{it}, and between α_i and ε_{it} (i.e., α_i is simply a constant term), we can stack each observation unit's data records at different time periods together, not distinguishing between cross section and time series. In that sense, we can run a regression over all the stacked data points using OLS, which is called a **pooled OLS regression**. This type of regression is the easiest to run, but is also subject to many types of errors.

Fixed Effects (FE) Modeling

An **FE model** specification regresses differences in Y_{it}'s (using their average $\bar{Y}_i$) on differences in X_{it}'s (using their average $\bar{X}_i$). OLS is used and any time-invariant explanatory variables (e.g., gender, and distance to CBD) are not included in the model (so their β cannot be estimated but the term α_i picks up all such fixed effects).

$$Y_{it} - \bar{Y}_i = \beta(X_{it} - \bar{X}_i) + (\varepsilon_{it} - \bar{\varepsilon}_i)$$

The "t" indicates the time period, and the "i" the observation. The ε does not include the time-invariant error (α_i). OLS estimation is used to estimate the β parameters for the mean-differenced variables.

First Difference (FD) Modeling

FD is similar to FE in that differences of Y_{it} and X_{it} are used, and the coefficients of time-invariant explanatory variables cannot be estimated. OLS is used to estimate the β parameters using the following FD model form (Equation 8.2):

$$Y_{it} - Y_{i,t-1} = \beta\left(X_{it} - X_{i,t-1}\right) + \left(\varepsilon_{it} - \varepsilon_{i,t-1}\right) \tag{8.2}$$

The FD model works well when two time periods are collected for each observation unit. However, analysts should be aware of some caveats when applying this method to situations involving more than two time-periods (T≥3). For T=3, we could subtract period 1 from 2, and period 2 from 3. Nevertheless, this estimator is inefficient because we could also subtract period 1 from 3 (which is not used). In addition, with more than two time-periods the problem of serially correlated $(\varepsilon_{it} - \varepsilon_{i,t-1})$ arises, resulting in inefficient OLS estimates because one OLS assumption is independence across the error terms. Thus standard errors will be biased. To remedy this we can use GLS or Huber-White sandwich estimators.

Random Effects (RE) Modeling

Unlike the FE model, the RE model requires the assumption that α_i is not correlated with any explanatory variables. The advantage of this assumption is that coefficients of both time-invariant (e.g., distance to city center and gender) and time-variant (e.g., population size and income) explanatory variables can be estimated (Equation 8.3).

$$Y_{it} = \beta_0 + \beta_1 X_{it} + \alpha_i + \varepsilon_{it} \tag{8.3}$$

The RE model assumes that α_i and ε_{it} (both uncorrelated with any of the explanatory variables) are uncorrelated and normally distributed, $\alpha_i \sim N(0, \sigma_\alpha^2)$ and $\varepsilon_{it} \sim N(0, \sigma_\varepsilon^2)$. Let μ_{it} indicate their sum; in other words, $\mu_{it} = \alpha_i + \varepsilon_{it}$. The variance-covariance matrix for each observation's μ_{it} (with a total *T* time periods) is expressed as the variance-covariance matrix in Equation 8.4:

$$= \Omega_{T\times T} = \begin{bmatrix} \sigma_\alpha^2 + \sigma_\varepsilon^2 & \sigma_\alpha^2 & \cdots & \sigma_\alpha^2 \\ \sigma_\alpha^2 & \sigma_\alpha^2 + \sigma_\varepsilon^2 & \cdots & \sigma_\alpha^2 \\ \vdots & \vdots & \ddots & \vdots \\ \sigma_\alpha^2 & \cdots & \sigma_\alpha^2 & \sigma_\alpha^2 + \sigma_\varepsilon^2 \end{bmatrix} \tag{8.4}$$

Because the variance structure is unknown (thanks to unknown parameters σ_α^2 and σ_ε^2), FGLS is used to estimate the linear regression model shown in Equation 8.4.

FGLS works by first estimating $\hat{\sigma}_\varepsilon^2$ and $\hat{\sigma}_\alpha^2$ from the linear regression equation used by FE model (or the "within-effect" model) and from the "between-effect" model respectively, and then regress transformed Y_{it} on transformed X_{it} using OLS.

Resource

Greene's 2008 study (listed in References) provides details of the FGLS procedure.

8.6 Discrete Choice Models and Estimation Methods

Discrete choice models are used extensively in transportation to predict discrete outcomes such as a traveler's mode or destination choice, a business or household's location decision, and a property's land use status. Discrete choice regression models may use cross-sectional or panel data, but must lead to a discrete (as opposed to a continuous) response variable (*Y*). Table 8.6 presents the different logit modeling options. Categories of discrete choices are assigned an integer value.

MLE is the most commonly used method for estimating β parameters for any of the discrete choice models. These models are regularly used to predict the *probability* of a decision maker *i* choosing an alternative *j* based on unobserved (latent) utility values (U_{ij}). Possible explanatory variables include characteristics of the decision-maker and the attributes of the available choices. Because the model most likely does not include all factors affecting the decision, and observation units with the same *X*'s can make very different choices both in the long run and short run, there is latent heterogeneity in the model, via random utility error term,ε_{ij}, where *i* indexes observation unit and *j* indexes alternative (e.g., car vs. bus vs. walk/bike modes). Thus, discrete choice models are also called **random utility models** (RUM), thanks to the error term in the utility function to account for unobserved factors, $U_{ij} = \beta x_{ij} + \varepsilon_{ij}$.

Discrete Choice Model	Functional Form Foundation
Logit	**Logistic function**
Binary logit	Response variables are bi-variate (e.g., 0 or 1; Red vs. Blue)
Multinomial logit	Response involves more than 2 alternatives (e.g., car vs. bus vs. walk/bike)
Nested logit	Response involves more than 2 alternatives and some alternatives are more related than others (reflected by a nest structure)
Probit	**Cumulative normal function**
Binary probit	Bi-variate response
Multinomial probit	Response involves more than 2 outcomes

Table 8.6: Discrete Choice Models

Because multinomial and nested logit models are used extensively in transportation, the remainder of the discussion focuses on those important specifications and estimations.

Figure 8.7 presents the shape of the logistic function for the probability of choosing one alternative versus another (A vs. B). The sigmoidal curve is $\rho = \frac{e^u}{1+e^u}$ and shows mixed choice in the middle, dominance of alternative A in the high-up region, and another outcome (alternative B) in the lower region. The y-axis shows the probability of choosing the second alternative, B, where both alternatives are competitive because their different costs or levels of attractiveness are similar.

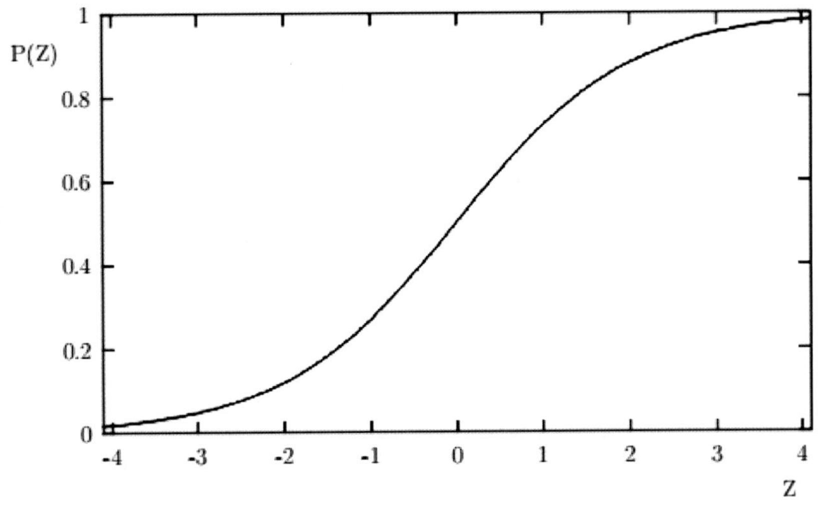

Figure 8.7: The Logistic Curve (Source: Cramer 2002)

The logistic function relates the probability of selecting one of several discrete alternatives as a function of explanatory variables. From the P(x) equation for a logistic function, algebraic rearrangement of the equation creates a functional form similar to linear regression for the transformed shares of the discrete alternatives (Figure 8.8). The error term in the systematic utility is assumed to follow a Gumbel distribution, which leads to a logit function with no error term thanks to mathematical properties associated with Gumbel distribution.

The logistic function

$$p(x) = \frac{e^{\beta_0 + \beta_i x_i \ldots}}{1 + e^{\beta_0 + \beta_i x_i \ldots}}$$

↓

The logistic function rearranged

$$\frac{p(x)}{1 - p(x)} = e^{\beta_0 + \beta_i x_i \ldots}$$

↓

Natural log applied to both sides of equation

$$ln\left(\frac{p(x)}{1 - p(x)}\right) = ln\left(e^{\beta_0 + \beta_i x_i \ldots}\right)$$

↓

The logistic regression equation

$$ln\left(\frac{p(x)}{1 - p(x)}\right) = \beta_0 + \beta_i x_i \ldots$$

Figure 8.8: Linearizing the Logistic Function

Resource

Koppelman and Bhat's 2006 USDOT instruction manual on multinomial and nested logit models explores those models with transportation examples.

Mixed logit models are a highly flexible extension of the conventional logit model and now enjoy wide application thanks to advances in computational techniques. This model specification permits parameters in the observed utility term to be a function of other parameters. For example, the probability for person n choosing alternative i is expressed as

$$P_{ni} = \int \frac{e^{x_{ni}\beta}}{\sum_j e^{x_{nj}\beta}} \cdot f(\beta|\theta)d\beta$$

where x_{ni} is the observed explanatory variable for person n, including both alternative specific variables and person specific variables with β the corresponding parameter coefficients. The parameter of interest under mixed logit case is the vector parameter θ rather than β. Mixed logit model circumvents disturbing issues such as independence from irrelevant alternatives (IIA), as the odds for person n choosing alternative i versus s is dependent not only on factors associated with these two alternatives but also other alternatives. We may also consider a mixed logit model if the population contains M segments, in which the probability of person n choosing alternative i is expressed as $P_{ni} = \sum_{m=1}^{M} \frac{e^{x_{ni}\beta_m}}{\sum_j e^{x_{nj}\beta_m}} \cdot f(\beta_m)$ where $f(\beta_m)$ is the multivariate density for the parameter vector β_m.

Resource

Mixed logit models are explored in Dr. Kenneth Train's online book, *Discrete Choice Methods for Simulation*.

The Bayesian hierarchical model can also be used to tackle discrete choice problems. In a hierarchical model scheme, observable outcomes are modeled conditionally on a set of parameter, which is assumed to have a probability distribution with further parameters, also known as *hyperparameters*. In practice, the non-hierarchical model tends to fit large datasets poorly when the number of parameters is insufficient, but overfit the datasets with a large number of parameters. Thanks to the added structure in hierarchical model, such a dilemma can be avoided.

Examples of how to assign the values for the different forms of logit models are provided in Figure 8.9.

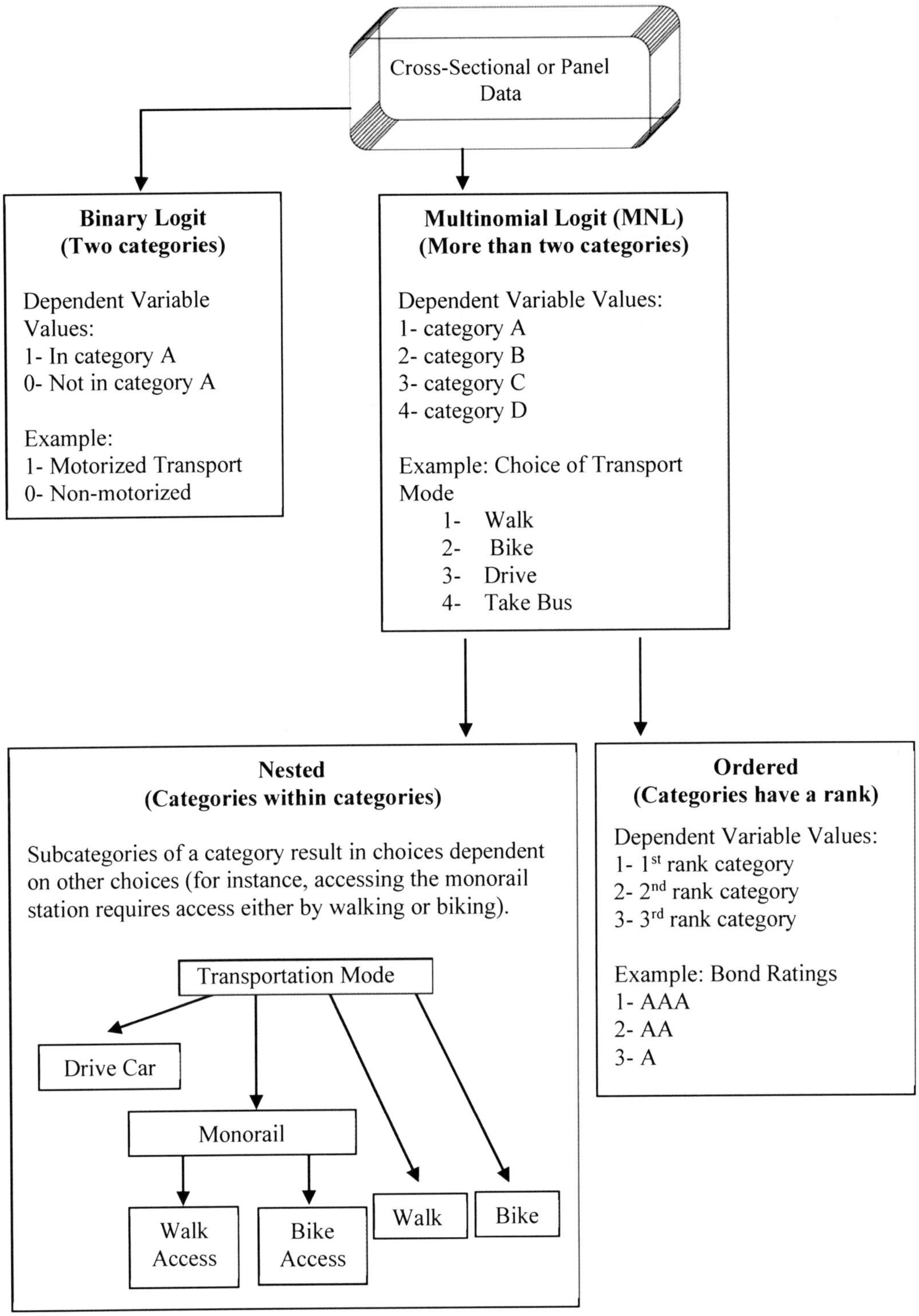

Figure 8.9: Logit Modeling Options

8.7 Time-Series Modeling

Time-series data can be used to model the impact of historical data on economic and other variables (such as population, GDP, and traffic counts) and forecast future values of those variables.

Time-series modeling is similar to cross-sectional data modeling in that the time series is like a sample, but instead of being for a fixed population, it is for a random process evolving over time (a stochastic process).

Several modeling options are available for time-series data (Figure 8.10). **Univariate models** are designed for forecasting one variable, whereas **multivariate models** can forecast for more than one variable simultaneously. Most β parameter estimates are made using MLE, so whereas the other sections of this chapter focused on the different parameter estimation methods, this section on time-series modeling gives an overview of the different functional forms of time-series models.

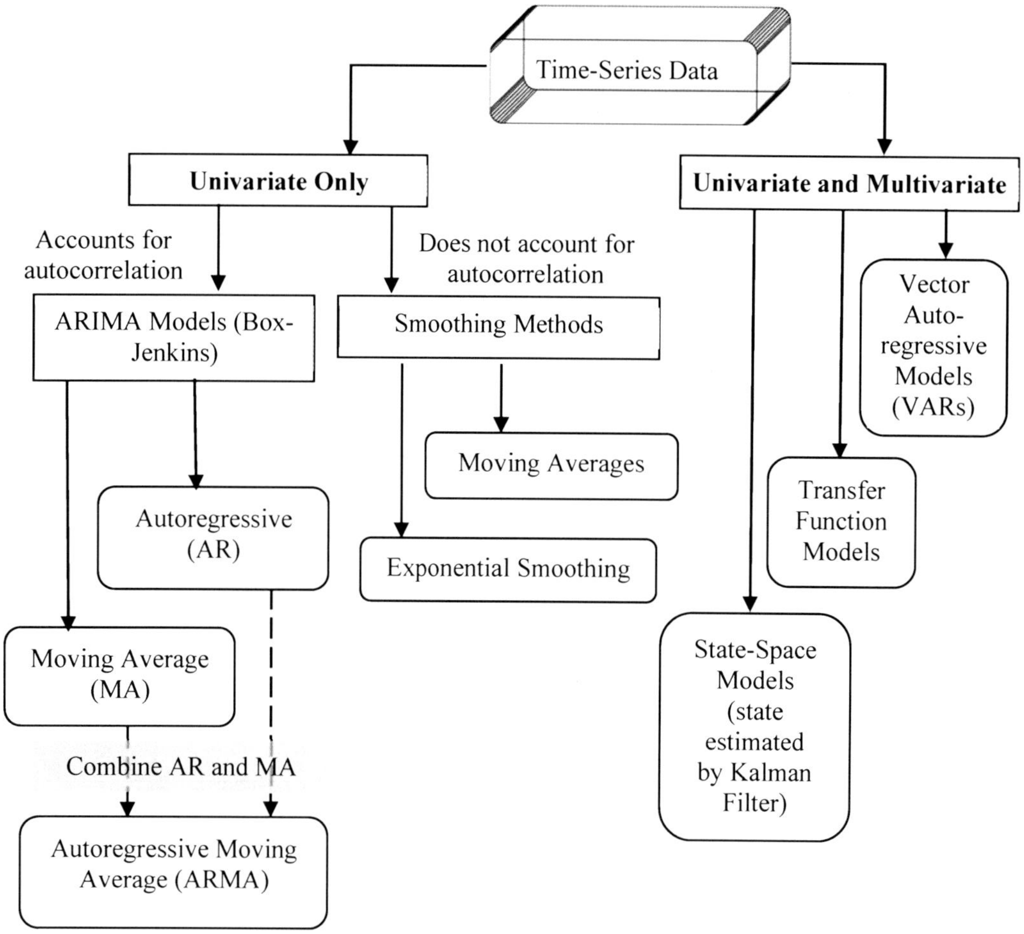

Figure 8.10: Time-Series Modeling Options

Single Response Variable Settings

ARIMA Models (Box-Jenkins)

ARIMA stands for **autoregressive integrated moving average** and is a base model for the autoregressive, moving average, autoregressive moving average, and ARIMA models that all account for autocorrelation but differ in functional form.

Autoregressive (AR) models take the following form that assumes the value of Y at time *t* depends on the value of *Y* from a previous time period. The *p* indicates the number of time periods used in the AR functional form (Equation 8.5). The β parameters are generally estimated by MLE.

$$Y_t = \beta_0 + \beta_1 Y_{t-1} + \beta_2 Y_{t-2} + \cdots + \beta_p Y_{t-p} + \varepsilon_t \qquad (8.5)$$

AR models are univariate in that they use a single variable observed over time. The dependent variable is the value of the variable in the future time period of interest, and all of the explanatory variables are the previous values of the dependent variable.

Moving average (MA) models differ from AR in that MA models use the errors from previous time periods as explanatory variables instead of the values of *Y* from previous time periods, with the resulting functional form (Equation 8. 6) and β parameter estimated by MLE. The subscript "*q*" denotes the total number of time periods included.

$$Y_t = \beta_0 + \varepsilon_t - \beta_1 \varepsilon_{t-1} - \beta_2 \varepsilon_{t-2} - \cdots - \beta_p \varepsilon_{t-q} \qquad (8.6)$$

The **ARMA (autoregressive moving average) model** simply combines the right-hand sides of both the AR and MA equations into one long equation to solve for Y_t. Either MLE or one of two LS methods—conditional or unconditional—can be used to estimate the ARMA parameters.

AR, MA, and ARMA models all assume the time series are stationary, meaning Y_t has constant mean, variance, and covariance with other Y_t for other time periods (because it is time invariant, the mean, variance, and covariance are the same regardless of the time period). However, as might be expected from dynamic trends such as traffic counts, population, spending, or sales over time, the time series can trend upward or downward, resulting in a non-constant mean and non-constant (co)variance for the *Y* over time.

To maintain the condition of time invariance of the mean and (co)variance of *Y*, the time-series data can be differenced. *Differencing* involves subtracting the value of Y_t observation from value of Y_{t-k} observation, where *k* could be 1 for a linear trend, 2 for a quadratic trend, or more for seasonal trends. The general functional form of the ARIMA model consists of Y*, the difference between Y_t and Y_{t-k} observations:

$$Y_t = \beta_0 + \beta_1 Y_{t-1}^* + \beta_2 Y_{t-2}^* + \cdots + \beta_p Y_{t-p}^* + \varepsilon_t$$

Smoothing Methods

Smoothing methods are not really econometric approaches because these forecasting methods are concerned only with one variable, do not consider explanatory variables, and do not involve estimation of parameters using econometric techniques. Instead, they are essentially extrapolation methods, categorized into two types. **Simple moving averages** assigns equal weights to past values of the time series to determine the moving average. **Exponential smoothing** assigns the most recent time-series data points with larger weights relative to the past.

Multivariate Response Variable Settings

The multivariate time-series models are complex. Their inclusion in the Figure 8.10 flowchart is to show that forecasting for more than one variable at a time is possible. In addition, it should be noted that econometricians developed multivariate Box-Jenkins methods.

Resource

See Washington at al. (2003), Gujarati (1995), and Kennedy (1998) for more on multivariate models.

8.8 Computer Programs for Econometric Applications

Numerous software programscan be used to estimate some or most econometric models. For example, TransCAD is the first and only GIS software designed specifically for use in the field of transportation; it stores, displays, manages, and analyzes transportation data. It is an especially powerful tool for planning and travel demand modeling. It's modeling capabilities include trip generation, trip distribution, mode split modeling, traffic assignment, and other network processes (see http://www.caliper.com/TCTravelDemand.htm.) For example, the mode choice module can be used to forecast the shares of trips made by mode via multinomial and nested logit models.

MS Excel's Analysis Toolpak is also a powerful tool for some basic analysis and modeling procedures. It includes ANOVA (analysis of variance), Pearson correlation analysis, two-sample F-Tests to compare two population variances, histograms, moving average forecasts, and OLS regression.

See Chapter 6 for specifics on using Excel's functions.

Other useful statistical software packages are R, SPSS, and SAS. R is anopen-source software for statistical computing and graphics (http://www.r-project.org/). It compiles and runs on a wide variety of UNIX platforms, Windows, and Mac OS. It is also known for better performance for optimization problems. SPSS (Statistical Package for the Social Sciences), as its name suggests, is among the most widely used software packages for statistical analysis in social science. It's base capabilities include descriptive statistics, bivariate statistics, linear regression, factor analysis, and cluster analysis. One competing software package is SAS (Statistical Analysis System). In addition to base modeling capabilities (e.g., descriptive statistics), SAS offers a number of components with separate licenses for more sophisticated analysis such as econometrics, time series, and data mining.

8.9 Statistical Significance and Prediction

After selecting a model (the functional form) and estimating the parameters, the next step is to determine if the parameters, and their associated variables, are statistically significant. This section briefly describes how analysts determine statistical significance.

The overall goal of econometric regression analysis is to extract from the data the associations between the response and explanatory variables to determine which explanatory variables influence the response variable. Explanatory variables that do not contribute to determining the response variable should not be a part of the model. Therefore, any estimated parameter that lies close to zero ($\beta_i = 0$) for an explanatory variable indicates that the variable is not significant in the prediction or forecast of the response variable.

Results of parameter estimations indicate statistical significance by reporting either or both of the estimate's p-values or t-statistics.

To determine the significance with p-values, researchers compare the p-value with a significance level set by the researcher, usually 0.01, 0.05, or 0.10. Those levels are selected based on how confident the analyst wants to be in their determination of significance. For instance, to be 99% confident (in a statistical sense), the analyst would select a significance level of 0.01 (1.00 – 0.99 = 0.01) and 0.05 for 95% confidence (1.0 – 0.95).

A p-value less than the analyst-specified significance level means the estimated parameter is far enough away from the hypothesized mean of 0 that the analyst can reject the hypothesis that the parameter is not significant ($\beta_i = 0$). In other words, a small p-value means the variable is statistically significant and should be included in the model. For instance, a p-value less than 0.01 would mean the estimated β_i parameter is more than two standard deviations away from the hypothesized value of 0. The p-value is an estimate of the probability that a different data sample could result in a β parameter estimate even further than the hypothesis that $\beta_i = 0$.

An alternative approach is to calculate a t-statistic for the standard hypothesis that the parameter estimate is equal to zero ($\beta_i = 0$). The t-statistic divides the estimated parameter by the standard error of the parameter (a regression model output):

$$t - statistic = \frac{\hat{\beta}_i - 0}{SE_{\hat{\beta}_i}}$$

If the t-statistic exceeds an analyst-specified t-critical value (also selected based on the level of confidence), the β_i is sufficiently far from 0 that it is considered statistically significant.

In summary, if the p-value is less than an analyst-specified significance level, or the t-statistic is greater than the analyst-specified t-critical value, the hypothesis

Resource

Any statistics textbook will explain the theory and mechanics of determining statistical significance (e.g., Devore's *Probability and Statistics for Engineering and the Sciences*, used in sophomore-level college statistics courses).

that the estimated parameter is equal to zero is rejected, and the parameter is considered statistically significant and retained in the regression model.

Practical significance is also an important factor to consider when interpreting parameter estimates. In very large samples (e.g., n>10,000), almost all variables will be statistically significant. Practical significance measures can help decipher which variables are most influential among statistically significant variables. Practical significance has to do with whether changes in an explanatory variable will have a noticeable effect on the response variable, while statistical significance addresses the question whether the estimated parameters are statistically different from zero.

Chapter 10's case studies provide examples of statistical-significance analysis.

Elasticities are often used to evaluate practical significance. These are expressed as $E_{y_i,x_{ik}} = \frac{\partial y_i}{\partial x_{ik}} \cdot \frac{x_{ik}}{y_i}$, where y_i is the response variable for observation i, x_{ik} is the kth covariate, ∂ represents partial derivative, and E_{y,x_k} measures the percentage change in the response per 1% change in the kth covariate for observation i. Elasticities offer a quick way to judge the overall and relative practical significance of a variable. Some parameter estimates can appear small in an absolute sense due to the unit of measure (such as when annual income is measured in \$1 units). But when calculated in percentages, a 1% increase in household income can have a significant impact on VMT. An elasticity estimate will showcase this when the small absolute β may be misleading.

We can compute an averaged elasticity by taking the average over all the $E_{y_i,x_{ik}}$'s across the sample records, or substituting the averaged values of x_{ik} and y_i into the equation shown previously (i.e., $E_{\bar{y},\bar{x}_k} = \frac{\partial \bar{y}}{\partial \bar{x}_k} \cdot \frac{\bar{x}_k}{\bar{y}}$). Such elasticity equations tend to hold for continuous X's. For indicator variable (e.g., x = 1 if female and x = 0 if male), analysts often simply substitute the two values (0 and 1) separately into the regression equation and compute the difference between the two predicted y values (because a person cannot be 1% more female, but rather simply male or female). The result measures changes in the response variable when x_k changes from 0 to 1 (or from 1 to 0).

Other methods are available for quantifying impact of covariates. Analysts can, for example, simply examine the effect of one standard deviation change in all covariates. In the case of Poisson rate or negative binomial models with exponential link functions (to ensure non-negative rates, like crash rates or traffic counts per hour), **incidence rate ratios** (**IRR**s) are often tabulated. IRRs are the exponential function of each parameter ($e^{\beta i}$), and convey the ratio of rates (rate of incidence) after and before the associated covariate (x_i) is increased by one unit. Econometrics boasts a tremendous toolbox for evaluating regression results, with many approaches depending on the specification used.

8.10 Summary

The field of econometrics contains a wide range of models and estimation methods and is actively refining and expanding the econometric tool set. Therefore, this chapter did not aim to provide encyclopedic coverage of econometric models, but instead sought to

- Place into context the role of econometrics (specifically regression) in answering transportation economic problems.
- Describe in general the steps taken for econometric analysis (e.g., selecting data, models, and parameter estimation methods and testing for statistical significance).
- Differentiate between the types of data sets and note how the type affects the choice of model and estimation method.
- Explain why multiple methods are used for estimating parameters (e.g., error correlations and variances) and provide guidance on finding the model and estimation method that works best for the data and the question at hand.

The diversity of the models and parameter estimation methods yields some crossover of methods and applications. For instance, MLE is used to estimate parameters for logit and linear cross-sectional regression models. Discrete choice models use cross-sectional data as well, and in more advanced models, panel data.

Econometrics is used extensively in transportation economics, and econometric tools are instrumental in anticipating travel demands and crash counts, the analysis of costs and benefits (Chapter 1 and 6), economic impacts (Chapter 7), and pricing (Chapter 2). For more details on the models and parameter estimation methods presented, see the econometric textbooks listed in Section 8.1 or consult this chapter's References.

8.11 An In-Depth Look

Systems of Simultaneous Equations

Simultaneous equations systems (**SES**) host a set of linked equations, with one or more explanatory variables determined (as response variables) in another equation or equations, and/or with shared parameters across equations (such as in systems of demand equations for different goods and services, all derived from the same underlying utility or preference function). All SES have contemporaneous correlation or autocorrelation. A typical economic example of contemporaneous correlation within simultaneous equations consists of a consumption function (C), which depends on income X, and an income equation, that equals consumption plus investment I (Kennedy 1998, p. 157):

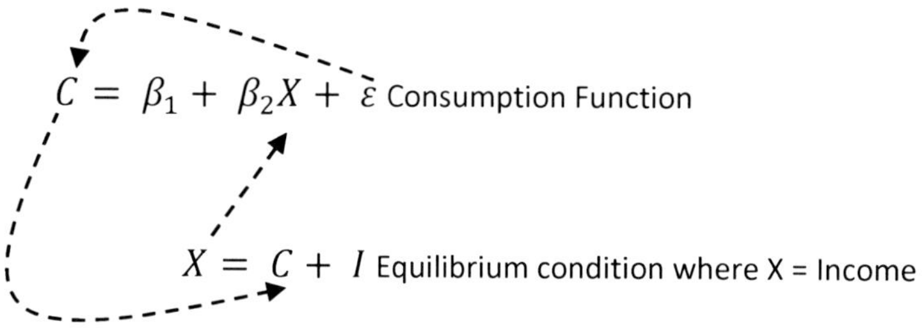

If the error term (ε) in the consumption function increases, due to some unobserved effects (such as optimism or government stimulus), then consumption (C) increases. The increased consumption causes income to increase in the equilibrium condition and simultaneously in the consumption function. Because the X in the consumption function is correlated with the error term via the equilibrium condition, X and the error term are contemporaneously correlated. This is an example of a single-equation with endogenous right-hand variables.

Figure 8.11 summarizes the parameter estimation options for systems of equations. Brief discussions of each method follow.

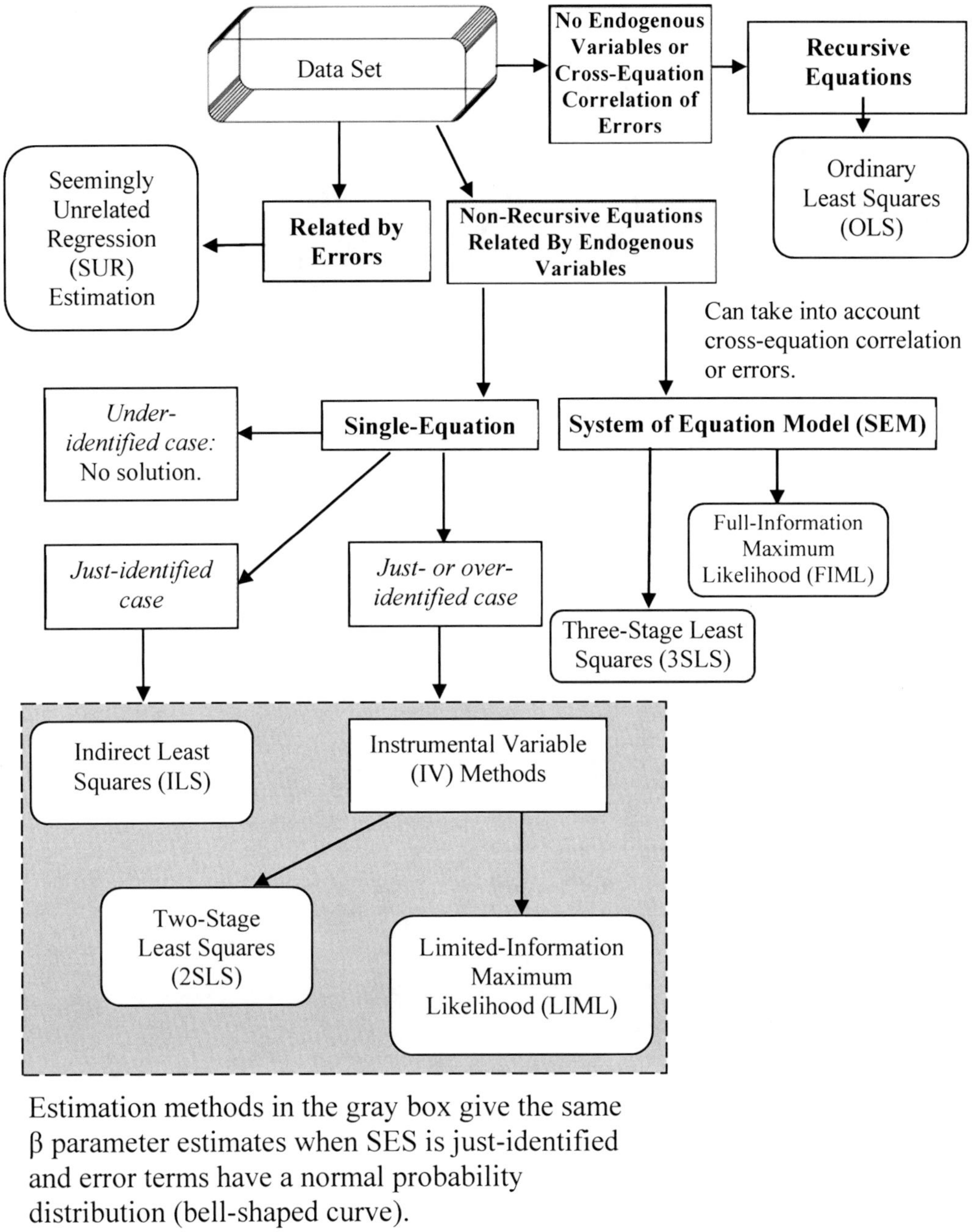

Figure 8.11: Parameter Estimation Methods for Systems of Equations

Recursive and Non-Recursive SES

If the SES consists of a series of equations related in one direction, meaning "the equations can be ordered such that the first endogenous variable is determined only by exogenous variables, the second determined only by the first endogenous variable and exogenous variables, the third by only the first two endogenous variables and exogenous variables, and so forth" (Kennedy 1998, p. 169), then the SES is **recursive** and OLS may be used to estimate the β parameters.

Non-recursive SES may be estimated either by the single-equation estimation methods that estimate β parameters for each equation separately, or by the systems—equation estimation methods that take into account all the restrictions imposed by the interrelated equations to estimate the β parameters simultaneously and the cross-equation contemporaneous correlation of the errors.

Seemingly Unrelated Regression (SUR)

Some SEMs are not interrelated by sharing a variable, but are related by correlation between each equation's error terms only. The autocorrelation with the errors is handled with an estimation method called **seemingly unrelated regression (SUR)**, used specifically for these related-by-errors-only equations. The following is an example of a pair of SUR equations:

$$\left.\begin{aligned} Z &= \beta_1 + \beta_2 X + \varepsilon \\ Y &= C + \beta_3 T + \mu \end{aligned}\right\}$$

Seemingly unrelated: The equations for Z and Y are related by correlation between errors (ε and μ) only.

SUR combines the set of equations related only by correlation among error terms into essentially one large equation. Correlation(s) among errors are estimated (from the estimation of the seemingly unrelated equations by OLS), and β parameters are estimated using FGLS estimation method described earlier.

Concept Example: SUR

In the above equation, *Z* represents the truck VMT on a highway and *Y* represents the passenger car VMT. The explanatory variables used to predict *Z* and *Y* are different. Explanatory variable *X* represents the travel time and T represents the travel cost. The error terms, however, are correlated because variables missing from the model—such as preferences for a particular route (because of scenery, for example)—affect both *Z* and *Y*.

Single-Equation Estimation Methods

Single-equation estimation methods are further divided into whether the systems of equations are under-, just- or over-identified. **Just-identified SES** results in unique β parameter estimates because for each parameter there is only one equation to solve. **Over-identified SES** allows for estimation of parameters, but they are not unique because two or more equations provide two or more different estimates for a parameter. **Under-identified SES** does not allow for parameter estimation because the reduced-form equation does not result in an equation to solve for each parameter. The use of reduced-form models can potentially result in too little or too much information, preventing parameter estimation.

Indirect Least Squares (ILS)

The **ILS method** first estimates parameters using OLS for the reduced-form equations, each consisting of one endogenous variable as the response variable and all of the exogenous variables in the system of equations as explanatory variables.

A reduced-form equation is created for each endogenous variable. The resulting reduced-form parameter estimates are then used to estimate the β parameters of primary interest (hence the term "indirect"). ILS is used for just-identified SEMs, because over-identified SEMs can be problematic for ILS.

Instrumental Variable (IV) Methods

Figure 8.11 showed two models subtypes for **IV methods**.

Two-Stage Least Squares (2SLS)

2SLS estimates the β parameters in two stages. In the first stage, the endogenous variables are regressed (made the response variable) only on the exogenous variables as explanatory variables, as in the first step of ILS, to form a reduced-form regression equation. However, unlike ILS, the regression equation is then used to provide estimated values of the endogenous variables. The new list of values is assigned to a new variable that serves as the IV in the next stage. During the second stage, the β parameters are estimated, using the new variable created from the first stage as an IV for the endogenous variable in the equation.

Limited-Information Maximum Likelihood (LIML)

As mentioned before, estimation methods fall under two general categories of LS or maximum likelihood. **LIML** uses MLE estimation to find the reduced-form parameter estimates. The reduced-form regression equation is then used to find the estimated values of the endogenous variable. As with 2SLS, the new variable is then used as an IV for the endogenous variable in the original equation to find the β parameter estimates.

Systems of Equation Models (SEM)

As described in Section 8.4, **SEM** refers to a mathematical model consisting of a set of interrelated equations that must be solved simultaneously. Figure 8.11 showed two model subtypes.

The large economic models described in Chapter 7 consist of one or more SEMs.

Three-Stage Least Squares (3SLS)

3SLS starts off as in 2SLS, but has an added step. This additional step allows analysts to estimate β parameters by considering all the equations simultaneously to account for the interrelation of the equations and the autocorrelation. The first step obtains parameter estimates for the reduced-form equations (discussed under 2SLS). The second step determines the correlation between errors in the equations, and the third step uses these estimated correlations to find the β parameters using FGLS as discussed earlier in this chapter.

Full-Information Maximum Likelihood (FIML)

FIML is just like LIML, except that correlation across equations is also accounted for.

Transportation Application: Discrete Choice Modeling

Koppelman and Bhat (2006) provided a simple example of a nested logit work-trip mode choice model using San Francisco Bay Area Travel Survey data. Four nested structures were considered: **motorized (M) alternatives**, including Drive Alone (DA), Shared Ride 2 (SR2), Shared Ride 3+ (SR3+), and Transit (TRN); **private automobile (P) alternatives** containing DA, SR2, and SR3+; **shared ride (S) alternative**; and **non-motorized (NM) alternatives** comprising Bike (BIK) and Walk (WLK), as shown in Figure 8.12.

The logsum parameter (also known as *nesting coefficient* or *dissimilarity parameter*) is greater than 1 for Model P (i.e., the model with private automobile modes clustered in the same nest), which implies that the assumed nest structure is not supported by the data and thus inappropriate. The NM model (for non-motorized nest) cannot reject the multinomial logit model at any reasonable confidence level, as reflected by the small Chi-square test statistics in Table 8.6. By contrast, Models M and S successfully reject the multinomial logit model in a statistically significant way. In other words, the trips recorded tend to exhibit strong correlation among motorized modes (including DA, SR2, SR3+, and TRN) and among shared-ride modes (i.e., SR2 and SR3+).

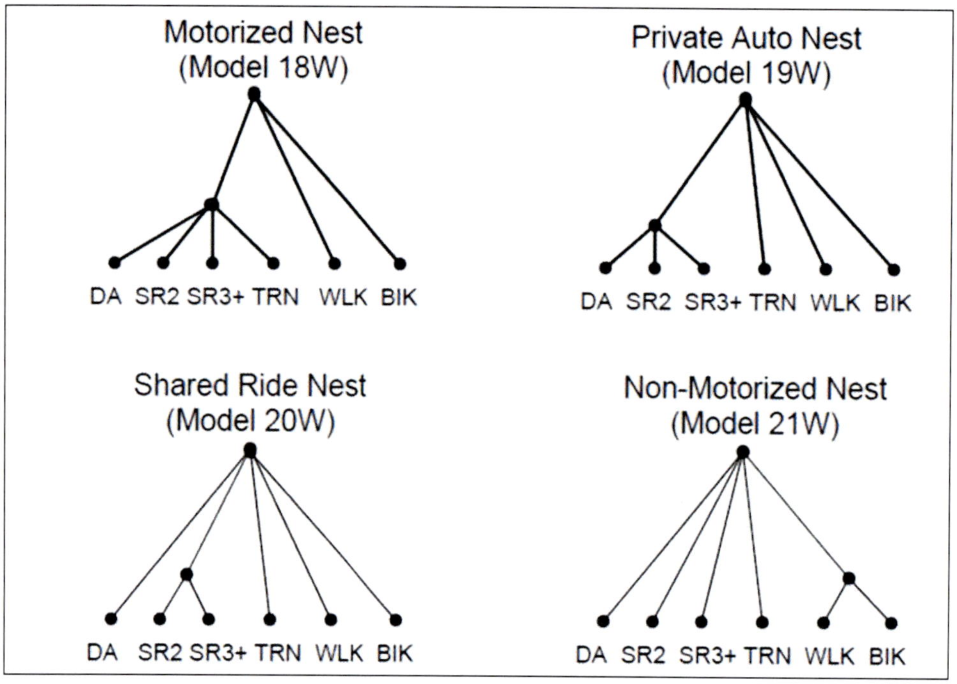

Figure 8.12: Single Nest Models (Source: Koppelman and Bhat 2006)

Variable	Model 17W	Model 18W	Model 19W	Model 20W	Model 21W
Nest	None	Motorized (M)	Private Auto (P)	Shared Ride (S)	Non-Motorized (NM)
Travel Cost by Income (1990 cents per 1000 1990 $'s)	-0.0524 (-5.0)	-0.0388 (-5.1)	-0.0607 (-9.3)	-0.0455 (-4.6)	-0.0519 (-6.3)
Travel Time (minutes)					
Motorized Modes Only	-0.0202 (-5.3)	-0.0146 (-4.8)	-0.0202 (-5.6)	-0.0206 (-5.7)	-0.0199 (-5.2)
Non-Motorized Modes Only	-0.0454 (-7.9)	-0.0462 (-8.4)	-0.0462 (-8.2)	-0.0452 (-7.7)	-0.0454 (-8.6)
OVT by Distance (mi.) Motorized Modes	-0.133 (-6.8)	-0.112 (-5.9)	-0.136 (-7.1)	-0.134 (-6.9)	-0.135 (-8.2)
Household Income (1,000's of 1990 dollars)					
All Private Vehicle Modes	0.00	0.00	0.00	0.00	0.00
Transit	-0.0053 (-2.7)	-0.0039 (-3.5)	-0.0045 (-2.4)	-0.0054 (-2.8)	-0.0053 (-2.8)
Bike	-0.0086 (-1.7)	-0.0095 (-2.1)	-0.0075 (-1.7)	-0.0089 (-2.0)	-0.0092 (-2.8)
Walk	-0.0060 (-1.9)	-0.0066 (-2.3)	-0.0050 (-1.7)	-0.0062 (-2.1)	-0.0056 (-2.1)
Vehicles per Worker					
Drive Alone (base)					
Shared Ride (any)	-0.317 (-4.8)	-0.225 (-4.7)	-0.511 (-5.8)	-0.315 (-5.2)	-0.317 (-5.4)
Transit	-0.946 (-8.0)	-0.704 (-5.7)	-0.872 (-8.0)	-0.938 (-9.0)	-0.947 (-9.1)
Bike	-0.702 (-2.7)	-0.742 (-3.4)	-0.611 (-2.8)	-0.703 (-3.2)	-0.693 (-4.1)
Walk	-0.722 (-4.3)	-0.772 (-5.4)	-0.614 (-4.2)	-0.724 (-5.0)	-0.714 (-5.3)
Dummy Variable for Destination in CBD					
Drive Alone (base)	0.00	0.00	0.00	0.00	0.00
Shared Ride 2	0.260 (2.1)	0.192 (2.8)	0.398 (3.8)	0.396 (3.6)	0.260 (2.1)
Shared Ride 3+	1.07 (5.6)	0.778 (4.9)	1.59 (6.5)	0.641 (4.6)	1.07 (5.5)
Transit	1.31 (7.9)	0.918 (4.9)	1.37 (7.8)	1.32 (7.7)	1.31 (7.3)
Bike	0.489 (1.4)	0.478 (1.3)	0.524 (1.5)	0.501 (1.4)	0.414 (1.5)
Walk	0.102 (0.4)	0.113 (0.5)	0.117 (0.5)	0.114 (0.5)	0.104 (0.5)
Empl. Density - Work Zone (employees / square mile)					
Drive Alone (base)	0.00	0.00	0.00	0.00	0.00
Shared Ride 2	0.0016 (4.0)	0.0011 (4.3)	0.0025 (8.1)	0.0019 (5.5)	0.0016 (4.3)
Shared Ride 3+	0.0023 (5.0)	0.0016 (4.6)	0.0036 (7.7)	0.0019 (5.5)	0.0023 (5.0)
Transit	0.0031 (8.7)	0.0022 (5.3)	0.0035 (10.2)	0.0032 (9.0)	0.0031 (9.1)
Bike	0.0019 (1.6)	0.0014 (1.2)	0.0023 (1.9)	0.0020 (1.6)	0.0022 (2.4)
Walk	0.0029 (3.9)	0.0023 (3.0)	0.0033 (4.5)	0.0030 (3.8)	0.0028 (3.6)
Variable	Model 17W	Model 18W	Model 19W	Model 20W	Model 21W
Nest	None	Motorized (M)	Private Auto (P)	Shared Ride (S)	Non-Motorized (NM)
Constants					
Drive Alone (base)	0.00	0.00	0.00	0.00	0.00
Shared Ride 2	-1.81 (-17.0)	-1.32 (-5.9)	-2.57 (-12.1)	-1.68 (-17.4)	-1.81 (-18.5)
Shared Ride 3+	-3.43 (-22.6)	-2.50 (-5.9)	-4.96 (-11.9)	-2.21 (-8.8)	-3.43 (-23.0)
Transit	-0.685 (-2.8)	-0.404 (-2.6)	-0.847 (-3.4)	-0.680 (-2.8)	-0.682 (-2.9)
Bike	-1.63 (-3.8)	-1.38 (-3.5)	-1.82 (-4.7)	-1.62 (-4.2)	-1.44 (-4.9)
Walk	0.0682 (0.2)	0.334 (0.9)	-0.116 (-0.3)	0.0732 (0.2)	0.0791 (0.2)
Nesting Coefficients / Dissimilarity Parameters					
Non-Motorized Nest					0.766 (-2.1)
Shared Ride Nest				0.329 (-5.4)	
Automobile Nest			1.47 (4.0)		
Motorized Nest		0.723 (-2.3)			
Log-likelihood at Zero	-7309.601	-7309.601	-7309.601	-7309.601	-7309.601
Log-likelihood at Constants	-4132.916	-4132.916	-4132.916	-4132.916	-4132.916
Log-likelihood at Convergence	-3444.185	-3442.315	-3435.996	-3442.415	-3443.554
Rho Squared w.r.t Zero	0.5288	0.5291	0.5299	0.5291	0.5289
Rho Squared w.r.t. Constants	0.1666	0.1671	0.1686	0.1671	0.1668
Chi-Squared vs. MNL		3.74	16.38	3.54	1.26
Rejection significance		.053	.000	.060	.261

Table 8.6: Parameter Estimates for Single Nest Work Trip Models under Different Nest Structures (Source: Koppelman and Bhat 2006)

Example

System of Equations Modeling

Brownstone and Fang (2009) employed a system of five regression equations with correlated error terms to analyze vehicle ownership and utilization. This study analyzed the effect of local population density on household vehicle ownership. They drew a randomly selected sample population of 5,863 from the 2001 National Household Travel Survey. Their five dependent variables were number of passenger cars, number of light-duty trucks, mileage on cars, mileage on trucks, and housing density at the census tract level. Density at the census tract level served as an endogenous variable with metropolitan statistical area (MSA) density serving as the IV. The various socioeconomic, land use, and transportation factors used as regressors are shown in Table 8.7.

The behavior for each household is coded by five equations, among which two equations of car and truck counts are modeled as bi-variate ordered probit, two equations of car and truck miles travelled are modeled as censored Tobit model, and the last equation has the natural log of residential density on the tract level on the left hand side (i.e., the dependent variable). This system of equations was modeled under the Bayesian paradigm with the results shown in Table 8.8 (total number of draws = 20,000 and burn-in = 2000). Unsurprisingly, a 1% increase in the average MSA residential density is associated with .57% rise in the residential density on the census tract level. Household size is positively associated with number of trucks and truck utilization but negatively associated with car ownership and utilization. Income is estimated to exert a positive impact on vehicle ownership and utilization in a statistically significant way. In addition, accessibility to transit (represented as an indicator variable of whether an MSA has rail) tends to reduce the purchase and operation of trucks. More importantly, accounting for endogeneity based on the correlation matrix of structural errors is necessary.

In summary, residential density has a negligible effect on car ownership and use but reduces truck purchases and distances in statistically significant ways.

Variables	National		Subsample	
	Mean	(Std.)	Mean	(Std.)
Observations	25,057		5,863	
Housing units/sq. mile (block)	1397	(1505)	1448	(1523)
Population/sq. mile (block)	3638	(4657)	3793	(4760)
Employment/sq. mile (tract)	1306	(1472)	1353	(1492)
Housing units/sq. mile (tract)	1217	(1367)	1244	(1363)
Population/sq. mile (tract)	3102	(4051)	3227	(4168)
Number of adults	1.91	(0.70)	1.89	(.69)
Number of children	.65	(1.05)	.61	(1.02)
Highest education achieved high school	30.6%		30.5%	
Highest education achieved bachelor	37.8 %		38.1%	
Youngest child under 6	14.6%		14.1%	
Youngest child between 6 and 15	17.2%		16.3%	
Youngest child between 15 and 21	5.9%		5.6%	
MSA has rail	22.1%		22.7%	
Resides in urban area (tract)	75.3%		77.1%	
If annual household income is between 20k and 30k	12.4%		12.8%	
If annual household income is between 30k and 50k	23%		22.8%	
If annual household income is between 50k and 75k	17.9%		18.2%	
If annual household income is between 75k and 100k	11%		10.6%	
If annual household income is greater than 100k	12%		11.2%	
If the household owns home	80.1%		79.4%	

Table 8.7: Descriptive Statistics for Vehicle Ownership and Utilization Model (Source: Brownstone and Fang 2009

Variable	Coefficient				
	number of cars	number of trucks	annual avrg car miles (in 1,000)	annual avrg truck miles (in 1,000)	Log of block density
log(block density)	0.0375	-0.1969	0.0342	-3.2304	-
	(0.0433)	(0.0455)	(0.4929)	(0.6602)	-
Number of bikes	-0.0273	0.1093	-0.1293	1.2140	0.0138
	(0.0130)	(0.0130)	(0.1480)	(0.2097)	(0.0127)
Household size	-0.1204	0.0980	-1.1827	2.1654	-0.0317
	(0.0270)	(0.0274)	(0.3115)	(0.4278)	(0.0262)
Number of adults	0.3239	0.1671	3.5415	1.5293	-0.0113
	(0.0346)	(0.0358)	(0.4002)	(0.5610)	(0.0336)
Urban	-0.0039	0.1747	-1.0355	3.1176	2.4098
	(0.1250)	(0.1298)	(1.4218)	(1.9134)	(0.0385)
Income between 20k and 30k	0.1255	0.3805	1.1598	5.5918	-0.0032
	(0.0561)	(0.0614)	(0.6343)	(0.9864)	(0.0532)
Income between 30k and 50k	0.1554	0.5828	2.4567	8.7760	-0.0686
	(0.0501)	(0.0556)	(0.5693)	(0.8782)	(0.0483)
Income between 50k and 75k	0.1347	0.7135	2.7229	11.8910	-0.1108
	(0.0553)	(0.0603)	(0.6334)	(0.9540)	(0.0539)
Income between 75k and 100k	0.3262	0.6780	4.2178	11.4340	-0.1700
	(0.0655)	(0.0697)	(0.7414)	(1.1015)	(0.0641)
Income greater than 100k	0.2539	0.7526	3.9113	12.8280	-0.3294
	(0.0660)	(0.0700)	(0.7490)	(1.1065)	(0.0646)
Income data missing	0.2381	0.2795	0.6552	3.7614	-0.1050
	(0.0650)	(0.0731)	(0.7459)	(1.1589)	(0.0631)
Owns home	0.0675	0.3937	-0.4018	3.3768	-0.3576
	(0.0423)	(0.0458)	(0.4828)	(0.7257)	(0.0372)
MSA has rail	0.0598	-0.1962	0.2095	-2.0256	-0.0203
	(0.0421)	(0.0449)	(0.4758)	(0.7046)	(0.0413)
Highest education: high school	0.1008	-0.0022	1.1975	0.6450	0.0217
	(0.0385)	(0.0402)	(0.4415)	(0.6449)	(0.0375)
Highest education: Bachelor	0.2265	-0.1654	2.5117	-1.1363	0.1622
	(0.0421)	(0.0441)	(0.4815)	(0.7033)	(0.0403)
Youngest child under 6	0.1033	0.1264	2.4547	2.1375	-0.0254
	(0.0711)	(0.0730)	(0.8176)	(1.1478)	(0.0695)
Youngest child between 6 and 15	0.1197	0.0873	2.1364	1.3270	-0.0418
	(0.0634)	(0.0649)	(0.7299)	(1.0186)	(0.0619)
Youngest child between 15 and 21	0.0779	-0.1235	2.0036	0.4597	-0.0193
	(0.0683)	(0.0717)	(0.7839)	(1.1416)	(0.0685)
log(average MSA Density)	-	-	-	-	0.5743
	-	-	-	-	(0.0244)

Notes: The base groups are households with income below 20k, do not own home, are high school dropout, have no children, and live in rural area. Posterior standard deviations are reported in parentheses;

Table 8.8: Parameter Estimates for Vehicle Ownership and Utilization Model (Source: Brownstone and Fang 2009)

8.12 References

Anseline, L. (1988) Spatial Econometrics: Methods and Models. Norwell, MA, Kluwer Academic Publishers.

Ben-Akiva, M. & Lerman, S. (1985) Discrete Choice Analysis: Theory and Application to Travel Demand. Cambridge, Massachusetts: The MIT Press.

Best, N., Cowles, M., & Vines, S. (1995) CODA: Manual version 0.30. URL: http://www.mrc-bsu.cam.ac.uk/bugs/.

Brownstone, D. & Fang, H. (2009) A Vehicle Ownership and Utilization Choice Model with Endogenous Residential Density. University of California Transportation Center, Working Papers. URL:http://www.escholarship.org/uc/item/2hc4h6h5.pdf;origin=repeccitec. Accessed: 8/07/2011.

Cameron, A. C. & Trivedi, P. K. (2010) Microeconometrics Using Stata. College Station, Texas: Stata Press.

Cramer, J. S. (2002) The Origins of Logistic Regression. Tinbergen Institute Discussion Paper.

Dargay, J. M. (2001) The Effect of Income on Car Ownership: Evidence of Asymmetry. Transportation Research Part A: Policy and Practice, 35(9), 807–821.

Eisele, W. L. & Frawley, W.E. (2000) A Methodology for Determining Economic Impacts of Raised Medians: Final Project Results. Texas Transportation Institute, Texas A&M University.

Greene, W. H. (2008) Econometric Analysis (6th ed). Upper Saddle River, New Jersey: Prentice Hall.

Gujarati, D. N. (1995) Basic Econometrics (3rd ed.). New York: McGraw-Hill, Inc.

Kennedy, P. (1998) A Guide to Econometrics. Cambridge, Massachusetts: The MIT Press.

Koop, G. (2003) Bayesian Econometrics. West Sussex, England: Wiley.

Koppelman, F. & Bhat, C. (2006) A Self Instructing Course in Mode Choice Modeling: Multinomial and Nested Logit Models. U.S. Department of Transportation.

Kweon, Y. & Kockelman, K. (2005) The Safety Effects of Speed Limit Changes: Use of Panel Models, Including Speed, Use, and Design Variables. *Transportation Research Record* 1908: 148–158.

Mannering, F. & Winston, C. (1985) A Dynamic Empirical Analysis of Household Vehicle Ownership and Utilization. *RAND Journal of Economics*. 16(2): 215–236.

McCausland, B. & Stevens, J. (2001) Bayesian Analysis, Computation and Communication: The PC MATLAB Version of the BACC Software. Available at http://www.econ.umn.edu/~bacc/bacc2001.

Srinivasan, S. & Kockelman, K. (2002) The Impacts of Bypasses on Small- and Medium-Sized Communties: An Econometric Analysis. *Journal of Transportation and Statistics* 5(1), 57–69.

Xiong, X. & Kockelman, K. (2012) The Cost of Right-of-Way Acquisition: Recognizing the Impact of Condemnation via a Switching Regression Model. Forthcoming in *Transportation Research Record.*

Wang, X. & Kockelman, K.(2007) Specification and Estimation of a Spatially and Temporally Autocorrelated Seemingly Unrelated Regression Model. *Transportation* 34(3), 281–300.

Washington, S. P., Karlaftis, M.G., & Mannering, F.L. (2003) Statistical and Econometric Methods for Transportation Data Analysis. Boca Raton: Chapman & Hall/CRC

Wooldridge, J. M. (2002). Econometric Analysis of Cross Section and Panel Data. Massachusetts: the MIT Press

Chapter 9. Data Sets

A data set is a collection of variable values from actual observations obtained by surveying a sample of observational units in a larger population (of households, businesses, states, residential parcels, bridges, etc.). Data can be used to anticipate future traffic conditions, forecast pavement deterioration, understand the economic impacts of transportation policies, analyze crash severity as a function of vehicle and roadway design features, make a case for added funding, appreciate public opinion, and so forth.

As discussed in Chapter 1, all private and public benefits and costs should be considered when evaluating transportation projects and policies. To anticipate long-run project impacts, future conditions (such as rainfall, pavement cracking, operating costs, and truck counts) can be forecasted using appropriate model equations based on relevant data sets. Model predictions are based on input data, and the accuracy of data inputs tends to play a critical role in the quality of model outputs.

Following are some example transportation applications and their associated data sets, as used in transportation planning and decision making.

1. **Policy Evaluation:** In NCHRP Report 12-23, Kockelman et al. (2006) analyzed observed speed and crash data across distinct U.S. regions to anticipate the outcome of relaxing speed limits on high-speed roadways. They used the nation's Fatality Analysis Reporting

System and General Estimates System data sets, along with Washington State's Highway Safety Information System (HSIS) records (for roadway design attributes), Washington DOT traffic count and speed data by time of day, and Southern California traffic count, speed, and crash data. They estimated that speed limit increases are followed by a less-than-commensurate increase in actual speeds (at a ratio of about 10 to 3), and slightly higher overall crash rates (e.g., 3% increase for a 10 mi/h speed limit increase). But fatal crash rates rose substantially (e.g., 24% increase for a 10 mi/h speed limit increase). While their data sets were solid, issues of concern for the analyses are that higher speed limits are often associated with higher design standards, and crashes are rare events.

2. **Environmental Impacts:** Dallman et al. (2011) directly sampled diesel trucks' exhaust plumes at the Port of Oakland and found that California's new emissions control requirements for drayage trucks had caused the fleet's black carbon and oxides of nitrogen to fall sharply, by 40% or more, in the Port program's first year of vehicle replacements and retrofits.

3. **Economic Impacts:** Srinivasan and Kockelman (2002) used economic and population census data for small towns across Texas, along with traffic count data, to analyze the sales impacts of relief routes or bypasses on gas stations, restaurants, services, and total retail sales. They found that bypassed cities lost sales in all four industry categories, with gasoline stations most affected and service receipts least affected. Lower impacts were observed in cities with high traffic counts, and those with less traffic diversion, as expected. The march of time, access controls on the new facility, proximity to a large urban area, and other attributes also diminished impacts.

As transportation planning and design interests shift toward more sustainable and cost-effective investments, new and more comprehensive data collection effects are greatly needed. Fortunately, new technologies are paving the way in multiple areas: BlueTooth devices are facilitating low-cost trip-table generation (for travel demand modeling), smartphones are facilitating longer-term and lower-burden travel surveys, smaller and better GPS devices are allowing for detailed vehicle tracking, and GIS databases and satellite images are mapping land use and network details with greater spatial resolution.

The following table lists examples of existing U.S. data sets that are useful for economic and other analyses of transportation questions. The list shows databases that are primarily applicable to transportation planning applications first, followed by those more clearly economic in nature, then those related to safety, environmental issues, and infrastructure. Of course, many analyses require a marriage of multiple data sets (e.g., anticipating crash losses as a function of network design details, weather conditions, traffic flows by vehicle type and time of day, and local land use conditions). It is hoped that this short listing of 28 meaningful data sources will allow readers to become sufficiently familiar with the great variety of available databases so that they can put these and related data sets into practice, for a variety of valuable analyses.

References

Dallman, T. Harley, R., & Kirchstetter, T. (2011) Effects of Diesel Particle Filter Retrofits and Accelerated Fleet Turnover on Drayage Truck Emissions at the Port of Oakland. *Environmental Science and Technology* 45(24), 10773–10779.

Kockelman, K. (2006) Safety Impacts and Other Implications of Raised Speed Limits on High-Speed Roads. *National Cooperative Highway Research* Program Report, 17–23.

Srinivasan, S. & Kockelman, K. (2002) The Impacts of Bypasses on Small- and Medium-Sized Communties: An Econometric Analysis. *Journal of Transportation and Statistics* 5(1), 57–69.

Data set tables appear on the following nine pages.

Examples of Existing U.S. Data Sets

Data Set Name	Source	Description	Availability & Frequency	Indicators	Why Useful	Reference
Common Application: Transportation Planning						
Longitudinal Employment & Household Dynamics (LEHD)	U.S. Census Bureau	LEHD uses modern statistical & computing techniques to combine federal & state administrative data on employers & employees with core Census Bureau censuses & surveys.	• In the Winter 2000, Florida & Illinois states. • As of September 2003, 27 states (including Texas) became official LEHD state partners. • The most recent was released in May 2009.	**1.** Origin/ Destination employee numbers from household to place of employment. **2.** Information on the characteristics of workers by block residence • Number of workers living on each block • Proportion of workers earning low, medium, or high annual wages • Mean annual wages **3.** Information on the characteristics of businesses by block • Mean quarterly pay per worker • Industries operating on each block (SIC, Standard Industrial Classification division)	Potentially an alternate/ additional source of place of work & flow data.	U.S. Census Bureau, Longitudinal Employer-Household Dynamics. Available at http://lehd.did.census.gov/led/about-us/FAQ.html#lehd Last accessed on 5/3/2011.
Quarterly Workforce Indicators (QWI)	Bureau of Labor Statistics (BLS)	A set of economic indicators such as employment, job creation, wages, & worker turnover. These data can be queried by different levels of geography including state, county, metro, & workforce investment area, as well as by detailed industry, gender, & age of workers.	From 2001. Issued quarterly.	**1.** Total employment **2.** Net job flows **3.** Job creation **4.** New hires **5.** Separations **6.** Turnover **7.** Average monthly earnings **8.** Average new hire earnings	To determine transportation needs of different areas (for example, to designate+19 Transportation Reinvestment Zones).	U.S. Census Bureau, Longitudinal Employer-Household Dynamics. Available at http://lehd.did.census.gov/led/datatools/qwiapp.html Last accessed on 5/3/2011.

Data Set Name	Source	Description	Availability & Frequency	Indicators	Why Useful	Reference
Metropolitan Travel Survey Archive	The Bureau of Transportation Statistics (BTS), FHWA, & University of Minnesota	To store, preserve, & make publicly available travel surveys conducted by metropolitan areas, states, & localities. For Texas, household travel of Tyler & Smith, Longview & Gregg, & Laredo & Webb Counties are available.	For Texas: **1.** Laredo 2002 **2.** Tyler 2003 **3.** Longview 2003 Issued annually.	**1.** Demographics **2.** Number of daily trips by mode. **3.** Number of daily trips by destination purpose. **4.** Daily trip rates. **5.** Trip durations.	It provides information about travel preferences or change in travel behavior of people, over a period of time, across the population.	University of Minnesota, Metropolitan Travel Survey Archive. Available at http://www.surveyarchive.org/index.html
National Household Travel Survey (NHTS)	BTS & FHWA	Contains comprehensive data on long distance & local travel, & transportation patterns in the United States.	The 2009 is the latest, while 2001, 1995, 1990, 1983, 1977, & 1969 are previous versions. Issued periodically.	**1.** Purpose of the trip (work, shopping, social, etc.) **2.** Means of transportation (car, walk, bus, subway, etc.) **3.** Trip duration **4.** Time of day/day of week **5.** Demographic, geographic & economic data for analysis purposes	**1.** To quantify travel behavior **2.** To analyze changes in travel characteristics over time **3.** To relate travel behavior to the demographics of the travel **4.** To study the relationship of demographic & travel over time **5.** To plan new investments	http://www.bts.gov/programs/national_household_travel_survey/
Commodity Flow Survey (CFS)	Partnership between BTS & the U.S. Census Bureau.	Provides data on the movement of goods in the United States, including information on commodities shipped, their value, weight, & mode of transportation, as well as the origins & destinations of shipments.	1993, 1997, 2002, & 2007 are available. Conducted every 5 years.	**1.** Shipment value **2.** Shipment weight **3.** Mode of transportation **4.** Origin & destination of shipments **5.** Business establishments in mining, manufacturing, wholesale trade, selected retail industries & selected auxiliary establishments (e.g., warehouses) of in-scope multiunit companies	To estimate distance distributions & origin-destination flows by commodity type, mode, shipment size, & value	http://www.bts.gov/publications/commodity_flow_survey/

Data Set Name	Source	Description	Availability & Frequency	Indicators	Why Useful	Reference
Freight Analysis Framework (FAF)	This dataset is developed by FHWA from variety of sources	It includes a comprehensive data about freight movement among states & major metropolitans by all modes of transportation.	From 1997. Updated every 5 years.	**1.** Commodity **2.** Origin-destination **3.** Routing **4.** Shipment **5.** Transportation	It forecasts future freight flows among regions & assigns those flows to the transportation network.	http://ops.fhwa.dot.gov/freight/freight_analysis/faf/index.htm
American Community Survey (ACS)	U.S. Census Bureau	It consists of U.S. population & household characteristics.	From 2000. Issued annually, providing 1, 3, and 5 year estimates.	**1.** Sex by age **2.** Total population **3.** Movers between regions **4.** Means of transportation to work **5.** Aggregate travel time to work (in minutes) of workers **6.** Household type (including living alone) by relationship for the population in Households **7.** Much more	Regular demographic, housing, social, & economic data for comparison across states and communities.	McGuckin, N & Ruiter, E. (2007). A Guidebook Using American Community Survey Data for Transportation Planning. National Academy Press, Washington DC. Available at http://onlinepubs.trb.org/onlinepubs/nchrp/nchrp_rpt_588.pdf
Census Transportation Planning Package (CTPP)	AASHTO, State DOTs, U.S. Census Bureau, FHWA, BTS, Federal Transit Administration (FTA), & Transportation Research Board (TRB)	Contains summary tabulations available for traffic analysis zones (TAZs) that have been defined by state & regional transportation agencies.	From 1990. Issued every 10 years.	Information on commute characteristics & socio-economic data from Census 2000	To provide a wide range of data for transportation planning activities at the state & local level.	USDOT, RITA: http://www.fhwa.dot.gov/planning/census_issues/ctpp/contacts/index.cfm

Data Set Name	Source	Description	Availability & Frequency	Indicators	Why Useful	Reference
ACS Public Use Microdata Sample (PUMS)	U.S. Census Bureau	A sample of the actual responses to the ACS & include most population & household characteristics.	From 2000. Issued annually, providing 1, 3, and 5 year estimates.	**1.** Industry **2.** Occupation **3.** Place of work **4.** Workers by educational attainment by means of transportation **5.** Average age of workers by means of transportation **6.** Discrete choice models (e.g., household vehicle & worker level models) **7.** Persons by race/ethnicity by disability status by poverty status	Based on its indicators, useful data for transportation planners.	U.S. Census Bureau, American Community Survey, PUMS Data. Available at http://www.census.gov/acs/www/data_documentation/pums_data/
TexasAhead	Texas Comptroller of Public Accounts	It consists detailed data for Texas cities & counties such as labor market, consumer demand, financial markets, energy, gross state product & income	Varies.	**1.** Monthly unemployment **2.** Nonfarm Employment by industry **3.** Texas Gross State Product **4.** Texas Personal Income & Real Personal Income **5.** Sources of Personal Income in Texas & Per Capita Personal Income **6.** Texas Consumer Price Index **7.** Retail Sales **8.** Short-term & long-term Interest Rates	It gives an insight into economic indicators in Texas for planning	Texas Comptroller of Public Accounts, TexasAhead. Available at http://www.texasahead.org/
Highway Congestion (Urban Mobility Study)	Texas Transportation Institute	To provide information on trends in levels of mobility & congestion of 68 urban areas across the U.S. from available traffic data.	From 1982. Issued annually.	**1.** Driver-hours of delay per year **2.** Traffic rate index **3.** Wasted fuel per year **4.** Congestion cost per year **5.** Annual highway congestion cost **6.** Roadway congestion index	To estimate cost of delay.	http://mobility.tamu.edu/ums/

Data Set Name	Source	Description	Availability & Frequency	Indicators	Why Useful	Reference
Employment by Industry & Occupation	BLS	The information includes estimates of employment, hourly wages, & annual wages for 770 detailed occupations & for 22 major occupational groups.	From 1988. Issued annually.	**1.** Total employment by occupation **2.** Wages by occupation	To determine transportation needs of different areas. Also, to designate areas as Transportation Reinvestment Zone	http://www.bls.gov/oes/
Common Application: Economic Impacts						
Automobile Driving Costs	American Automobile Association (AAA)	To estimate annual driving costs for five different kind of vehicles, a compact, mid-sized, large, light truck car, & mini-van.	From 1992. Issued annually.	**1.** Estimated cost of gas & oil, maintenance & tires per mile **2.** Estimated cost of insurance, license, registration, taxes, & depreciation per year	Total cost per mile is given for vehicles driven 10,000, 15,000, & 25,000 annually.	http://newsroom.aaa.com/wp-content/ uploads/2012/04/YourDrivingCosts 2012.pdf
Automobile Driving Cost vs. Public Transit Cost	APTA	To estimate annual driving cost & public transit costs for commuters in the U.S according to vehicle size/type & miles driven per year.	From 1990. Frequency varies.	1) Automobile driving costs by class of vehicle (small, mid-size, SUV, etc.) & estimated miles per year 2) Public transit costs for different fare levels & with various surcharges.	To understand the difference between automobile driving cost & public transit cost.	http://www.apta.com/resources/statistics/Pages/TransitFareDatabase.aspx
Gross Domestic Product (GDP) by Industry	Bureau of Economic Analysis	Data on the nationwide GDP for transportation industries.	From 1947. Issued annually.	**1.** Contributions to gross domestic product (GDP) **2.** U.S. gross domestic demand (GDD) attributed to transportation-related final demand. **3.** U.S. gross domestic product (GDP) attributed to for-hire transportation services.	General economic importance of transportation, & by mode.	http://www.transtats.bts.gov/DatabaseInfo.asp?DB_ID=275&DB_Name=Gross%20Domestic%20Product%20%28GDP%29%20by%20Industry&Link=http%3A//www.bea.gov/industry/index.htm&DB_URL=

Data Set Name	Source	Description	Availability & Frequency	Indicators	Why Useful	Reference
Consumer Price Index (CPI)	BLS	It produces monthly data on changes in the prices paid by urban consumers for a representative basket of goods & services.	A few indexes are available from as far back as 1913. Issued monthly.	It covers following services: **1.** Food & beverages **2.** Housing **3.** Apparel **4.** Transportation **5.** Medical care **6.** Recreation **7.** Education & communication **8.** Other goods & services	For adjusting dollar values at different points in time to consistent, comparable values.	United States Department of Labor, BLS, Consumer Price Index. Available at http://www.bls.gov/cpi/home.htm
U.S. Retail Gasoline Prices	The Energy Information Administration	Gasoline price per gallon by grade, by region, & by formulation in the U.S. Retail Gasoline prices database	First year of data varies depending on formulation & grade. Issued weekly.	**1.** Motor fuel prices **2.** Regional retail motor fuel prices **3.** World crude oil prices	Produce weighted average price estimates at the city, state, regional, & national levels.	http://www.transtats.bts.gov/DatabaseInfo.asp?DB_ID=540&DB_Short_Name=Gasoline%20Prices&DB_Name=U.S.%20Retail%20Gasoline%20Prices&Link=http%3A
Common Application: Safety						
Fatality Analysis Reporting System (FARS)	The National Highway Traffic Safety Administration (NHTSA)	Data on all fatal motor vehicle crashes in the United States.	From 1991. Issued annually.	**1.** Motor vehicle fatal crashes by posted speed limit **2.** Motor vehicle safety data **3.** Motorcycle rider safety data **4.** Occupant fatalities by vehicle type & non-occupant fatalities **5.** Occupant & non-motorist fatalities in crashes by number of vehicles & alcohol involvement (AI) **6.** Passenger car occupant safety data **7.** Transportation fatalities by mode **8.** Truck occupant safety data	In order to improve traffic safety & determine cost of motor vehicle crashes.	http://www.nhtsa.gov/FARS

Data Set Name	Source	Description	Availability & Frequency	Indicators	Why Useful	Reference
Highway Safety Information System (HSIS)	FHWA	It contains accident, roadway inventory, & traffic volume data for California, Illinois, Maine, Michigan, Minnesota, North Carolina, Utah, & Washington.	From 1991. Issued annually.	**1.** Time **2.** Environment **3.** Accident **4.** Vehicle information **5.** Driver information **6.** Occupant information **7.** Roadway elements **8.** Pedestrian/Bicyclist information	It is useful to study current highway safety issues.	FHWA, Highway Safety Information System (HSIS). Available at http://www.hsisinfo.org//index.cfm
NASS General Estimates System (GES)	NHTSA	A sample of all police-reported motor vehicle crashes. Help identify safety problem areas, provide a basis for regulatory & consumer initiatives,& form the basis for Cost-Benefit Analysis of traffic safety initiatives.	From 1988. Issued annually.	**1.** Crash injury type **2.** Crash type **3.** Involving a young driver **4.** Day of week **5.** Interstate **6.** Intersection **7.** Involving a pedestrian **8.** Involving speeding **9.** Plus much more	Useful in traffic analysis by NHTSA & DOT agencies, determining changing needs of the traffic community safety, & answering motor vehicle safety questions from Congress & the public agencies.	NHTSA, NASS General Estimates System. Available at: http://www.nhtsa.gov/Data/National+Automotive+Sampling+System+(NASS)/NASS+General+Estimates+System Last accessed on 5/15/11.
Motor Carrier Management Information System (MCMIS)—Crash File	The Federal Motor Carrier Safety Administration (FMCSA)	Data from State police crash reports involving drivers & vehicles of motor carriers operating in the U.S.	From 1993. Issued annually.	**1.** Distribution of transportation fatalities by mode **2.** Injured persons by transportation mode **3.** Median age of automobiles & trucks in operation in the U.S. **4.** Transportation accidents by mode **5.** Transportation fatalities by mode **6.** Truck occupant safety data **7.** Truck profile **8.** U.S. automobile & truck fleets by use	To estimate cost of motor vehicle crashes.	http://ai.volpe.dot.gov/default.aspx

Common Application: Environmental Impacts

Data Set Name	Source	Description	Availability & Frequency	Indicators	Why Useful	Reference
U.S. Greenhouse Gas Emissions & Sinks	The U.S. Environmental Protection Agency	Annual inventory of U.S. greenhouse gas emissions & sinks.	From 1990. Issued annually.	**1.** Greenhouse gas emissions by type, mode, & industrial sector **2.** Greenhouse gas emissions by mass & by mass weighted by the each gas's potential to cause global warming	To estimate environmental impact of building or expanding new facilities.	http://epa.gov/climatechange/emissions/usinventoryreport.html
Common Application: Infrastructure Conditions						
Highway Performance Monitoring System (HPMS)	FHWA, state highway agencies, local governments, & metropolitan planning organizations	To provide data about the extent, condition, performance, use, & operating characteristics of U.S. highways.	From 1993. Issued annually.	**1.** Condition of U.S. roadways by functional system **2.** Public road & street mileage in the United States by type of surface **3.** Roadway vehicle-miles traveled (VMT) & VMT per lane-mile by functional class **4.** System mileage within the United States **5.** U.S. vehicle-miles	For assessing highway system conditions & performance, investment decisions.	http://www.fhwa.dot.gov/policyinformation/hpms.cfm
Motor Vehicle Output	The Bureau of Economic Analysis	Data on domestic consumer expenditures, private fixed investment, government fixed investment, & exports for automobiles & light trucks.	From 1967. Issued quarterly.	**1.** Real inventories & sales **2.** Personal consumption expenditures **3.** Government current receipts & expenditures **4.** Gross private domestic investment & capital transfers **5.** Motor vehicle output	To anticipate vehicle sales & use, fleet turnover, energy, expenditures, emissions, etc.	http://www.bea.gov/index.htm
Vehicle Miles Traveled (VMT)	FHWA	Contains vehicle miles traveled data, by roadway type and vehicle type.	From 1980. Issued either annually or monthly.	**1.** Roadway VMT & VMT per lane-mile by functional class **2.** U.S. passenger-miles **3.** U.S. vehicle-miles	To estimate condition of roadways & bridges caused by traffic.	http://www.bts.gov/publications/national_transportation_statistics/#chapter_1
Common Applications (Multiple Uses): Transportation Planning & Infrastructure Conditions						

Data Set Name	Source	Description	Availability & Frequency	Indicators	Why Useful	Reference
National Transportation Statistics	BTS	Includes physical components, safety record, economic performance, the human & natural environments, current traffic operations, crashes & injuries by mode, price and purchase statistics, & energy use of U.S. transportation system.	From 1960. Publication frequency varies; currently updated quarterly	**1.** Transportation system **2.** Transportation safety **3.** Transportation & the economy **4.** Transportation, energy & the environment	Gives general information about transportation system, for trend analysis, safety & cost understanding, investment & policy decisions.	http://www.bts.gov/publications/national_transportation_statistics/
National Transit Database (NTD)	APTA	Data on transit systems: extent, operations, ridership, & costs.	Issued annually.	Capital & operating funds, operator wages, stations provided, vehicle mileage, fares, energy use, taxes & more.	For assessing transit investments, provision, & operating changes.	http://www.apta.com/resources/statistics/Pages/NTDDataTables.aspx
Vehicle Inventory & Use Survey (VIUS)	U.S. Census Bureau	Provides data on the physical & operational characteristics of the U.S. private & commercial truck populations.	From 1963 until 2002. Issued every 5 years.	**1.** Trucks, truck miles, & average annual miles for all trucks **2.** Trucks by vehicle size **3.** Trucks by truck type **4.** Truck miles by vehicle size **5.** Truck miles by truck type **6.** Truck miles distribution by operational characteristics	Inventory analysis, cost allocation, safety/risk assessment, environmental impact estimates, and so forth.	U.S. Census Bureau, Vehicle Inventory & Use Survey. Available at http://www.census.gov/svsd/www/vius/products.html

Other data sets of interest: American Housing Survey, Residential Energy Consumption Survey, Motor Vehicle Occupant Safety Survey, and more.

Chapter 10. Case Studies

The multiple topics introduced in this Reference are all interconnected. Many come into play simultaneously. This chapter highlights a variety of specific contexts that demonstrate this interconnectedness. These transportation case studies involve both qualitative and quantitative analyses for comprehensive assessment of economic impacts and indicators:

1. Conducting cost-benefit assessments of network improvements
2. Quantifying the sales losses associated with highway relief routes
3. Forecasting the traveler welfare effects of different tolling policies
4. Anticipating right-of-way acquisition costs

In addition to the case studies, also presented is an introduction to Transportation Project Impact Case Studies (T-PICs), a collection of case studies tracking the economic impacts of highway projects collected by the U.S. Strategic Highway Research Program. The use of T-PICS to anticipate potential economic impacts of highway projects is discussed after the four case studies.

Case Study 1: Cost-Benefit Evaluation of Network Improvements

As discussed in Chapter 6, competing investment opportunities and policies can be compared via *cost-benefit analysis* (CBA). CBA seeks to identify the most cost-effective opportunities, by monetizing the impacts of network investments, road pricing, and other transportation projects, in order to prioritize them in the face of limited budgets. This section first describes a New Jersey DOT CBA for network improvement options, and then discusses the use of a comprehensive Project Evaluation Toolkit (PET) for Texas applications.

Evaluating Network Improvement Opportunities in New Jersey

The NJDOT used CBA to place a value on the expected impacts of a State Route 18 (SR 18) improvement project on the Piscataway, NJ economy. The proposed project was a 2.5-mile extension of SR 18 at its northern terminus, connecting it to IH 287. CBA was required to compete for a USDOT TIGER II grant. The corridor's repair costs were based on expected resurfacing and maintenance expenses, and its economic effects were the monetized values of travel time and delay savings. The TIGER grant's livability considerations provide equations for monetizing a project's air and noise pollution impacts, and its safety considerations provide equations for quantifying changes in crash costs. The equations are used to monetize all six impact types: vehicle operating, congestion, crash, air pollution, noise, and maintenance.

Assuming a capacity increase of 120% (a high-end estimate), the NJDOT's CBA suggested that over a 25-year span (with a decidedly low annual discount rate of 2.8%), the project benefits would equal $6.82 billion of cost savings over the base case. In contrast, costs were estimated to be just $49.7 million, returning a benefit-cost (B/C) ratio estimate of a striking 138.

Uncertainty is always associated with the various inputs, particularly as a project's precise capacity impacts are difficult to estimate. A basic sensitivity analysis was performed by varying two key inputs: the value that can be placed on the travelers' time—known as *value of travel time* (VOTT) as described in Chapter 1—and capacity increase. The single VOTT varied from $7.90 to $27.20 per hour for passenger cars (with truckers' VOTT held fixed at just $19.90 per hour) and the capacity increase ranged from 25% to 120%. The resulting B/C ratios ranged from 108 to 207, and NJDOT used these ratios to defend the long-term economic viability of the proposed project in their pursuit of the TIGER grant monies.

A Project Evaluation Toolkit (PET)

To allow for a wide range of network improvement scenarios when evaluating competing opportunities, University of Texas researchers developed a user-friendly and fast-running Project Evaluation Toolkit (PET) to compute B/C ratios, internal rates of return, emissions totals, toll revenues, and other indicators of interest. New roads, variable tolls, capacity addition, speed harmonization, and other system changes can be compared against one another and against a no-build alternative to guide early project selections within a city or regional network. The Toolkit anticipates near- and long-term project impacts, including traveler welfare, multiple emissions species, crash counts by severity, travel time reliability effects, and toll revenues. PET generates these estimates by modeling traffic pattern changes (using utility-maximizing logit model specifications, as discussed in the Chapter 8). It also relies on the rule of half (Chapter 2); monetized values of reliability, emissions, and crash changes (Chapter 1); and optimization and sensitivity analysis techniques (Chapter 6). Designed to be an easy-to-use, quick-response tool, PET enables decision-makers to optimally allocate scarce resources.

Fagnant et al. (2011) demonstrated the use of PET through two Austin applications:

(1) Capacity expansion of US 290 between US 183 and SH 130 (with three alternative scenarios: grade-separated freeway upgrade, grade-separated flat-rate tollway, and grade-separated variable-rate tolling by time of day).

(2) Travel demand management along IH 35 between US 183 and 15th Street (including $1 toll, $2 toll, and lane-removal strategies).

The results of the first case study (US 290's expansion) show favorable B/C ratio estimates for all three scenarios: 14.0 for the freeway upgrade, 6.5 for the tollway, and 3.9 with variable tolling. The freeway upgrade produced the highest traveler welfare impacts (and, to a lesser degree, travel-time reliability benefits), while crashes and emissions were predicted to fall under all three scenarios (although VMTs rose, thanks to improved travel conditions). Both tolling scenarios resulted in very favorable revenues for project financing, with projected internal rates of return (IRRs) estimated at 23% for the flat-rate tollway and 29% for variable-rate tolling, by time of day.

Fagnant and Kockelman (2012) also conducted a sensitivity analysis (as described in Chapter 6) to evaluate the impacts of 28 sets of unknown model inputs (like travelers' values of time, agency valuation of emissions and lives, and link-performance parameters). Their 600 simulation runs relied on Monte Carlo random draws (from lognormal distributions of inputs), resulting in a wide distribution of potential outcomes (including some negative B/C ratios) and average B/C values that were slightly less than the single-run, average-input solutions. The sensitivity results also highlighted the fact that project outcomes (B/C ratios, IRRs, and other economic indicators) were best when system-wide capacity values were lowered—thus raising the benefits of added capacity.

Fagnant et al.'s (2011) second case study (travel demand management along IH 35) showed how many travelers re-routed their trips to city streets, rather than forego most existing trips or switching to less congested times of day. This behavior resulted in higher VMT values overall, causing even more delays to travelers, worsening reliability, and raising crash counts and emissions. For instance, in the $2 toll scenario's initial year, traffic in the surrounding links increased by 40%, while traffic on IH 35 fell by 22%, resulting in a net system-wide VMT increase of 1.3%, or 70.5 million annual VMT. **Economic techniques embodied in tools like PET help policymakers, planners, engineers, and other stakeholders quickly and rigorously evaluate a variety of transportation investment and policy options to anticipate the best scenario (and avoid the worst).**

Case Study 2: The Economic Impacts of Bypasses

Bypasses or "relief routes" are built to improve traffic flow around a city's central business district and often impact local and regional economic development, land use, and city growth patterns. Yeh et al. (1998) estimated that bypasses reduced average downtown, main-route traffic by 72% in small communities (populations under 2,000). Such rerouting of passerby traffic away from existing, downtown businesses can noticeably reduce local business sales, while causing shifts in regional economies through changes in access, travel times, travel costs, and safety.

Bypass effects can be both positive and negative. They are largely designed to save time by allowing through travelers to avoid speed reductions and intersection stops in a city or town center. They may spur development that caters to longer-distance travelers, from nearby towns and other parts of the region, including big-box retail, automobile dealers, department stores, and hospitals. In the broadest sense, lowering transportation costs should decrease production costs, thereby increasing profits, sales, and regional employment levels.

Many bypasses are also pursued in order to reduce congestion, traffic noise, and downtown crashes (involving both pedestrians and vehicles). A Kansas study by Babcock and Davalos (2004) noted that many local businesses, formerly downtown, opened new locations along Kansas's relief routes, along with entirely new establishments that target a wider, out-of-town market. Further, this finding is complemented by that of Thompson et al. (2001), who observed that 57% of businesses along Kentucky bypasses were from the retail sector, compared to 31% of businesses in downtown areas.

Negative impacts to local economies vary in scale and over time, and can depend on city size, businesses present, through flows, truck shares, proximity to the bypassed site, and proximity to a metropolitan area. Some bypassed businesses may fail in the face of lost traffic. Thompson et al. (2001, p. 22) found that "the average vacancy rate in the downtown area of communities with a bypass was 18.4% versus 10.9% in similar communities without a bypass" and that 7.6% of bypass businesses had previously been located in the bypassed, downtown areas. Yeh et al. (1998) argued that small communities lose traffic and businesses to bypasses because they lack the economic diversity that will attract through travelers. Either way, business counts and traffic counts are not the same as business sales. The following section describes econometric models for traffic and sales volumes, before and after the installation of Texas relief routes.

Econometric Modeling to Assess the Economic Impacts of Relief Routes

In an effort to more rigorously and comprehensively assess the economic impacts of bypasses on small and mid-size communities, researchers Srinivasan and Kockelman (2002) examined sales data between 1954 and 1992 across 42 Texas locations (23 with bypasses and 19 without bypasses). All towns had populations between 2,500 and 50,000, and the sales data were by industry, as obtained every 5 years in the U.S. Economic Census. Additional demographic data, such as unemployment rates, income per capita, the share of elderly in the population, and the ratio of median household income to average household size, were also obtained for each community via the U.S. Population Census.

Per-capita sales data were obtained for the following four industry categories: all retail establishments, gasoline service stations, eating and drinking establishments, and services. These categories were used as the dependent or response variable, as defined in Chapter 8. To reflect the repeat observation of the same community (nine times each) in the nearly four decades of data, Srinivasan and Kockelman (2002) relied on random-effects models, one for each industrial sector's per-capita sales. An unobserved error component or "random effect" term was included for each community. Because the cities were observed at the same points in time, another form of correlation was also permitted: where sales observed in the same year had the same unobserved error-related term. Moreover, because unobserved city-specific factors impacting one industrial sector are likely to impact the other three sectors, a system of seemingly unrelated regression (SUR) equations was used, and estimated using feasible generalized least squares techniques (as described in Chapter 8). The relatively sophisticated model specification allows analysts to improve parameter estimates by reducing uncertainty of final estimates (as measured by standard errors). The final coefficient estimates from this correlated structure of equations and observations (following a process of stepwise removal of statistically insignificant factors) are shown in Table 2.

As the results show, several variables were helpful in explaining the sectors' per-capita sales. For example, per-capita sales at the state level (STATE SALES PERCAP) helped control or account for global trends in that specific industry's sales levels over time (reflecting recessions and expansionary periods), and thus this variable enters with a positive coefficient (because the state-level sales generally move in synch with the city-level sales values for that same industry category). Inclusion of a year variable (YEAR) for each data point/observation helped account for time-related trends (with YEAR 1982 serving as an indicator variable, due to the fact that the 1982 economic census only provided sales data for establishments with payrolls). The share of population that is elderly and the unemployment rate of the town were only useful in predicting overall retail sectors, and were estimated to have a positive effect on that sector's sales, at least locally; they were not statistically significant in the other equations. As expected, per-capita traffic levels along the original route, upstream of the bypass point (TOT TRAFFIC PERCAP), represented a major positive factor for sales in all four sectors, and the share of traffic splitting onto the bypass had a negative effect.

Table 2: Coefficient Estimates

Per Capita Sales	*Total Retail*	*Gasoline Service Stations*	*Eating & Drinking Establish-ments*	*Service Industries*
Variable	**Coefficient Estimate**			
CONSTANT	5.34E+05	3.81E+04	1.50E+04	9.36E+04
STATE SALES PERCAP	1.52E+00	3.01E+00	6.97E–01	6.51E–01
YEAR	–2.78E+02	–1.99E+01	–7.81E+00	–4.85E+01
YEAR 1982		–3.92E+02	–1.16E+02	3.18E+01
ELDERLY SHARE	1.09E+02			
UNEMP RATE	1.21E+02			
INCOME PERCAPITA	6.49E–01		2.59E–02	1.53E–01
LARGECITY POP/DIST	1.14E–01		1.44E–02	3.14E–02
TOT TRAFFIC PERCAP	2.80E+03	2.97E+02	1.66E+02	3.97E+02
RELIEF ROUTE INDIC.	5.35E+03	–8.83E+01	1.77E+02	6.54E+02
NUM YEARS SQUARED				–1.52E+00
TRAFFIC SPLIT	–1.72E+04	–7.57E+02	–6.75E+02	–1.51E+03
Adj. R^2 (fit statistic)	0.59	0.31	0.59	0.65

The presence of the bypass (as indicated by the binary variable RELIEF ROUTE INDIC.) has a positive coefficient, as it enters the equations along with the last two variables in the table: number of years (squared) since the bypass was added and the split in traffic along the new route, both of which have negative impacts, offsetting this indicator's ostensibly positive effect. It is important to recognize this interplay across competing and complementary variables when interpreting results. The model results also indicated that the longer a relief route has been in service (indicated by NUM YEARS SQ), the lower the predicted per-capita service-industry sales. These results allow the computation of the critical split in traffic where introduction of a bypass and the resulting traffic diversion will reduce a town's overall sales (per capita). The model yielded the following estimated amounts by which sales will fall: Srinivasan and Kockelman (2002) report this as 31% for retail sales, 26% for eating and drinking establishments, and 43% for service industries. On average, gasoline service station sales were negatively impacted regardless, according to the model results, making gas sales the most economically impacted of the sectors studied. Such results highlight the industry types that are more sensitive to traffic diversion, presumably based on their customer base.

Interestingly, the ratio of the population of the nearest central city (of a metropolitan statistical area) to its distance from the bypassed town (i.e., LARGECITY POP/DIST) was estimated to have a positive effect on per-capita sales. This result implies that the closer a bypassed community is to a large city, or the more populated that city, the lesser the effect of a bypass, presumably thanks to a broader base of region-familiar, potential customers and/or less reliance of travelers on town businesses before the relief route's introduction.

As Srinivasan and Kockelman note, "the better a relief route works from a traffic standpoint, the greater its adverse impact on local per capita sales" (2002, p. 68). This serves a cautionary reminder that communities are sensitive to bypasses and altering traffic flows in such a major way can be expected to impact economies in significant and unique ways. However, higher volume locations (for example, those with five average annual daily traffic (AADT) per capita) were much less affected than others. Related surveys of businesses and other key stakeholders, per Handy et al.'s (2001) report, indicated that many affected individuals appreciated the accompanying loss in heavy-duty truck traffic through their town centers, suggesting that noise, emissions, safety, and pavement damage benefits may more than offset sales-dollar effects. **In general, such comprehensive investigations highlight the complexity and wide variety of economic impacts that transportation decisions can have, while quantifying their magnitudes. Census data, statistical rigor, and thoughtful model specification, coupled with some qualitative analysis, can provide a wealth of understanding for more sound transportation decision-making.**

Case Study 3: The Economic Impacts of Congestion Pricing

As discussed in Chapter 2, the variable-rate tolling of roadways to moderate congestion—a policy called *congestion pricing*—can inequitably impact lower-income travelers. Kalmanje and Kockelman (2004) proposed credit-based congestion pricing (CBCP) to help address such equity issues and improve all travelers' net benefits, by providing certain drivers in a region (such as vehicle owners, workers, or all adult residents) with a monthly toll credit. Those who exhaust their credits early in the month simply pay out of pocket to continue using congested roadways, while more budget-conscious travelers—who conserve their credits by driving at off-peak times of day, carpooling, or using alternative modes of transportation—end each month with extra credits. These can be applied toward other goods (such as transit passes) or sold to fellow travelers. Gulipalli and Kockelman (2008) estimated joint destination-mode choice models to examine the traffic, air-quality, traveler-welfare, and administrative costs impacts of such a policy's implementation across the Dallas-Fort Worth (DFW) networks.

They simulated three traffic scenarios in order to assess CBCP's economic impacts: the status quo (including existing flat-rate tolls); marginal cost pricing (MCP) on freeways (14% of the region's coded lane miles); and MCP on all road network links. As described in Chapter 2, *marginal cost* is the value of total travel time that new drivers add to those traveling behind/with them. Nested logit models (as described in Chapter 8) for destination and mode choices were used to simultaneously allocate/predict trip ends and modes, using DFW's 1996 household survey data and Dallas Area Rapid Transit (DART) on-board survey data. In all, 4,874 destination zones, 4 modes (drive alone, shared ride, transit, and walk/bike), 3 trip purposes (home-based work, home-based non-work, and non-home-based), and 6 traveler classes (based on income and vehicle ownership) were tracked using TransCAD, in conjunction with a background commercial-trip table.

As utility-maximizing decision-makers, many travelers avoided higher-toll facilities and higher-toll times of day, with only slight changes in mode and destination choices, while enjoying lower travel times on previously congested facilities. System-wide VMT was predicted to fall by 7 to 8% following the implementation of MCP, with average trip length falling from 9.63 to 9.09 miles. In the long run, with work-trip destinations (job locations) allowed to vary for all workers, VMT at volume-to-capacity ratios above 1.0 was predicted to fall 73% in the MCP-on-freeways-only scenario and 81% in the MCP-on-all-roads scenario, suggesting dramatic improvements in what were previously bottleneck/traffic-queued locations. The fall in VMT and congestion also resulted in air quality benefits, with 5 to 7% emissions reductions, compared to the status quo.

Gulipalli and Kockelman (2008) also estimated traveler welfare impacts via a measure of *consumer surplus* (which is the benefits to travelers beyond what they pay to access destinations of interest, in time and money, as described in Chapter 2). The net benefits of MCP were calculated as the differences in probability-weighted choice utilities for all trips, normalized by each traveler class's marginal utility of money (MU$), in order to arrive at units of dollars. A travel budget of $15 per month per eligible traveler also was added. This amount was derived from congestion-pricing toll revenues that well exceeded system-administration costs in the case of tolling congested freeways (rather than applying such technology to all DFW roadways). Average net benefits per traveler per day were estimated to be highest (at +40¢) for travelers residing near the central business districts, and lowest (at -30¢) for those in the northwest DFW region (west of Carrollton) and in the south (between Cleburne and Waxahachie). In the losing regions, low-income groups were estimated to benefit more than medium- and high-income groups (due to less extensive pre-policy trip-making). The authors noted that travelers from neighborhoods least likely to benefit from such a policy could be given larger monthly CBCP budgets, at least for some period of time, to even out the spatial distribution of welfare changes, thereby addressing, to some extent, questions of environmental justice.

Assuming 140 new toll plazas (1 toll plaza for every 3 centerline miles of congested or nearly congested freeway), 700 electronic toll readers, 350 high-speed cameras, and transponders for every DFW adult yields an estimated start-up cost of $48.3 million (or $11 per DFW resident) to implement such a pricing scheme. Maintenance and operation costs were estimated to be $260,000 million per year based on a $100,000 per year per lane-mile cost, or $54 annual cost per DFW resident (using FHWA values from past electronic tolling applications). Applying a capital recovery factor or discount rate assumption of 6% over the 10-year toll-collection system, the net present cost (per DFW adult) was estimated to be $55. This is small when compared to the estimated $440 "lost" each year in freeway-related congestion costs per resident. Moreover, transponder costs have fallen dramatically (with Texas TollTags now costing under $1 each, rather than the $10-per-transponder assumption in the original estimate) and Texas has fully automated much of its toll collection process, resulting in lower start-up and administration/operations costs. **CBCP encourages travelers to budget their travel based on congestion. Revenue redistribution, which can aid in bridging the gap of user benefits of different income groups, has played a key role in increasing the political acceptability of congestion pricing.**

Case Study 4: Right-of-Way Acquisition Costs

Project construction and corridor expansions often require right-of-way (ROW) purchases, which are a key expense for most DOTs. These ROW costs are often difficult to estimate in advance and can represent a public relations challenge. Being able to predict ROW expenses helps planners estimate total costs and secure project financing, but the estimation process is fraught with challenge and mis-predictions. Project administrators often claim that they lack adequate information on parcels, lack time to perform an in-depth analysis, and must often provide estimates years before properties are acquired, resulting in long-run under-estimation biases. To expedite cost estimation and reduce uncertainties in ROW budgeting, basic least-squares regression models can be quickly developed, using existing DOT data sets (and MS Excel or other software, as described in Chapter 8).

Researchers Heiner and Kockelman (2005) used such models to predict property values and anticipate ROW acquisition costs along Texas corridors in Austin, DFW, Houston, and San Antonio. To get a sense of entire-property purchase prices, they used commercial and residential transactions data, along with detailed parcel information, from CoStar (a real estate information provider) and the Travis County Appraisal District (TCAD). For payments to property owners for whole and partial acquisitions, they relied on TxDOT's ROW Information System, or ROWIS, supplemented by spatial details available in parcel maps. In the case of the latter, acquisition sizes were interacted with parcel indicators, to reflect the fact that twice the level of acquisition should cost (approximately) twice as much, everything else constant. The resulting equation (for each distinct data point or acquisition) is as follows:

$$TOTALCOST = \beta_0 + SF_{acquired} \sum_i (\beta_{i,land}\, X_{i,land}) + SF_{improvement} \sum_j (\beta_{j,land}\, X_{j,land}) + SF_{remainder} \sum_k (\beta_{k,land}\, X_{k,land}) + \varepsilon$$

where $SF_{acquired}$ represents the square footage of acquired land, $SF_{improvement}$ is square footage of taken improvements (i.e., structures built on the parcel), and $SF_{remainder}$ is the size of remainder/non-acquired area (where applicable). $X_{i,land}$ is a vector of explanatory variables indicating land use type, number of driveways, frontage length, parcel shape, year of acquisition, location, and whether the site is a corner property (as those tend to be more expensive, thanks to better customer access). Similarly, $X_{j,imp}$ variables describe the structure type, age, and condition; and $X_{k,dam}$ values characterize the parcel's remainder (e.g., any loss in frontage, whether the parcel's shape changed, and ratio of remainder area to original parcel size). As explained in Chapter 8, the error term ε accounts for unobserved attributes (such as soil quality, vegetative cover, and strength of local schools). Because these error terms tend to get larger (from under- or over-prediction) on larger, more complex, and more expensive properties, a second model was used in tandem to reflect scatter in the data in the sizes of the ε values. As described in Chapter 8, *feasible generalized least squares* (FGLS) regression (two stages of least-squares modeling) can be applied to reflect such heteroskedasticity and improve parameter estimates in the main model equation (which are the β terms shown above).

The study found that retail property costs along expanding corridors were the most consistent in value and easiest to predict with their economic equation. As expected, urban properties, especially those in downtown locations, cost much more than others, all other features constant. Multi-story office buildings were valued the highest (per square foot) and industrial use properties the lowest, everything else constant. Using TCAD data, properties in "excellent" condition were forecast to fetch \$22 more per square foot (of improved space) than a comparable property in "fair" condition. Using CoStar's commercial sales data for all Texas metropolitan regions yielded a similar result: a property in "excellent" condition was estimated to be worth nearly \$28 more per square foot than a property in "fair" condition. CoStar data also provided information on parking spots, which were estimated to add roughly \$6,000 (each) to a (commercial) property's value.

These types of linear regression models are helpful in predicting ROW costs using property attributes, while also identifying properties that are most challenging to assess. For example, a veterinary hospital site's value was significantly over-predicted by the Heiner and Kockelman model application across 10 commercial properties in Fort Bend County, Texas.

In addition to acquisition costs paid directly to owners, overall project costs and project duration can be significantly impacted by the negotiation or condemnation process that DOTs go through on each property. For example, Xiong and Kockelman's (2012) analysis of 2008 through 2011 ROWIS data suggest that condemned properties resulted in 78% higher acquisition costs for TxDOT (equivalent to an added \$14.7 per square foot) than negotiated properties, and 51% greater price uncertainty (measured as variance in the pricing model's error term). Moreover, the process of condemnation has been estimated to add an average of 8 months to parcel acquisition duration (versus negotiation), due to court trials in place of administrative settlement. Unfortunately, lost time is lost money (and sometimes reputation). Such currency can take the form of contractor costs, congestion delays due to later capacity expansion, weakened public support, and so forth. **Econometric models help transportation planners, engineers, and policymakers make better decisions about how much to budget for a future or near-term project, which properties to purchase early, which to evaluate more deeply, and which to work hardest to acquire through negotiated settlements.**

Anticipating the Economic Impacts of Highway Projects Using T-PICS

Crucial in anticipating potential project impacts is examining past, similar projects' costs and benefits. As a part of a research initiative by the U.S. Strategic Highway Research Program, the EDR Group and ICF Consulting assembled a large collection of economic analyses from already implemented transportation projects in an easy-to-access format called Transportation Project Impact Case Studies (T-PICS), available at http://transportationforcommunities.com/t-pics/.

T-PICS contains two main tools: Case Search and My Project Tool. Case Search allows users to search the database by attributes such as project type, region, motivation, class level, economic distress, keyword, or a combination of all six. This specificity allows the user to narrow the results to an existing project that is similar to a newly proposed transportation improvement. The T-PICS results page displays basic descriptions of each project, and a user can also choose to compare multiple projects.

The second important function of T-PICS, My Project, can be used to roughly predict the economic impact of a project. Due to the pure volume of entries in this database, finding and analyzing every project that relates to a newly proposed design is impractical. Instead, based on input criteria, this tool finds all projects that are similar, summarizes their economic impacts, and scales the data to the specific criteria of the proposed project. This analysis can be customized even further by specifying certain area characteristics, such as land use policies, business climate, and infrastructure type(s).

Summary

This chapter's case studies illustrate how the economic concepts introduced in this Reference weave together and can be used to enhance transportation planning, investments, and policymaking. For example, road pricing and improvement decisions along a specific corridor can have positive and negative impacts across an entire network, requiring an appreciation of latent demand, complete traveler welfare changes, travel time reliability, and sensitivity analysis to reflect analyst uncertainty.

These case studies also illustrate the value of statistical analysis, highlighting the flexibility and practical application of several econometric models using Census, land use, and agency data sets to quantify relationships. The methods of analysis developed in this Reference are powerful tools for mining existing and new data sets while exploiting a variety of practical methods. Such concepts and tools can be essential in anticipating project and policy impacts, thereby enhancing DOT decision-making, even in the presence of uncertainty.

References

Babcock, M. & Davalos, J. (2004) Case Studies of the Economic Impact of Highway Bypasses in Kansas. Kansas Department of Transportation, Kansas State University. Available online at http://www.ksdot.org/PublicLib/doccontent.dll?LibraryName=PublicDocs^dt00mx38&SystemType=2&LogonId=a97e035c62020933902b470f45be164e&DocId=003687790

Bina, M. & Kockelman, K. (2009) Location Choice vis-a-vis Transportation: The Case of Recent Home Buyers. *The Expanding Sphere of Travel Behaviour Research*, Emerald Publishers.

Fagnant, D., Kockelman, K., & Xie, C. (2011) Anticipating Roadway Expansion and Tolling Impacts: A Toolkit for Abstracted Networks. Presented at the 90th Annual Meeting of the Transportation Research Board, January 2011, Washington D.C.

Fagnant, D. & Kockelman, K. (2012) Transportation Project Outcomes Under Uncertainty: An Examination of Benefit-Cost Ratios and Other Impacts. Proceedings of the 91st Annual Meeting of the Transportation Research Board, and forthcoming in *Transportation Research Record.*

Gulipalli, P. & Kockelman, K. (2008) Credit-Based Congestion Pricing: A Dallas-Fort Worth Application. *Transport Policy* 15(1), 23–32.

Handy, S., Kockelman, K., Kubly, S., Srinivasan, S., Jarrett, J., Oden, M., & Mahmassani, H. (2001) The Impacts of Highway Relief Routes on Small Towns in Texas. Research Report 1843. Center for Transportation Research, The University of Texas at Austin.

Heiner, J. & Kockelman, K. (2005) The Costs of Right of Way Acquisition: Methods and Models for Estimation. Journal of Transportation Engineering, 131(3).

Kalmanje, S. & Kockelman, K. (2004) Credit-Based Congestion Pricing: Travel, Land Value, and Welfare Impacts. *Transportation Research Record* 1864, 45–53.

Kockelman, K., Fagnant, D., & Xie, C. (2011) PET: A Toolkit for Anticipating and Comprehensively Evaluating Transportation Project Impacts. Available online at: http://www.caee.utexas.edu/prof/kockelman/PET_Website/homepage.htm.

Le, T. (2009). Improving Right-of-Way Acquisition in Highway Projects through Scope Definition and Management of Inherent Factors. Ph.D. dissertation. The University of Texas at Austin.

Rutgers Intelligent Transportation Systems Laboratory (RITS) (2010) Cost/Benefit Analysis of NJDOT Route 18/Hoes Lane Improvement Project. Available online at: http://www.state.nj.us/transportation/works/studies/rt18_i287connect/pdf/cb_analysis.pdf

Srinivasan, S. and Kockelman, K. (2002) The Impacts of Bypasses on Small- and Medium-Sized Communities: an Econometric Analysis. University of Texas at Austin. Available online at http://www.ce.utexas.edu/prof/kockelman/public_html/TRB01Bypasses.pdf.

Ten Siethoff, B. and Kockelman, K. (2002) Property Values and Highway Expansions: An Investigation in Timing, Size, Location, and Use Effects. *Transportation Research Record No. 180*. TRB. National Academy Press, Washington, D.C.

Thompson, E., Miller, J., & Roenker, J. (2001) The Impact of a New Bypass Route on the Local Economy and Quality of Life. Kentucky Transportation Center, University of Kentucky. Available online at http://www.ktc.uky.edu/Reports/KTC_01_10_SPR_219_00_2I.pdf.

Vadali, S. & Sohn, C. (2001) Using a Geographic Information System to Track Changes in Spatially Segregated Location Premiums. *Transportation Research Record* 1768,180–192. TRB. Washington, D.C.

Williams, K. & Frey, R. (2004) Corridor Preservation: Best Practices for Local Government. *Transportation Research Record No. 1895*, 156–162.

Xiong, X. & Kockelman, K. (2012). The Cost of Right-of-Way Acquisition: Recognizing the Impact of Condemnation via a Switching Regression Model. Proceedings of the 91st Annual Meeting of the Transportation Research Board in Washington D.C., and forthcoming in *Transportation Research Record.*

Yeh, D., Gannon, M., & Leon, D. (1998) The Economic Impacts of Highway Bypasses on Communities. Wisconsin Department of Transportation, Economic Planning and Development, Madison, Wisconsin. Available online at http://www.dot.wisconsin.gov/library/publications/topic/plans/bypass.pdf.

Index

S

T

U

V

W

Made in United States
North Haven, CT
24 November 2024

60904891R00177